T0206364

Optical Polymer Waveguides

Jörg Franke · Ludger Overmeyer ·
Norbert Lindlein · Karlheinz Bock ·
Stefan Kaierle · Oliver Suttmann ·
Klaus-Jürgen Wolter
Editors

Optical Polymer Waveguides

From the Design to the Final 3D-Opto Mechatronic Integrated Device

Editors
Jörg Franke
Lehrstuhl für Fertigungsautomatisierung
und Produktionssystematik
Friedrich-Alexander Universität
Erlangen-Nürnberg
Erlangen, Germany

Norbert Lindlein
Institut für Optik, Information und
Photonik, Friedrich-Alexander
Universität Erlangen-Nürnberg
Erlangen, Germany

Stefan Kaierle
Laser Zentrum Hannover e.V.
Hannover, Germany

Klaus-Jürgen Wolter
Institut für Aufbau- und Verbindungstechnik
der Elektronik, Technische Universität
Dresden, Dresden, Germany

Ludger Overmeyer
Institut für Transport- und
Automatisierungstechnik
Gottfried Wilhelm Leibniz
Universität Hannover, Garbsen, Germany

Karlheinz Bock
Institut für Aufbau- und Verbindungstechnik
der Elektronik, Technische Universität
Dresden, Dresden, Germany

Oliver Suttmann
Laser Zentrum Hannover e.V., Coherent
LaserSystems GmbH & Co. KG
Hannover, Germany

ISBN 978-3-030-92856-8 ISBN 978-3-030-92854-4 (eBook)
https://doi.org/10.1007/978-3-030-92854-4

Responsible Editor: Alexander Grün
This Springer imprint is published by the registered company Springer Nature Switzerland AG
The registered company address is: Gewerbestrasse 11, 6330 Cham, Switzerland

Preface

To keep pace with the high demand of bandwidth to transmit increasing amounts of data even in real time, many applications are heading from metal-based transmission lines to optical waveguides. As optical glass fiber technologies are the backbone of our digital infrastructure covering long-haul transmission tasks, polymer optical fibers are used for short-range transmission. By providing a multitude of advantages like less weight, higher energy efficiency at increasing data rates and smaller dimensions of devices, the demand for new developments in optical assembly and interconnection technology is increasing. This book presents a novel approach to short-range optical communication based on the functionalization of three-dimensional structures with polymer optical waveguides to build up an optical transmission line.

This research report covers methods and technologies for the design, manufacturing and simulation of three-dimensional optically functionalized mechatronic components (3D-opto-MID). This includes

- computer-aided methods for the design and simulation of 3D-opto-MIDs
- a printing process to condition the substrates for the subsequent application of optical waveguides
- the technology for manufacturing optical waveguides on 3D-formed surfaces via Aerosol Jet Printing in order to build up polymer optical waveguides, which are able to transmit signals on a three-dimensional component
- a passive concept of coupling for a subsequent division of the optical signals along the optical path of the optical waveguides during field mounting
- and a concept of coupling for the direct connection of optoelectronic converter components to the optical waveguides.

Hence the book addresses scientists, engineers, students and interested non-experts to expand their knowledge with a new approach for the realization of optical networks and opens up new solutions to implement optically integrated and highly functionalized devices.

The research results presented in this book were obtained within the interdisciplinary research group Optical Design and Interconnection Technology for Assembly-Integrated Bus Systems (FOR 1660 OPTAVER) funded by the German Research Foundation (DFG). The work was carried out by a group of talented researchers, who are all experts in their scientific fields and have successfully utilized their abilities and ideas to make a long-lasting impact on the key technology field of short-range optical transmission. We are very proud to be able to connect the design and simulation as part of the CAD-CAM chain to design optically functionalized components with the actual manufacturing processes. This optical integration increases the functional density of these devices by replacing dedicated hardware like polymer optical fibers and allows us to present a holistic approach to 3D-opto-MID for the first time.

On behalf of all researchers contributing to this book, I would like to thank the DFG for the opportunity to conduct this highly innovative research. With the conducted investigations, our research group OPTAVER was able to answer how polymer optical waveguides can be applied to three-dimensional structural components. This extends the functionality of mechatronic integrated devices (MID) by optical means, which opens up a multitude of new applications in the field of optical transmission and sensing.

Erlangen
February 2022

Professor Dr.-Ing. Jörg Franke

Contents

Current Development in the Field of Optical Short-Range Interconnects

1

Lukas Lorenz and Karlheinz Bock

1.1 Advantages of Optical Communication

The current development in the field of information and data transmission is driven by a constantly growing amount of data generated worldwide. Current predictions range from 33 Zettabytes (ZB) in 2018 to 175 ZB in 2025 [1]. The reason for this enormous growth is the increasing network of devices that traditionally have no connection to a network. In this case, we speak of the Internet of Things (IoT). This affects all areas: mobile systems, transportation, sensor technology, cloud applications, medical devices and big data systems, to name just a few [2].

The consequence of the enormous increase in generated data is the demand for increasingly higher data transfer rates to handle, transmit and store these large amounts of data. While current CMOS circuits reach higher and higher speeds, the electrical connections can hardly support them [3]. Therefore, parallel point-to-point connections with enormous space requirements are becoming increasingly popular (e.g., Peripheral Component Interconnect (PCI) Express in home computers) [4, 5]. In the near future, traditional electrical connections will reach their limits. Here photonics offers promising alternatives. This is particularly evident when comparing energy efficiency and space requirements [3], as shown in Table 1.1. These values illustrate the advantages of optical over electrical connections. Furthermore, photonic systems are insensitive to electromagnetic interference and can therefore be used in electromagnetic compatibility (EMC) critical areas.

L. Lorenz (✉) · K. Bock
Institut für Aufbau- und Verbindungstechnik der Elektronik, Technische Universität Dresden, Dresden, Germany
e-mail: lukas.lorenz@tu-dresden.de

K. Bock
e-mail: karlheinz.bock@tu-dresden.de

J. Franke et al. (eds.), *Optical Polymer Waveguides*,
https://doi.org/10.1007/978-3-030-92854-4_1

Table 1.1 Comparison between electrical and optical data transmission

	Electrical	Optical	Conclusion
Energy efficiency at 20 Gbit/s [3, 6]	(30...40) mW/Gbit/s	17.5 mW/Gbit/s	At high data rates, optical transmission is more energy efficient.
Space requirement [6]	3.2 mm²	0.6 mm²	ICs connected with optics need less space than the electrical pendants
Footprint for a 10 Gbit/s connector with 144 channels [7]	1620 mm²	236 mm²	Optical channels can have a smaller pitch than electrical ones

According to prognoses, the advantages in these aspects will lead to optical multimode short-range interconnects to be the backbone of upcoming IoT and Industry 4.0 applications [8]. Furthermore, they will be decisive for the further development of these systems [2]. An example of such a system is shown in Fig. 1.1. Emerging autonomous driving applications require the processing and dissemination of large amounts of data (up to 4 TB per hour [9]). This could be done efficiently, quickly and with little space requirements in the vehicle via an optical bus system.

A look at the existing solutions for optical interconnects and bus systems reveals that components for optical integration are well developed [4]. For both transmitters [11, 12] and receivers [13, 14], established technologies can be used. In addition, in the field of classical packaging and interconnection technology, solutions are already available and optical systems are established:

- Telecommunications (transmission between locations)
- Data communication (transmission between individual computers)
- Computer communication (transmission within a computer).

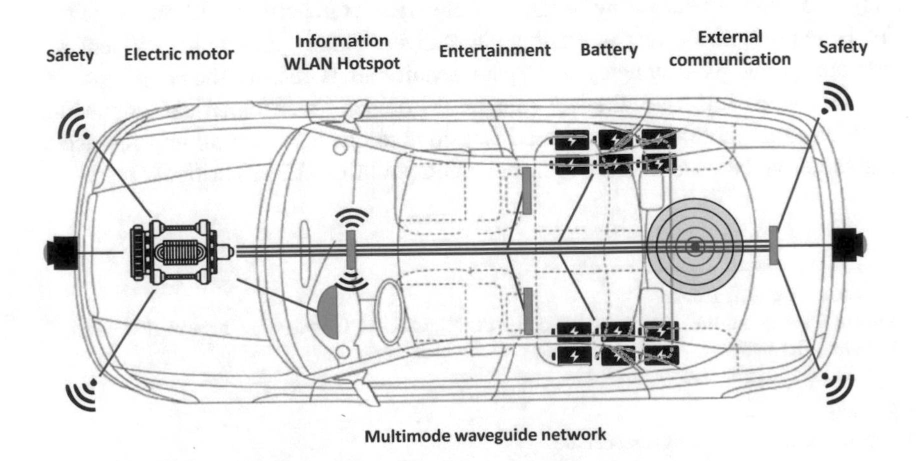

Fig. 1.1 Optical interconnects as backbone of the communication system in a car [10]

For longer transmission distances in the kilometer and longer meter range (telecommunications), optical transmission is already standard for many years. Optical solutions are also established in the area of active optical cables for rack-to-rack links in data centers (data communication) [15] and at board level (computer communication) [16, 17]. Furthermore, approaches for optical connections on the chip are already known (computer communication) [18].

However, the solutions for data communication or computer communication are all designed for use in classic IT environments, i.e., in server farms, on printed circuit boards (PCB), etc. For the emerging applications in the area of Internet of Things and Industry 4.0, it is necessary to carry out a paradigm shift and discuss new solutions. For this purpose, Lorenz [10, 19] focussed on a newly emerging application area: the device communication. This area includes applications outside the classical computer architecture. Although these applications have a connection to the Internet, the focus is on the connection and networking of modules in closed, embedded systems with connection lengths between 0.1 m and 100 m. These connections do not use classical cabling or wiring on a substrate, but embedded links in the structures—just like the system itself. The new demand for device communication also results in challenges in the design of 3D opto-mechatronic integrated devices (3D-Opto-MID [20]). The three most important challenges, which are addressed in this book, are summarized in Table 1.2.

An example of an application is illustrated in Fig. 1.2. It shows a sensor network in an airplane wing, where several modules are connected to an embedded waveguide via bus coupling. Since this is a lightning strike and EMC-critical environment, optical systems are particularly preferred.

Another example, depicted in Fig. 1.3, shows an optical strain sensor for car batteries. Since there is a need for galvanic isolation, optical sensors or at least optical connections are necessary. The waveguides are either directly applied on the battery housing (e.g., additively) or flexible foils are fixed, following the three-dimensional shape.

Table 1.2 Demands and challenges induced by upcoming applications in the device communication [10] and the chapters they are addressed in

Demand	Technical challenge	Associated chapter
Device integrated, large waveguide networks (e.g. in cars or airplanes) on structural elements	Fabrication of waveguides on 3D surfaces, which are not length limited	4 (Conditioning of Substrates) 5 (Waveguide Manufacturing)
Bus coupling of 3D-Opto-MID modules in large optical networks to avoid a huge cable harness	Coupling of waveguides without interruption, as well as asymmetric coupling ratios depending on the coupling direction	6 (Coupling)
3D-Opto-MID assemblies in compact design (e.g. as components in a sensor network)	Design, simulation and fabrication of optical networks in 3D with sufficient heat management	2 (Design) 3 (Simulation) 6 (Fabrication)

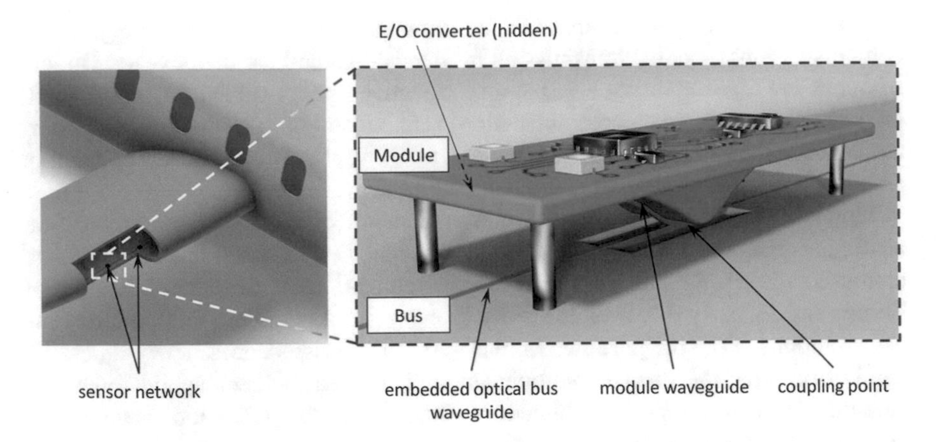

Fig. 1.2 Example application for a sensor network, which communicates via an embedded optical bus [10]

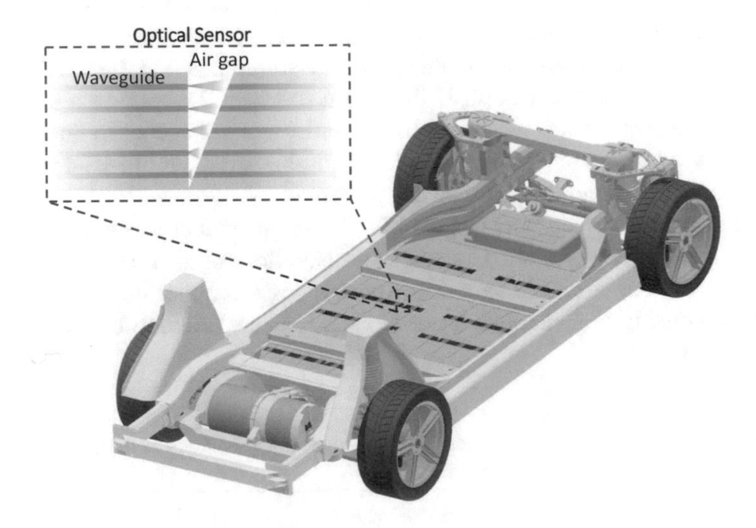

Fig. 1.3 Example for an optical strain sensor monitoring the battery health in an electric car

1.2 3D-Opto-MID for Optical Bus Systems

There are only inadequate solutions for the above-mentioned challenges. The following subchapter gives an introduction of optical short-range connections, optical bus systems and 3D-Opto-MID. The corresponding main chapters discuss then the detailed state of the art.

1.2.1 Short-range optical waveguides networks

As already described in Sect. 1.1, in the communication technology a differentiation is made between the individual connection levels, which essentially depend on the length of the optical transmission.

For telecommunications, which covers long-distance transmission, single mode fibers and wavelengths of $\lambda = 1300$ nm and $\lambda = 1550$ nm are used, which ensure low attenuation and high signal quality over long distances. However, due to the small core diameters of single mode fibers, packaging is very challenging because of the very small alignment tolerances.

Single mode waveguides, but planar light wave circuit (PLC), are also used in module and chip-level applications (computer communication) for interconnects over shortest distances. The integration of waveguides in silicon (Silicon Photonics) is of great importance here. In this area, there are already approaches for optical bus coupling via ring resonators [18, 21].

These areas—and thus single mode waveguides in general—are irrelevant for the device communication, as they deal with either extremely long (> 1 km) or extremely short (< 10 cm) transmission distances. For this reason, they will not be discussed in detail in this book.

The relevant range for the device communication and thus for this work can be limited to connection lengths of 0.1 m to 100 m. Multimode waveguides with larger core cross sections are suitable for this purpose. The advantages are less critical tolerance requirements, which simplifies production and assembly significantly. The disadvantages of multimode transmission (higher attenuation and mode dispersion), on the other hand, are not relevant for such short transmission distances.

Because multimode fibers have their first optical window at $\lambda = 850$ nm and the often used GaAs-based vertical-cavity surface-emitting lasers (VCSEL) have a great availability at this wavelength, 850 nm is established as the standard wavelength in optical short-range connections. Hence, the approaches presented in this book are based on this wavelength as well.

To achieve waveguide networks in and on structural elements, there are already solutions to embed fibers into carbon fiber-reinforced polymers (CFRP) [22, 23]. However, this is a very complex process. Furthermore, it cannot be used with other materials and the coupling to and from the fibers is very difficult. Hence, this technology is more suitable for structural health monitoring with fiber Bragg gratings than for the communication between several devices.

More suitable for the connection of different modules or devices are optical networks in printed circuit boards. These so-called electro-optical printed circuit boards (EOPCB) use embedded glass layers for the optical routing and out-of-the-plane coupling elements to connect different devices [16, 24]. The major disadvantage is the limitation to 2D. Although there are solutions for stacked EOPCBs, where a 3D light path between different layers is possible, there is still no possibility to realize bends or curves in the vertical direction [25, 26]. Furthermore, these technologies are limited to the PCB-environment with a high effort for the fabrication of the glass panel-based waveguides, which is not flexible enough for the device communication.

To increase the potential of optical short-range connections for the device communication, arbitrary waveguide networks on three-dimensional surfaces are needed. According to that, there is a demand for new manufacturing techniques for multimode waveguides.

1.2.2 Optical bus systems

Even though there is already research successes in the field of optical computers [27], the processing and storage of information are currently exclusively electrical. For this reason, optical transmission always requires electro-optical (e/o) and opto-electrical (o/e) conversion. Discrete semiconductor lasers (edge or surface emitters [28, 29]) are usually used for e/o conversion, but integrated lasers for silicon photonics are also available [30]. Either they are directly modulated or a separate modulator is added. The conversion of the received signals is usually done by photodiodes [32] and the corresponding processing circuit [10].

There are two theoretical options to connect more than one e/o module: parallel point-to-point interconnects or bus coupling. The first possibility is the connection of the modules/participants (e.g., sensors) to the base/control side (e.g., central processing unit) with several waveguides, i.e., every interconnection has its own transmission path. This has the advantage that, in the event of failure of a waveguide, only one module is affected. However, the following disadvantages are associated with that:

- Very complex hardware on the base side
- Design of a huge cable harness with numerous single waveguides
- High space requirements.

Due to the described limitations when using several e/o modules, the disadvantages outweigh the advantages of optical communication compared to the electrical pendants. Hence, optical transmission has not been established for device communication applications yet. A solution to this challenge would be the second possibility of transmitting a signal: via a bus system. According to the definition, a bus is a shared transmission path between several nodes, in which two nodes communicate while the rest remain silent [32].

For optical transmissions, a ring bus—as shown in Fig. 1.4—is preferred, since the direction of propagation of the light does not have to be changed. Hence, additional splitters/combiners for splitting into transmitting (Tx) and receiving (Rx) elements or light deflecting elements (mirrors, fiber Bragg gratings) are not required. An additional advantage of such an optical bus system is the possibility of using wavelength division multiplexing (WDM) [32] either to enable simultaneous communication between several nodes or to integrate several logical buses with different wavelengths into one physical bus. The disadvantage is that

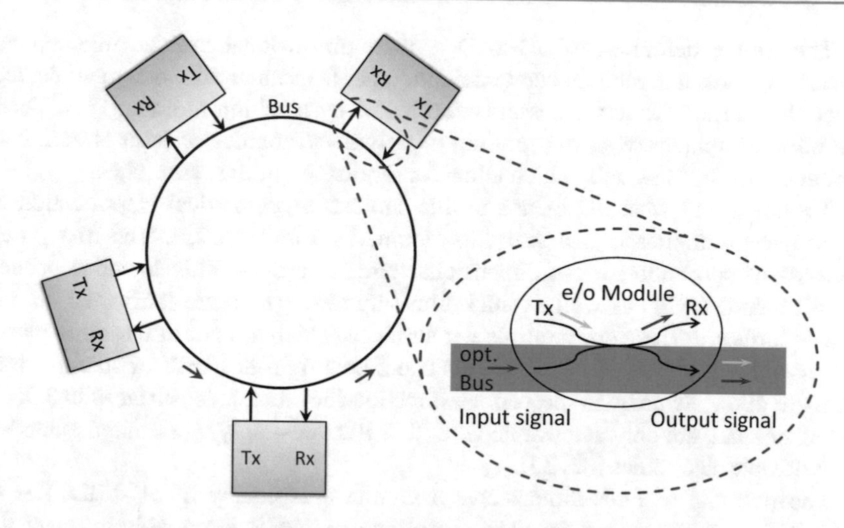

Fig. 1.4 Schematic of an optical ring bus and the required optical bus coupling [10]

all nodes in the system share the power budget of a single bus. All advantages and disadvantages of the optical ring bus are summarized in Table 1.3.

The fact that all modules share the power budget of one bus has led to the fact that no practical concept for optical bus coupling could yet be developed, as it will be discussed in Chap. 6. Hence, the possibilities for the transmission from one waveguide to another are still limited to butt coupling or grating couplers, which both need an interruption of the two coupling partners. On the other hand, parallel (bus)connections are standard in the electric domain.

1.2.3 Current development in the field of 3D-Opto-MID

As mentioned earlier, the integration of optical paths and elements into 3D-MID is getting more and more important. One example is the asymmetric optical bus coupler (AOBC) which will be described in Chap. 6. For that, a defined radius of one of the coupling waveguides is crucial, which is why a 3D assembly is necessary.

Table 1.3 Overview of the advantages and disadvantages of an optical ring bus

Advantages	Disadvantages
Simple cable harness (in ideal circumstances only one waveguide)	All modules share the power budget of the bus waveguide
Flexible reconfiguration of the network	The bus waveguide needs to be connected to its origin
In theory an arbitrary number of modules can be connected	
No loss of power if a module is disconnected	
Parallel use of the bus waveguide with WDM	

The former definition of 3D-MID is three-dimensional molded interconnect device. A newer and more accurate definition is 3D mechatronic integrated device, since the carriers are not necessarily fabricated by injection molding [33]. There are other techniques, e.g., 3D printing of polymers/ceramics or laser sintering of ceramic materials [34, 35], which allow for fabrication of 3D assemblies.

To understand what 3D means in this context, a geometrical classification of electronic assemblies is necessary, as summarized in Table 1.4. The first group consists of conventional two-dimensional circuit carriers. This includes printed circuit boards (PCB) as well as thick-film ceramics, which are fabricated on one planar surface. If there are multiple flat surfaces, which are not in the same plane, the assembly is classified as 2.5D. This could be a rigid-flex PCB or stacked rigid PCBs to allow elements in the vertical direction [36]. Freeform surfaces in 3D, on the other hand, are only achievable with 3D-MID and—only to a limited extend—with flexible electronics [33, 37].

The potential of three-dimensional assemblies, especially of 3D-MID, lies in the high geometric design freedom. According to that, it is possible to adapt the electronic parts to the geometry (coolers, alignment or functional structures, etc.), which allows for high flexibility and higher degrees of miniaturization due to higher package densities. Furthermore, for some applications it is crucial to obtain defined angles and/or radii between single parts, which is only possible with 3D-MID. [33]

However, because of a more difficult manufacturing compared to planar fabrication and assembly processes, 3D-MID stays behind standard techniques in terms of throughput. Furthermore, holistic design tools for mechanical, thermal, electrical and optical properties are not available for 3D-MID, which complicate the design process compared to standard PCB or ceramic assemblies. In addition, thermal management has to be considered when choosing the substrate material,

Table 1.4 Classification of packages according their spatial dimensions (adapted [33])

2D	2.5D	3D
Planar process surface	Multiple plane surfaces	Freeform surfaces
Standard PCBs, thick-film ceramics,	Rigid-flex PCBs, Stacked PCBs	3D-MID, flexible PCBs
+ High throughput + Matured technologies and design tools + Suitable for temperature critical applications - No functionality in vertical direction	+ High throughput + Flexion/Twisting possible + Functionality in vertical direction - Manual assembly of the single parts	+ Function integration + Three-dimensional design freedom + High degree of miniaturization - Difficult part assembly - No holistic design tools

which then influences the choice between PCB, thick film, flex PCB and 3D-MID. Hence, it always has to be carefully considered which requirements are made on the assembly and which production technology is then suitable and adequate for it.

To increase the potential of 3D-MID, [33] made several proposals:

- In order to address high current applications, thermally conductive materials are necessary.
- For optical applications, which need stable temperatures for LEDs and lasers, heat dissipation concepts have to be made.
- Improved 3D placement and interconnection as well as embedding technologies allow for higher throughputs and higher package densities (chip on MID).
- Holistic design tools are necessary to improve the R&D steps prior the fabrication.

1.3 New Approach for Additive Manufactured 3D-Opto-MID

For the upcoming applications in the device communication, the advantages of optical transmission were pointed out. However, in contrast to optical links on PCB or rack level, there is a lack of short-range connections on three-dimensional structural elements. This leads to the demands described in the previous subchapters. The goal of the presented work is to increase the competitiveness of optical short-range connections for the device communication compared to electrical pendants. Four main challenges could be identified:

- There is a demand for new waveguide manufacturing techniques for 3D.
- New coupling schemes are required to achieve optical bus systems.
- Classical 3D-MID needs to be extended by an optical functionality toward 3D-Opto-MID.
- For the development of such systems, a holistic design/simulation tool is necessary.

In this book, we want to address all of these issues in an interdisciplinary research. The question we want to answer is:

How is it possible to increase the competitiveness of optical short-range connections for the device communication on 3D structures in terms of waveguide design, simulation, fabrication, coupling and packaging?

This question is successively solved in an interdisciplinary research attempt (Fig. 1.5) throughout the single chapters of this book, beginning with the modeling and simulation followed by the conditioning of the substrate and manufacturing of the waveguides. At the end, a 3D-Opto-MID package for optical bus coupling is presented and the impact of the novel technology is evaluated.

To understand the decisions, made in the single chapters, a preview of the solution is necessary. To achieve maximum flexibility for arbitrary waveguide networks

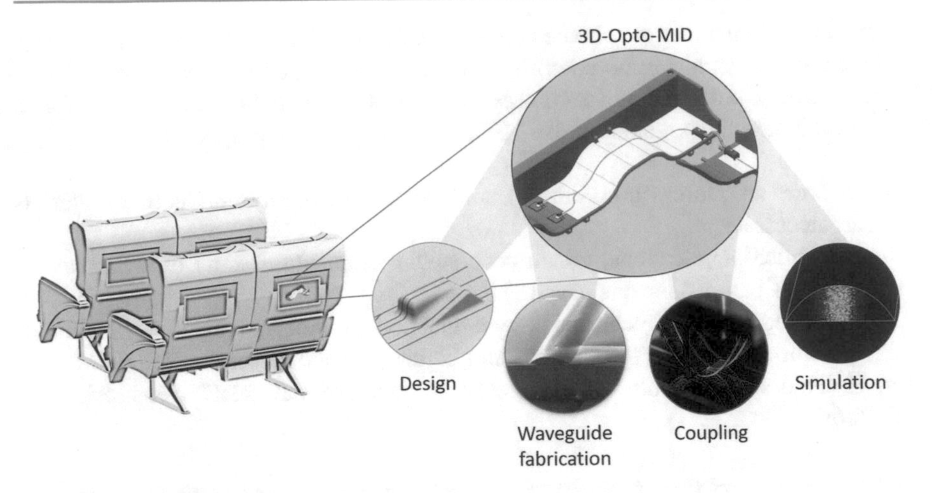

Fig. 1.5 Illustration of the interdisciplinary research topic of additively manufactured optical waveguides for 3D-Opto-MID by the example of a module for infotainment systems

on 3D surfaces, the core of the presented research approach is the additive manufacturing of multimodal waveguides using aerosol jet printing. It has a high potential for three-dimensional applications. The possibility to print waveguides on 3D surfaces brings optical short-range connections to a new level. Chapter 5 contains the detailed discussion of aerosol jet printing compared to the state of the art.

The main challenges of printed polymer waveguides are their unique cross section (circular segment) and the waviness along the waveguide. To control the latter one and to increase the possible aspect ratio of the core, conditioning lines are used, which influence the wetting behavior of the substrate related to the core material. This leads to an increased height of the circular segment at the same width. The application of these conditioning lines is discussed in Chap. 4. The whole approach is depicted in Fig. 1.6.

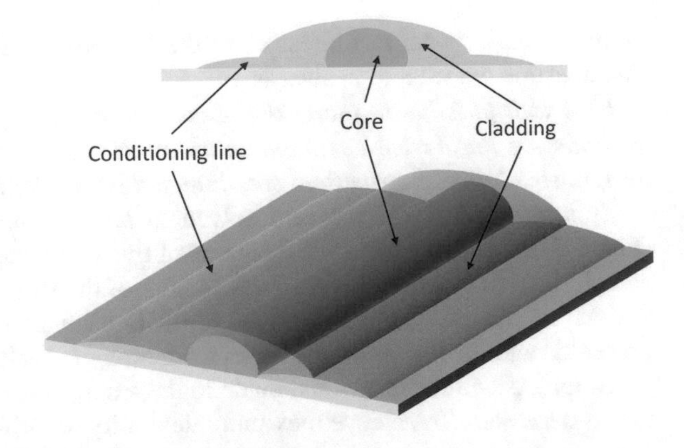

Fig. 1.6 Schematic of an aerosol jet printed waveguide with conditioning lines to increase the aspect ratio of the core

The coupling of these waveguides is another challenging task, which is described in detail in Chap. 6. The asymmetric optical bus coupling (AOBC), developed in this work, allows for interruption-free, bidirectional waveguide coupling, especially for photonic bus systems. Crucial for this coupling method is the bending of one of the coupling partners out of the plane (in vertical direction), which is why 3D substrates are necessary. Hence, classical 3D-MID is extended to include optical functions.

Furthermore, an all-new design tool is developed in Chap. 2, which allows for optical *and* electrical routing on 3D substrates including a standardized parameter pass to an optical simulation tool. It extends mechatronic data models by adding optical properties and functions. This creates a design that is adapted to the individual manufacturing processes. The simulation is especially developed for multimode waveguides with arbitrary cross sections and will be presented in Chap. 3.

Overall, the complete process flow for additively manufactured waveguides for 3D-Opto-MID is portrayed from the design, simulation toward the manufacturing to the final package and assembly.

We sincerely thank the Deutsche Forschungsgemeinschaft for funding the research group OPTAVER FOR 1660.

References

1. Reinsel, D., Gantz, G., Rydning, J.: „The Digitization of the World—From Edge to Core," IDC White Paper, Framingham, USA (2018)
2. Tekin, T.: Review of Packaging of Optoelectronic, Photonic, and MEMS Components. IEEE J. Sel. Top. Quantum Electron. **17**, 704–719 (2011)
3. Kam, D., Ritter, M., Beukema, T., Bulzacchelli, J., Pepeljugoski, P., Kwark, Y., Shan, L., Gu, X., Baks, C., John, R., Hougham, G., Schuster, C., Rimolo-Donadio, R., Wu, B.: Is 25 Gb/s On-Board Signaling Viable? IEEE Trans. Adv. Packag. **32**(2), 328–344 (2009)
4. MIT Microphotonics Center: "On Board Optical Interconnection Digest," (2013)
5. Rosenberg, P., Mathai, S., Sorin,W., McLaren, M., Straznicky, J., Panotopoulos, G., Warren, D., Morris, T., Tan, M.: „Low Cost, Injection Molded 120 Gbps Optical Backplane," J. Lightwave Technol. 590–596 (2012)
6. Pepeljugoski, P., Ritter, M., Kash, J., Doany, F., Schow, C., Kwark, Y., Shan, L., Kam, D., Gu, X., Baks, C.: „Comparison of bandwidth limits for on-card electrical and optical interconnects for 100 Gb/s and beyond," In: Proceedings SPIE 6897, Optoelectronic Integrated Circuits X, San Jose, CA, USA (2008)
7. Dangel, R., Berger, C., Beyeler, R., Dellmann, L., Gmür, M., Hamelin, R., Folkert, H., Lamprecht, T., Morf, T., Oggioni, S., Spreafico, M., Offrein, B.: Polymer-Waveguide-Based Board-Level Opitcal Interconnect Technology for Datacom Applications. IEEE Trans. Adv. Packag. **31**(4), 759–767 (2008)
8. International Technology Roadmap for Semiconductors 2.0: "Outside System Connectivity Edition 2015," (2015)
9. Hypermobil—Das Magazin zur Elektromobilität, Dezember (2016) [Online]. Available: http://hypermobil.de/daten-gigaybte-autonomes-fahren/
10. Lorenz, L.: Beiträge zur effizienten Kopplung von optischen Wellenleitern in der Gerätekommunikation. In: Bock, K., Wolter, K., Zerna, T. (Hrsg.) Dresdner Beiträge zur Aufbau- und Verbindungstechnik der Elektronik, TUDPress, Dresden, Germany (2018) ISBN-10: 3959081545, ISBN-13: 978-3959081542

11. Larsson, A., Gustavsson, J., Westbergh, P., Haglund, E., Haglund, E., Simpanen, E.: „High-Speed VCSELs for Datacom," In: 42nd European Conference on Optical Communication (ECOC), Düsseldorf, Deutschland (2016)
12. Philips GmbH U-L-M Photonics: „Single-mode & Polarization stable VCSEL ULM850-B2_PL-S0101U," (2013)
13. VI Systems GmbH: High Speed (up to 40Gbit/s) Multi-Mode Fiber-Coupled Photodetector Module (700–890nm). Deutschland, Berlin (2016)
14. Caillaud, C., Glastre, G., Lelarge, F., Brenot, R., Bellini, S., Paret, J., Drisse, O., Carpentier, D., Achouche, M.: Monolithic Integration of a Semiconductor Optical Amplifier and a High-Speed Photodiode With Low Polarization Dependence Loss. IEEE Photonics Technol. Lett. **24**(11), 897–899 (2012)
15. Charbonneau-Lefort, M., Yadlowsky, M.: "Optical Cables for Consumer Applications," J. Lightwave Technol. **33**, 872–877 (2015)
16. Brusberg, L., Whalley, S., Pitwon, R., Faridi, F., Schröder, H.: Large Optical Backplane With Embedded Graded-index Glass Waveguides and Fiber-Flex Termination. J. Lightwave Technol. **34**(10), 2540–2551 (2016)
17. Nieweglowski, K., Lorenz, L., Wolter, K.-J., Bock, K.: „Multichannel optical link based on polymer multimode waveguides for boardlevel interchip communication," In: Proceedings of IMAPS 20th European Microelectronics Packaging Conference (EMPC), Friedrichshafen, Germany (2015).
18. Sun, C., Wade, M., Lee, Y., Orcutt, S., Alloatti, L., Georgas, M., Waterman, A., Shainline, J., Avizienis, R., Lin, S., Moss, B., Kumar, R., Pavanello, F., Atabaki, A., Cook, H., Ou, A., Leu, J., Chen, Y., Asanovic, K., Ram, R., Popovic, M., Stojanovic, V.: „Single-chip microprocessor that communicates directly using light," Nat. **528**, 534–538 (2015)
19. Lorenz, L., Nieweglowski, K., Wolter, K.-J., Bock, K.: „Two-Stage Simulation for Coupling Schemes in the Device Communictaion using Ray Tracing and Beam Propagation Method, " in *IEEE Electronics System-Integration Technology Conference (ESTC)*. Dresden, Germany (2018)
20. Fanke, J., Härter, S.: "Transition of Molded Interconnect to Mechatronic Integrated Devices," In: 9th MID Congress, Fürth, Deutschland (2010)
21. Faralli, S., Gambini, F., Pintus, P., Scaffardi, M., Liboiron-Laboucer, O., Xiong, Y., Castoldi, P., Di Pasquale, F., Andriolli, N., Cerutti, L.: „Bidirectional Transmission in an Optical Network on Chip with Bus and Ring Topologies," IEEE Photonics J. **8**(2) (2016)
22. Teitelbaum, M., Yarlagadda, S., O'Brian, D., Wetzel, E., Goossen, K.: „Normal Incidence Free Space Optical Data Porting to Embedded Communication Links," 32–38 (2008)
23. Qiu, L., Goossen, K., Heider, D., O'Brian, J., Wetzel, E.: „Free-space input and output coupling to an embedded fiber optic strain sensor: dual-ended interrogation via transmission," Opt. Eng. (2011)
24. Schröder, H., Neitz, M., Schneider-Ramelow, M.: „Demonstration of glass-based photonic interposer for mid-board-optical engines and electrical-optical circuit board (EOCB) integration strategy," In: SPIE OPTO, Photonics West, San Francisco, CA, USA (2018)
25. Yoshimura, T., Miyazaki, M., Miyamoto, Y., Shimoda, N., Hori, A., Asama, K.: Three-Dimensional Optical Circuits Consisting of Waveguide Films and Optical Z-Connections. J. Lightwave Technol. **24**(11), 4345–4352 (2006)
26. Lee, W.-J., Hwang, S.H., Kim, M.J., Jung, E.J., An, J.B., Kim, G.W., Jeong, M.Y., Rho, B.S.: Multilayered 3-D Optical Circuit With Mirror-Embedded Waveguide Films. IEEE Photonics Technol. Lett. **24**(14), 1179–1181 (2012)
27. Dolev, S., Oltean, M.: Optical Supercomputing—4th International Workshop, OSC, Berlin. Springer Verlag, Heidelberg (2013)
28. Reider, G.: Photonik—Eine Einführung in die Grundlagen. Springer-Verlag, Wien (2005)
29. Eberlein, D., Glaser, W., Kutza, C., Labs, J., Manzke, C.: Lichtwellenleiter-Technik. Expert Verlag, Renningen (2006)

30. Tanaka, S., Jeong, S., Sekiguchi, S., Akiyama, T., Kurahashi, T., Tanaka, Y., Morito, K.: "Four-Wavelength silicon Hybrid Laser Array with Ring-Resonator Based Mirror for Efficient CWDM Transmitter," In: Optical Fiber Communication Conference and Exposition and the National Fiber Optic Engineers Conference (OFC/NFOEC), Anaheim, CA, USA (2013)
31. Fischer-Hirchert, U.H.: Photonic Packaging Sourcebook. Springer, Berlin Heidelberg (2015)
32. International Electrotechnical Commission (IEC), November (2013) [Online]. Available: http://www.electropedia.org/iev/iev.nsf/display?openform&ievref=351-56-10. Accessed Juni 2017
33. Goth, C., Kuhn, T.: MID Technology and Mechatronic Integration Potential. In: Franke, J. (ed.) Three-Dimensional Molded Interconnect Devices (3D-MID), pp. 1–22. Carl Hanser Verlag, München (2014)
34. Scheithauer, U., Schwarzer, E., Moritz, T., Michaelis, A.: Additive Manufacturing of Ceramic Heat Exchanger: Opportunities and Limits of the Lithography-Based Ceramic Manufacturing (LCM). J. Mater. Eng. Perform. **27**, 14–20 (2018)
35. Schubert, M., Friedrich, S., Bock, K., Wedekind, D., Zaunseder, S., Malberg, H.: "3D printed flexible substrate with pneumatic driven electrodes for health monitoring," In: 21st European Microelectronics and Packaging Conference (EMPC), Warsaw, Poland (2017)
36. Karras, J.T., Fuller, C.L., Carpenter, K.C., Buscicchio, A., McKeeby, D., Norman, C.J., Parcheta, C.E., Davydychev, I., Fearing, R.S.: „Pop-up Mars Rover with Textile-Enhanced Rigid-Flex PCB Body," In: IEEE International Conference on Robotics and Automation (ICRA), Singapore, Singapore (2017)
37. Schubert, M., Rebohle, L., Wang, Y., Fritzsch, M., Bock, K., Vinnichenko, M., Schumann, T., Bock, K.: "Evaluation of Nanoparticle Inks on Flexible and Stretchable Substrates for Biocompatible Application," In: 7th Electronic System-Integration Technology Conference (ESTC). Dresden, Germany (2018)
38. Soltani, M., Liu, Y., Zimmermann, A., Kulkarni, R., Barth, M., Groezinger, T.: "Experimental and computational study of array effects on LED thermal management on molded interconnect devices MID," In: 13th International Congress Molded Interconnect Devices. Würzburg, Germany (2018)

Computer-Aided Design of Electro-Optical Assemblies

2

Jochen Zeitler and Jörg Franke

Abstract

The integration of the different engineering disciplines in an integrated development procedure is the core of this chapter. For this purpose, the physical and technological principles of optical technologies must be considered. Product development methods of technical systems are a related field, since up to now, especially for 3D-Opto-MID, no adequate methodology exists. Analogies to mechatronic systems and their sub-processes do exist, but these must be evaluated and expanded with regard to the new optomechatronic components. A separate procedure that differs from conventional development methods is just as important as the challenges that must be placed on modeling systems, the designer and the production of the 3D-Opto-MID. From these defined requirements, it was possible to derive a concept for a 3D optomechatronics CAD software (OMCAD) which contains the essential steps for creating these products.

J. Zeitler (✉) · J. Franke
Institute for Factory Automation and Production Systems, Friedrich-Alexander-Universität Erlangen-Nürnberg, Erlangen, Germany
e-mail: jochen.zeitler@neotech-amt.com
J. Franke e-mail: joerg.franke@faps.fau.de

J. Franke et al. (eds.), *Optical Polymer Waveguides*,
https://doi.org/10.1007/978-3-030-92854-4_2

This chapter focusses on the methods and development of a software-based methodology for designing 3D-Opto-MID. The aim is to show a general concept for designing these cross-domain assemblies.

2.1 Demand on Software-Based Design Tools for Spatial Optoelectronics

The integration of electrical functions into mechanical components has been intensively researched over the last 20 years. In view of the potential offered by the integration of optics into electrical and mechatronic components, the research of a new class of interdisciplinary products is therefore obvious. Until now, optical interconnects have mostly been based on optical fibers or planar electro-optical circuit boards. Current research is focused on the complete integration of optical functions into three-dimensional and spatially complex components. These components are a combination of well-known MID technology (see Fig. 2.1) and the application of optical structures, so-called 3D-Opto-MID. To be able to manufacture these products, new processes and technologies are needed for stable and efficient production.

According to the production technology aspects, the planning process for stable systems and applications is also crucial. The use of computer methods to solve scientific and technical problems has already proven to be extremely effective in areas such as mechanical, electrical and mechatronic development.

In the planning process of new products and production plants, this class of computer-aided tools is referred to as engineering software. The same software is based on mathematical models containing both analytical and numerical methods. The goal of this engineering software is to capture one or more aspects of a real or planned system. Although modern computers have enormous computing power, they still rely on robust algorithms to keep the processing time low and reduce the number of faulty solutions. Knowledge-based engineering software increases user productivity, especially in the design process, by transforming data into explicit knowledge. This can be done in different ways: For example, geometric, electrical and optical data from manufacturing can be collected and used to define design

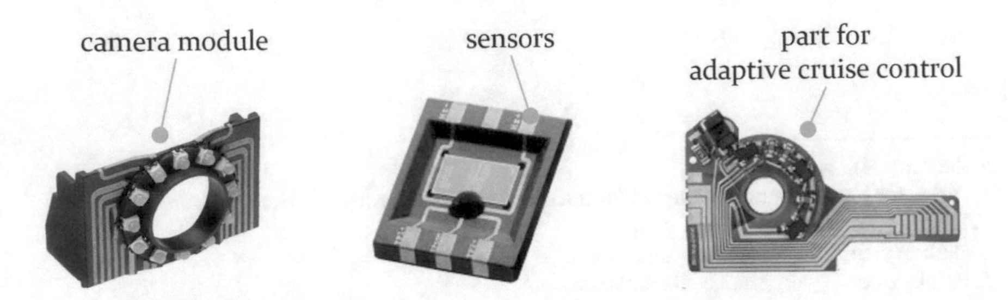

Fig. 2.1 Exemplary representation of a 3D-Opto-MID including its components [1]

rules. These help the product developers by uncovering undesirable developments in general. The more complex the system, the more difficult it is to realize such implicit arguments. On the other hand, in some cases, short notifications in the user interface with hints on problems are helpful to get attention to errors at an early stage [2]. An example can be found in the design of mechanical components and electronic circuits. These rely on a large portfolio of established computer-aided design (CAD) and electronic design automation (EDA) software tools. For example, printed circuit boards and production processes can be simulated extensively, or a wide variety of configurations of component placements can be tested to exploit the available space. In contrast, only a small number of software solutions are available for the relatively young technology of MIDs, which describes this integration of mechanics and electrics. The engineering discipline of optics, which is coming to the fore, further expands these challenges.

From this, the future need for tools can be derived that allow the production- and function-oriented design of spatial optomechatronic structures. These must combine functions from mechanical CAD (MCAD), electrical CAD (ECAD) and optical CAD (OCAD). Since there are no solutions available in this field so far, a fundamental research of these novel systems is necessary.

Against this background, the work of the research group OPTAVER is intended to contribute to the development of integrated optomechatronic assemblies. The goal is to realize optical conditions in interaction with the discipline of mechatronics in a holistic product development process for the development of optomechatronic assemblies (see also Fig. 2.2).

To this end, approaches to the design of such products must be demonstrated and evaluated using a computer-aided modeling system. The latter requires, as in a conventional CAD system, an appropriate graphic development environment, functional structures for the integration of electrical circuits as well as

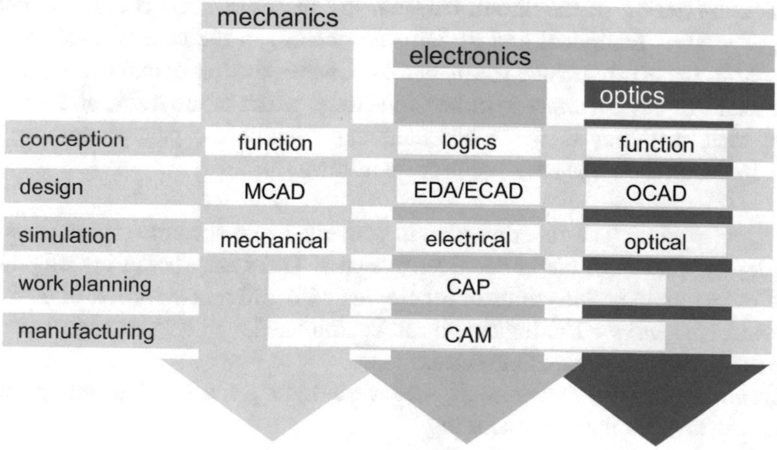

Fig. 2.2 General design process of electrical, optical and electro-optical circuits

functionalities for the design of optical networks. Therefore, it is necessary to identify relevant partial aspects and components to be realized in such a system. Based on these results, a procedure for optomechatronic product development will be derived. Physical and production-related influences play just as much a role here as the given conditions.

2.2 State of the Art in the Design of Electro-optical Circuits and MID

In the next subsections, development procedures for domain-specific and cross-domain systems will be explained. The focus here is primarily on electrical, MID and electro-optical assemblies.

2.2.1 General Procedures for Electronic and Electro-optical Assemblies

The basic purpose of a process chain for electronic and electro-optical product development is to transfer a functional idea into a working chip or circuit board using a reproducible process (the process chain). The result should be a functional system. Even if the design of simple photonic components can be done intuitively, a reproducible process supported by efficient software tools is of crucial importance. [3]

For both electronic and optical systems, the first step is to formulate basic concepts or ideas at a high level of abstraction. Afterward logical relations of different function modules can be captured in subsequent steps. In the literature, this first step is generally referred to as front-end design or schematic capture. [3]; [4]

Analogous to this, there is the back-end design or layouting, which essentially includes the design of the circuit carriers, the placement and routing of components as well as functional and design rule checks. In the case of electronic circuit boards, these can include the distances between the tracks and other important components or basic geometric dimensions (e.g., width of the track, size and spacing of vias, distances to the board contour). Finally, the post-processing takes place. In detail, the steps are structured as follows [3]:

- *Design capture:* The function idea is converted to a schematic plan consisting of logical blocks or hierarchical subsystems. There may be an investigation of different circuit architectures or topologies with different selection of devices.
- *Circuit simulation:* The logical circuit is simulated, and its parameters are optimized to make it work as intended.
- *Circuit layout:* The logic circuit is converted into a mask layout representation that can be used for manufacturing.

- *Verification:* The layout is checked for errors to ensure that it is compatible with the manufacturing process, and simulations are performed to ensure that the layout performs the desired function.
- *Manufacturing:* The generated layout data goes through a series of post-processing steps to convert it into the actual formats. The assembly is then produced on the basis of this data.
- *Testing:* The manufactured product is tested, and the results are compared to the original design. If necessary, design information is updated to improve the next generation of designs.

For better understanding, the next three sections will focus on the specific design processes for electronics, 3D-MID and electro-optics.

2.2.2 Electronics Design Process

Very large-scale integration (VLSI) describes the process of embedding a geometric chip layout from an abstract circuit description. This can be, for example, a netlist that is transferred to a physical layer such as silicon [5]; [6]. The final product is an integrated circuit (IC). At a higher level in the design process and on a smaller scale, the purpose of printed circuit board (PCB) design is to create a geometric layout by mounting VLSI components (e.g., chips) and laying traces on a printed circuit board. In the early years of semiconductor technology, designs were created manually on paper. Soon, however, with the increase in the number of transistors on a chip and improved semiconductor manufacturing processes, new automated tools became necessary to facilitate the design process—electronic design automation (EDA) systems. These tools are mainly used to generate new chips and circuits [5]. In VLSI and PCB design, there are fundamental elements that need to be optimized simultaneously: area, speed, power dissipation, design time and testability. Partitioning on different levels is a typical example that illustrates the decomposition of a system into small subsystems down to the smallest logical blocks. [6]

According to Gajski and Kuhn [7], there are three design areas, each with its own hierarchy:

- The first area is the behavioral domain, where design is described on a functional level by mathematical equations or Boolean algebra.
- The second area is the structural domain, which defines a circuit as the composition of subcircuits (e.g., transistors that form a NAND gate at the chip level or electronic components that perform a specific function at the PCB level).
- The third domain, the physical domain, provides information about the location of the circuit elements.

Figure 2.3 shows the process of designing electronic circuits on different levels for IC, multi-chip module (MCM) and PCB.

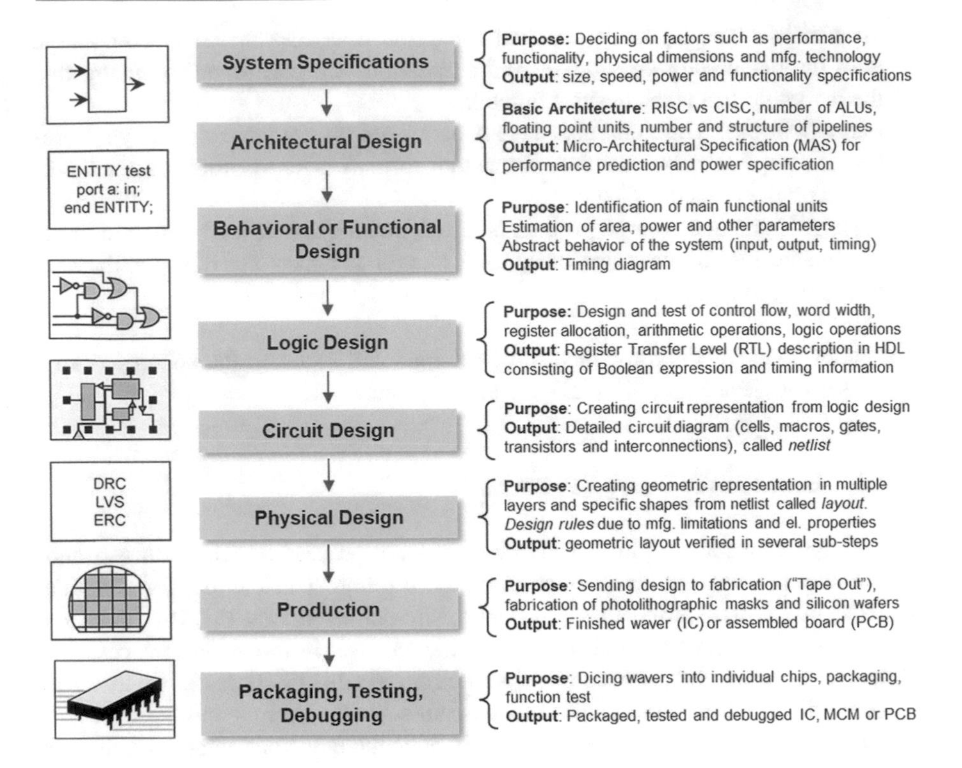

Fig. 2.3 Important steps of the digital design process with its purpose and output (based on [8])

The correctness of the created layouts can be checked by different verification methods. A practical approach is prototyping, where a working design is assembled from discrete components on breadboards. However, this becomes very complex with a large number of components. Simulation tools based on computer models analyze the output signal for a given input signal. As the layout with its internal states and possible input signals grows, simulation also reaches its limits.

2.2.3 Procedure for Spatial Electronic Assemblies (3D-MID)

3D-MID is a further development in contrast to the classic planar formwork carrier. MIDs are complex mechatronic systems, which not only focus on the electronic layout, but also require close cooperation between the mechanical and electronic development departments through function integration. However, many companies still do not use MID-specific systems for the development of new products. There are already guidelines in place, all of which pursue the goal of reducing the high complexity of development by means of a process model. In addition to the procedure according to PAHL/BEITZ [9], the guideline VDI 2206 [10] is very well known as a development methodology for mechatronic systems and plays a central role due to its universal interpretation.

Conventional MCAD and EDA systems cannot meet the requirements for the development of MIDs. In the case of EDA systems, this is mainly due to the fact that layout synthesis is only carried out for 2D layouts, and therefore, only the functionality of the logic design can be used sensibly for a potential MID tool. In contrast, MCAD systems offer no electronic functions and concentrate entirely on geometric modeling. If classic EDA and MCAD systems are now used to develop MIDs, this means that there must be a constant exchange of data between the MCAD system, which defines the installation space, and the EDA system, which is responsible for the layout design. At the same time, a change in one system means that this change must also be checked and secured in parallel in the other system (see also Fig. 2.4). These changes can cause costly and time-consuming iteration loops between the mechanical and electronic development departments. This is mainly due to possibly incompatible requirements, knowledge levels and file formats. In addition, the necessity to introduce unnecessarily large tolerances is an additional aspect to meet the requirements of the respective system. These mentioned points lead to unavoidable additional detail work.

2.2.4 Procedure in electro-optical design

In the field of electro-optical modeling of planar integrated systems, there are still no uniform procedures. In most cases, a system is designed based on an electrical

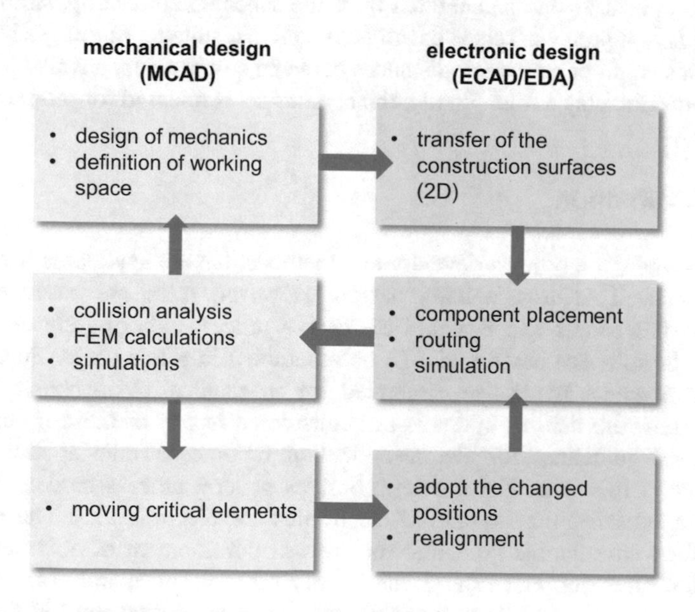

Fig. 2.4 Iterations between mechanical and electronic design [1]

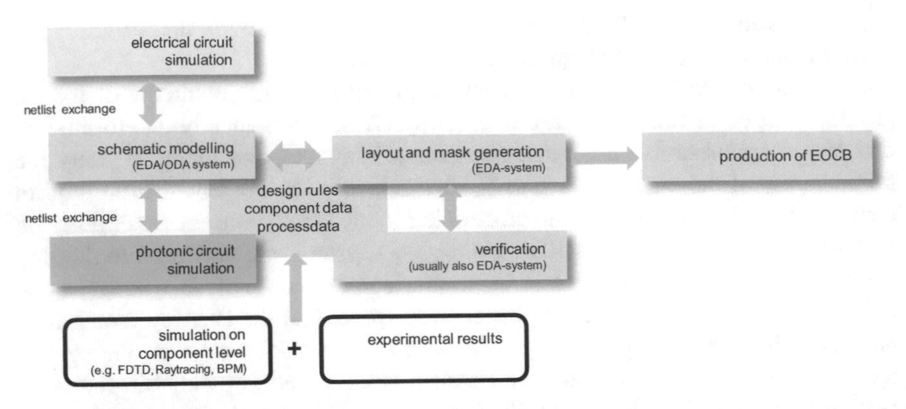

Fig. 2.5 Integrated design process of electro-optical circuit boards according to [11]

circuit board. Depending on the application, components or signal lines are replaced by optical intermediate layers or replacement components. Manufacturers of these systems generally proceed as follows (see also Fig. 2.5):

Based on electrical circuit diagrams/circuit simulations as well as photonic simulations, a schematic model is created in an EDA/ODA system, which is responsible for the logical wiring of the components. With an integrated editor, hierarchical blocks can be nested and linked together. The result is analogous to the electrical circuit, schematics, board layouts and parts lists.

Parallel simulations are performed on component level like raytracing, beam propagation or FDTD to ensure the functionality of the circuit. In a further step, an electro-optical layout is generated from the schematic modeling, similar to the electrical layout process. This is usually realized in a suitable EDA system. Design rules such as radii of curvature, distances between components, but also process or component data, play a role. Finally, the layout data is released for production.

2.2.5 Evaluation

As shown, there are product development methods that are applicable for domain-specific as well as cross-domain design. However, these are based either on sequential procedures which regard the design process either purely mechanical, electrical or software based and will be assembled in a late stage. Furthermore, these development models are intended for mechanical or mechatronic applications. Thus, the field of optics is not considered in any of these process models. General guidelines for electro-optical or optomechatronic applications are not known at this time. Nevertheless, besides general process models, there are concrete approaches for the field of electronics and electro-optics. These provide a basis for a transferable guideline for product development of optomechatronic assemblies. However, since the focus is only on electronics and optics development, and since the specific steps of digital design are not covered in the field of photonics development, a procedure for this must be figured out, which will be presented in detail later.

2.3 Need for Action for the Computer-aided Modeling of 3D-Opto-MID

Based on the technology description presented in the previous chapters, the product development methods and the basics of graphical engineering systems, a corresponding need for action for the modeling of 3D-Opto-MID components can be derived. In order to achieve this overall goal, a new methodology adapted to 3D-Opto-MID components is indispensable. This is accompanied by computer-aided modeling using a domain-spanning system. At the chair FAPS, the principle feasibility of domain-integrated systems has already been proven by various research projects. One example is the design tool MIDCAD [12]; [13]; [14]; [15] or the commercially available software NEXTRA by Mecadtron [16]; [17] for spatial electrical assemblies (3D-MID). However, these works can only serve as a basis for the design of optomechatronic assemblies, because the essential component of the optics is missing. In the following, general functional aspects as well as comprehensive aspects of a 3D optomechatronic CAD system, short OMCAD, shall be defined.

2.3.1 General Functional Requirements for a 3D Optomechatronic CAD System

This section describes general functional requirements that must be fulfilled by a 3D-Opto-MID design system.

One of the basic requirements is the realistic modeling of optical waveguides and components by means of a graphical user interface. Bypassing the limitations of specific engineering software, which is limited to either 2D or 3D assemblies, is a separate aspect that must be considered. This is especially important for the fact that such systems can usually only represent one or sometimes two engineering disciplines.

This situation is also accompanied by geometrical and topological inaccuracies, which result from frequent data transfers between the different software systems. These inaccuracies are a major problem since they affect the modeling of the optical transmission paths. The latter are important because they determine the quality of the signal to be transmitted. Examples for such inaccuracies are tiny and hardly or not visible gaps at surface boundaries, discontinuities and overlaps at the contact surfaces of solids, which are not explicitly recognized as errors by CAD systems [18]. A corresponding system must therefore be able to model circuit carriers as well as an exact description of the layout on them.

Furthermore, it is important for signal transmission to ensure a sufficiently defined passage of optical power for signal transmission. Therefore, the connection between 3D geometry design of optical assemblies and corresponding simulation methods is absolutely necessary. For this purpose, the different data representations of the geometry for simulation and design must be bridged, the logical circuit topology must be harmonized, and additional technological information (such as tolerance ranges, surface information, transition attenuation, material data) must be transmitted.

Since CAD data serves as the basis for subsequent process control, a design suitable for production is necessary. So-called design and manufacturing rules are indispensable for this. With the help of these guidelines, the manufacturability of the optomechatronic layout must be monitored automatically and continuously during the design process, and the designer must be informed immediately of any manufacturing problems.

A tool that meets all these requirements is therefore an optimal solution. However, further aspects for these requirements can be derived, which are necessary for 3D design, the integration of manufacturing engineering aspects or cross-disciplinary modeling. These will be considered in the following section.

2.3.2 General Aspects

Based on the requirements, comprehensive aspects from the areas of design, production and computer-aided systems must be taken into account. The layout design in particular is a special challenge, which is influenced by all of these areas. A layout, which is created directly in the 3D model, quickly reaches its limits with functions of established software. In particular, the high level of development of routing systems from the field of electronics, which are mainly found in the 2D area, cannot be directly transferred to a 3D problem. This is due, for example, to the complex surface structures of the circuit carriers. Thus, routing functionalities require the ability to deal with bodies of higher topological gender.

In the following, these aspects are explained in more detail. They are divided into the categories design, manufacturing and software systems.

2.3.2.1 Aspects of Design

Difficulties in the design arise with mechatronic products particularly with very small installation space such as digital cameras or mobile telephones, since the different assigned domains affect themselves mutually. By the use of optical components, it is to be expected that this problem intensifies additionally. Aspects for design questions can be roughly divided into restrictions regarding electrical/electro-optical circuit diagrams as well as mechanical design restrictions.

Mechanical design constraints mainly affect the circuitry of the 3D-Opto-MID. Thermal or mechanical stresses or strains can occur at certain points in different phases of its life cycle. Furthermore, holding and mounting surfaces have to be considered, which are called keep-out areas. These are areas where no further functional elements may be placed.

In general, multilayer layouts of complex 3D-Opto-MIDs are not yet possible for electrical/electro-optical design. Therefore, certain automated routing strategies (e.g., multilayer escape routing of ball grid arrays (BGAs)) are not applicable for the creation of the conductor layout, whether electrical or optical. Furthermore, through hole devices (THD) cannot be used. Also the placement of components is generally not possible on all surfaces, e.g., due to discontinuity or strong surface curvatures.

2.3.2.2 Aspects Resulting from Manufacturing

In order to ensure production-ready models, relevant aspects must already be considered during the modeling process. For the further processing of design data (post-processing), for example to create printing programs for aerosol jet printing of 3D-Opto-MID, it is necessary to rely on established and readable data formats.

In post-processing, it is usually absolutely necessary to support standardized CAD formats such as STEP [19] or IGES [20]. In addition, descriptions of surfaces, placed components and conductors must be exactly defined. This is also necessary to obtain exact print paths that guarantee the functionality of the finished component. In the production of waveguides, a wide variety of materials and tools can be used, which can influence the shape of the finished product and must also be defined in the software system. One of the resulting characteristics, for example, is undercuts that cannot be achieved by the machine kinematics.

2.3.2.3 Aspects of Software Systems

In the field of software systems, especially 3D CAD systems, aspects to be considered can be divided into geometrical and algorithms aspects.

For geometrical aspects, complex surfaces represent a challenge for the validation of optical components. Layouts that are planned on 3D bodies must have a precise analytical description. This must be taken into account so that they can be optically simulated and thus be validated. Especially for freeform surfaces, this aspect plays a role, since many CAD systems only approximate derived geometries.

Algorithmic aspects are especially important for the layout design. For example, algorithms that work on complex B-Rep surfaces are more computationally intensive than on a planar counterpart. For algorithms to be implemented, this means high implementation efforts, since a wide range of different surface types and volumes must be considered. Furthermore, sufficient computing power must be available, since the data structures of 3D bodies are many times more complicated and, in addition, 2D routing algorithms in their current form can only be parallelized to a limited extent.

2.3.3 Summary

In summary, requirements regarding the software system and aspects to be considered for integration can be divided. The essential requirements are listed in Table 2.1 below and describe the basic functionalities that such a system must fulfill.

1. Graphical user interface and visualization
2. Possibility of designing a circuit carrier and validating it with CAE software with regard to mechanical and thermal properties
3. Interfaces for the logical design and simulation of electronic and photonic circuits
4. Layout functionality for component placement and ladder routing
5. Analysis of passive components (e.g., waveguides) of the final circuit layout
6. Ability to perform design rule checks to ensure circuit integrity and function
7. Consideration of manufacturing engineering aspects in the early design process
8. Extraction of production drawings and NC data for various manufacturing processes.

Table 2.1 Functional requirements for a 3D-Opto-MID application

	aspects to be considered	
	electrical/electro-optical	mechanical
Design	▪ Netlist complexity is limited because 3D opto-MIDs are "single layer". ▪ Components must be placeable on spatially complex surfaces.	▪ Mechanical/thermal stress at specific locations on the product must be considered. ▪ Contact surfaces for handling and assembly must be taken into account.
	Post-processing	Manufacturing
Production	▪ Production Standardized formats (such as: STEP, IGES, STL, SAT) should be supported. ▪ Descriptions of curves and surfaces must be accurate. ▪ Geometry must be adapted to the requirements of post-processing tools	▪ Different manufacturing processes require different materials and have specific constraints on layout design. ▪ Constraints on wire density, wire thickness and spacing. ▪ Geometric constraints of the 3D circuit carrier (e.g. undercuts) due to machine kinematics.
	Geometrics	Algorithms
Softwaresystems	▪ Software systems algorithms must be adapted to handle a wide range of B-Rep volumes with many different surface types. ▪ Geometry derived from complex surfaces results in approximated geometry. ▪ Some surfaces in CAD systems can lead to problems with too large tolerances.	▪ Algorithms running on B-Rep surfaces are partially computationally intensive. ▪ Increasingly complex data structures required to process B-Rep shapes. ▪ Only limited parallelization possibilities with routing algorithms possible.

However, the biggest challenge is to create a robust software-based basis. This must be able to combine the different disciplines of optics, electronics and mechanics in a common design/modeling environment.

2.4 Concept of a Cross-domain Methodology for Optomechatronic Components

The relevant specific design processes presented previously represent different sub-domains of 3D-Opto-MID design. Since they mainly refer to planar assemblies and do not take into account the conditions of the spatial shape or do not integrate the domain of optics, none of the methods can be applied comprehensively.

For this reason, a procedure is necessary which sufficiently covers the domains of electronics, mechanics and optics. Figure 2.6 shows an example of a

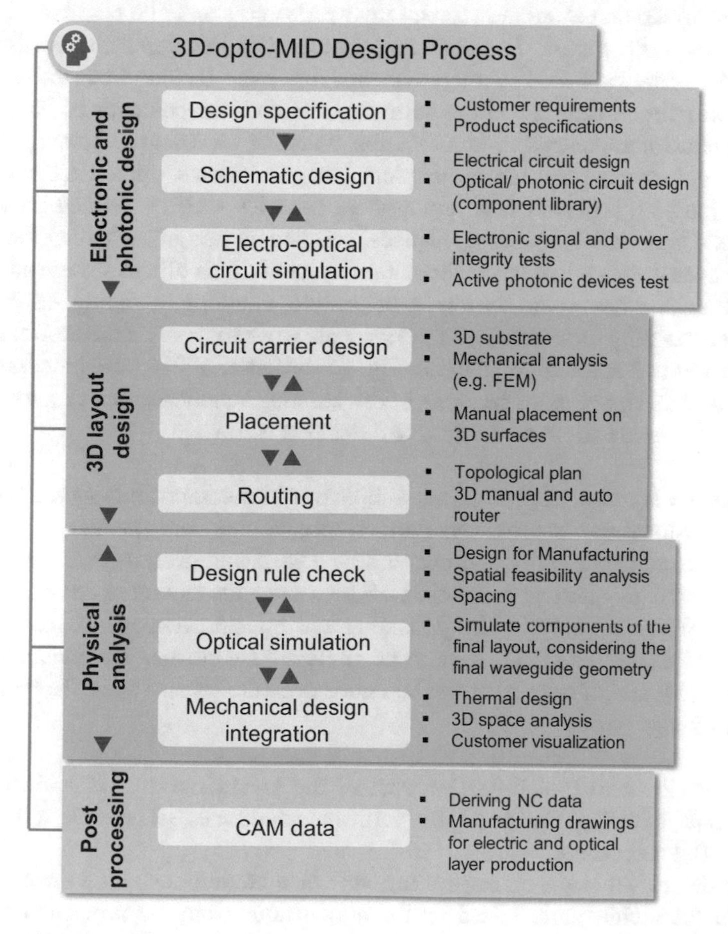

Fig. 2.6 Design procedure for integrated optomechatronic components based on [21] and extended in [22]; [23]

3D-Opto-MID product development process, which is based on [21] and has been further developed. In the example shown, this process concentrates primarily on the individual stages of computer-aided product development and refers to the requirements from the previous chapter. Analogous to electronic assemblies, electronics or optics design first takes place, which consists of the steps design specification, schematic design and electro-optical circuit simulation. These steps mainly comprise upstream planning processes. This includes the definition of the requirements or the product, the creation of the electronic or photonic circuit diagram as well as initial integrity tests to ensure the function. The main part of the 3D layout design and physical analysis includes the steps of carrier design, spatial placement of components and routing of waveguides/traces. In the physical analysis section, design rule checks are applied to ensure that the design is suitable for production, and optical simulation methods are used to ensure the quality of the waveguides. In the final post-processing stage, production drawings and associated CAM data are derived. Iteration loops between all involved steps should be possible.

A prerequisite for the implementation of such a development process is an appropriate computer-aided design system. This must include both the electrical and optical circuit design and the ability to simulate the behavior of passive and active optical components in combination with the electrical circuit. Active and passive components (e.g., lasers, couplers, ring modulators, photodetectors, waveguides) must be provided from component libraries with predefined parameters (geometry, material composition, physics, component design), whereby these have already been validated in other simulation systems. This allows designers to perform behavioral tests (e.g., for phase, dispersion, coupling, damping, etc.) for circuits incorporating these compact models. This work focuses on the sections of the presented flow that are highlighted in Fig. 2.6 because, unlike post-processing and logical design, they cannot be solved with existing functionalities of known software tools. These two process steps are described below:

- In the 3D layout design step, the ability to place components as well as routing functionalities is provided. Furthermore, design rule checks (DRCs) must ensure manufacturability and operability. The layout design has a direct influence on the geometrical parameters of the waveguide and other components.
- In the physical analysis, the parameterized optical components (e.g., curved waveguides) of the circuit have to be analyzed again (e.g., for attenuation and dispersion) and, if necessary, the layout or even the schematic has to be adapted accordingly.

The latter step creates a full 3D model of the layout during mechanical design integration, allowing designers to perform additional 3D spatial analysis or improve thermal design.

In order to set up a corresponding software environment, it is necessary to consider the architecture. Based on the conclusions from the previous chapter, a hybrid solution consisting of a plug-in system and an interface system is envisaged. The associated architecture is now presented conceptually in a layer model.

2.4.1 Considerations for the Design and Workflow of a Computer-aided 3D-Opto-MID System

For the basic structure, it makes sense to use a layer model due to the different tasks of the functions (see Fig. 2.7).

For the system to be implemented, the three layers can be broken down as follows:

- Design layer—responsible for basic design questions that need to be answered.
- Product model layer—necessary to exchange, collect and process data between different systems.
- Simulation layer—answers questions that cannot be solved in the design layer, in this case, the optical validation.

In the *design layer*, an MCAD system forms the core and provides a user interface, whereby the commands required for the design of the optomechatronic assemblies are assigned to a separate application environment. Theoretically, complex optical or even electrical circuits can be imported from existing OCAD/ODA or ECAD/EDA. It is intended to perform layout design and/or layout synthesis in the MCAD system. The commands for designing the circuit, routing the waveguides and placing optical components, are integrated via a plug-in solution. So is the design of optical foil substrates and the checking of design rules. The circuit carrier design is not explicitly listed, since it can be created with any CAD modeling tool and does not need to be implemented.

The simulation layer has the task of validating optical components. This requires corresponding descriptions that originally come from the MCAD system. These are, for example, surface descriptions, paths, light sources or materials

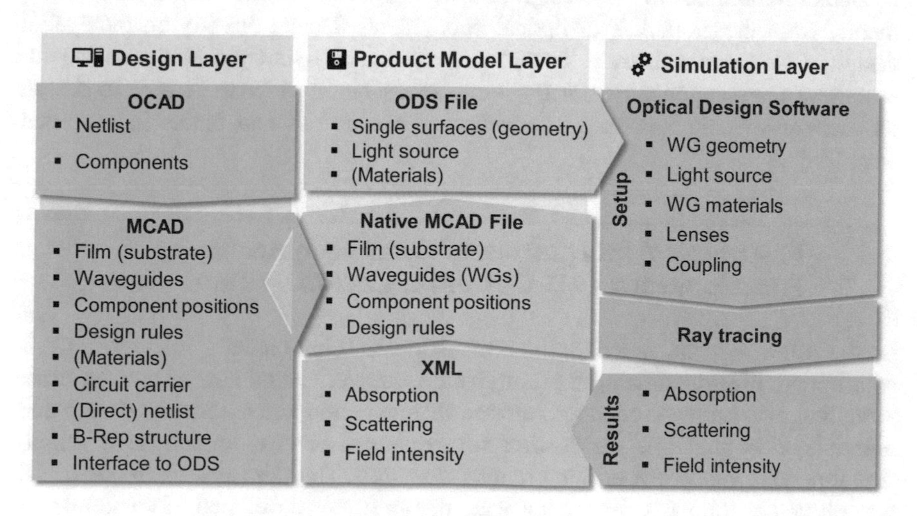

Fig. 2.7 Design, product model and simulation layer

of the optical conductors. Simulation algorithms can be started by configuring the optical simulation software accordingly. In the example used later, this refers mainly to a non-sequential ray tracing method. As a result, attenuation, scattering or intensity distribution can be transferred back to the design layer via the product model layer.

Visual feedback mechanisms or integrated design rules for the modeling process are also provided. In contrast to rules related to purely optical and material properties, the mostly geometric design rules result from the mechanical design. The interface to the optical simulation system reads the B-Rep structure of the waveguides and converts it into a data format readable by the optical simulation system. The visualization of the optical design rule checks (as a result of the optical simulation) is done in the MCAD system.

2.4.2 Integration of Optical Functionalities in a 3D-OMCAD System

In the following, the functionalities of an optomechatronically integrated CAD solution are described using a prototypical example—a OMCAD system (optomechatronical CAD system). Circuit carrier and substrate modeling, waveguide modeling, integration of components as well as automated design and manufacturing guideline verification are explained according to the procedure described in Sect. 2.5. Finally, implemented validation functions for optical testing are also explained. For the software demonstrator developed in the context of this work, the CAx system Siemens NX was selected as basis. As already explained, computer-aided design, also called computer-aided design (CAD), is used to capture and process geometric elements to create a three-dimensional model [24]. It is of particular importance for the design of spatial assemblies with electro-optical elements, since the properties of optical components depend directly on the spatial design of the circuit substrate. Thus, for the design of a 3D-Opto-MID, the consideration of the component geometry is indispensable in order to be able to design the domain-specific functions of mechanics, electronics and optics in a unified way. [25]

2.5 Prototypical Integration of Optomechatronic Functions into a 3D-Opto-MID Design System

In the following, the functionalities of an optomechatronically integrated CAD solution are illustrated using a prototypical example. Circuit carrier and substrate modeling, waveguide modeling, component integration and automated design and manufacturing guideline verification are explained. Finally, implemented validation functions for optical testing are also described. The CAx system Siemens NX was chosen as the basis for the software demonstrator developed in the course of this work. As previously stated, computer-aided design (CAD) is used to capture and process geometric elements to create a three-dimensional model [24]. It is of particular importance for the design of spatial assemblies with electro-optical elements, since the properties of optical components depend directly on the spatial design of the circuit carrier. For the design of a 3D-Opto-MID, the consideration of the component geometry is thus indispensable in order to be able to design the domain-specific functions of mechanics, electronics and optics in a unified manner.

If we look at the structures of the application from a programming perspective, it can be broken down into a three-level model. The three main elements—user interface, program logic and data types (see Fig. 2.8)—are interdependent. The user interface forms the interface to the operator (via menus, slides or direct commands). The program logic implements both the data handling and the algorithms, which in turn generate corresponding data types. The program logic thus forms the core of the application and is implemented largely isolated from the user interface according to the concept described in Sect. 2.4.

2.5.1 Circuit Carrier and Substrate Modeling

A crucial aspect in the modeling of printed optical waveguides is the selection of the materials enclosing the actual waveguide. A special feature of printed waveguides compared to POFs is that the enclosing materials consist of different materials due to the manufacturing process and the special geometric characteristics. The printed waveguide itself is located on a substrate material that is adapted to

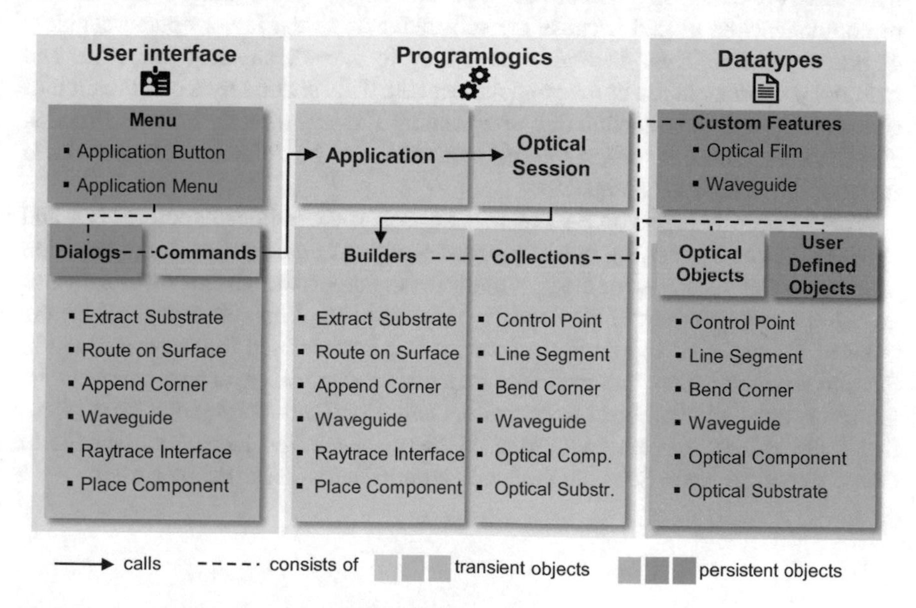

Fig. 2.8 Architecture of a 3D-Opto-MID design system

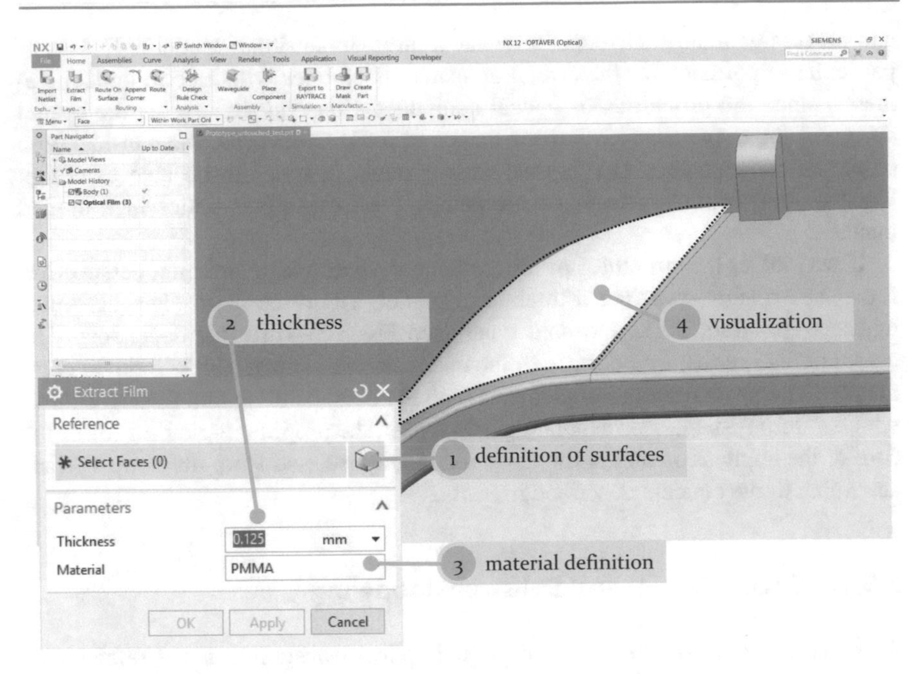

Fig. 2.9 Example for the definition of foil substrates

the spatial geometric characteristics of the circuit carriers, for example by thermoforming. In a further step, the optical conductors for 3D-Opto-MIDs are printed onto these optical layers. Without the film substrate, no waveguide can be created, or components be placed because the substrate acts as the lower optical cladding of the waveguide. Thus, it is also mandatory to deposit physical properties and material properties in the corresponding element. This is done by a combination of a user dialog and the selection of corresponding surfaces directly in the 3D model. The thickness and material of the foil substrates can be defined via a drop-down menu or direct input (see Fig. 2.9).

The function is based on the inherent offset surface builder feature of NX and creates surfaces by copying existing surfaces with a certain offset in the direction of their normal vector. After the user confirms the input, the surface feature creates an offset (corresponding to the defined thickness of the film substrate), adjusts the color of the surfaces, and finally converts it to a user-defined film feature listed in the part navigator. This automatically generates an optical layer that connects the feature to a native NX object to control its behavior. Only contiguous faces (faces that share at least one edge with one of the other selected faces) can be used to create an optical layer. Separate faces are automatically discarded by a filter.

2.5.2 Synthesis of Circuits

Another essential task to fulfill the tasks defined in Sect. 2.3 is the so-called circuit synthesis. In order to complete the optical circuit, components have to be placed. For the design, this means that components are placed either on defined connection points or at arbitrary positions on the circuit carrier.

However, this is preceded by a number of steps. Appropriate components must be integrated into the modeling system, and co-design aspects must be considered for the design of 3D optomechatronic components (3D visualization, netlists in 3D environment). The implementation of an interface between the different design systems is an integral part to link the (opto-)electronic circuit design phase with the mechanical and physical design phase (see Fig. 2.6 in Sect. 2.4). Therefore, the question of how interfaces between the systems can be realized within the framework of an OMCAD system is addressed first. To answer this question, the EDA system AUTODESK EAGLE [26] was used to generate schematic data (netlists) in .sch file format which was then transferred to the 3D-Opto-MID application. In general, regardless of the choice of ODA/EDA system, such netlists are created, which are stored in various manufacturer-specific formats. Although these differ in syntax depending on the system used, they are essentially similar in structure.

Each schematic file contains a structure that can be subdivided hierarchically according to the elements used. Such a structure will now be explained using the .sch format as an example.

As shown in Fig. 2.10, the .sch format contains, in addition to the library element (library), primarily three decisive main elements which represent different object groups: package, symbol and devicesets. The package element defines the geometric shape of the footprint as well as the contours of the component and the number of contacts. Thus, for electronic or optical components, footprints are defined as either THDs or surface mounted devices (SMDs). For SMDs, the size of the footprints is stored in x- and y-directions (represented by the SMD element for each contact), while for THDs, the hole depth is stored (represented by the pad element for each contact). The symbol elements (symbol) contain the schematic representation of the component, i.e., essentially the outlines/demarcations shown in the Schematic Editor. Furthermore, the pin elements are also included, which are uniquely defined by their names. The deviceset elements are used to define multiple components within a device (one symbol can represent multiple packages). The device element describes such a relationship. The part element represents a specific symbol-package combination, which basically represents a device in a deviceset.

The class element specifies the width of the conductors, the safety distance and the diameter for vias. Finally, the net element describes the relationships between part instances elements in the sheet element. Instance references to a part are defined in the part list. Each net element describes two connected parts, the connected connectors and the geometric segments that connect them.

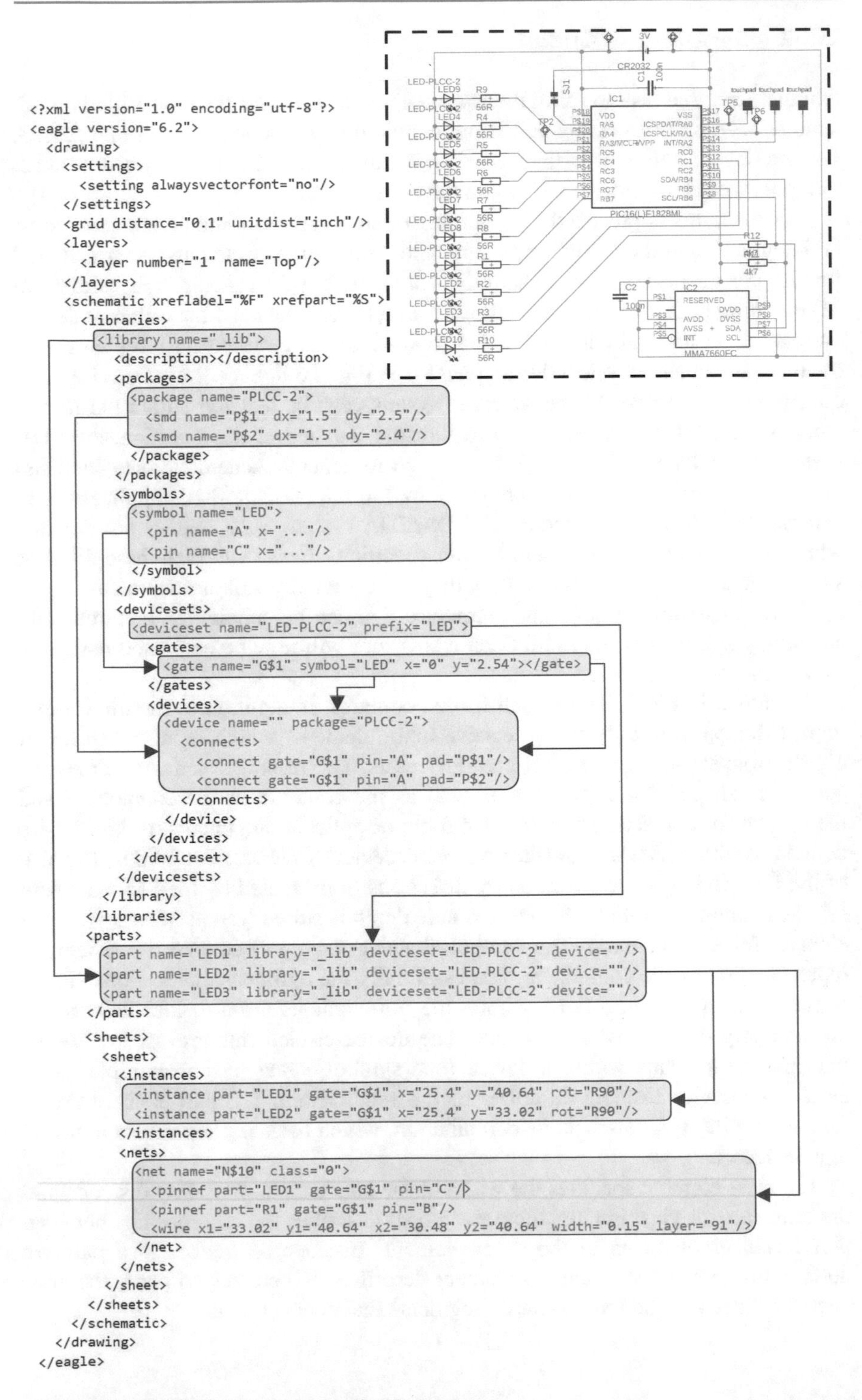

Fig. 2.10 Example for the structure of a schematic

Usually, the file format of an EDA system contains all important information about the components and nets of the assembly. However, components do not have 3D models by default since most EDA systems do not support display of 3D layouts. However, in an integrated CAD system for 3D-Opto-MIDs, the existence of the 3D model is mandatory for several reasons:

1. Due to the spatial nature of 3D-Opto-MIDs and the resulting shape of the substrates, spatial collision analysis with populated components is essential.
2. A 3D layout must be available to be able to reproduce manufacturing and assembly processes by means of CAM.
3. Exact mechanical and thermal analyses are only possible with exactly defined component sizes and materials.

This raises the question of integrating 3D models and their functionally relevant information in the form of component libraries. The task of a special 3D-Opto-MID component library is to provide netlist components for the layout phase including their 3D models. In general, 3D components can be generated in several ways:

1. Manual creation of solid models from component data sheets using a component generation tool or basic geometric modeling tools
2. Automated creation of solid models by a shape generator from the component information obtained during the import process (allows creation of standardized packages only)
3. Automated matching with open online libraries to download the 3D model and link it to the imported package (Fig. 2.11).

A third and final approach involves matching the imported components with an online library. For example, OCTOPART [27] can be included, a common parts library that can be accessed via an API to retrieve and download 3D models from a database.

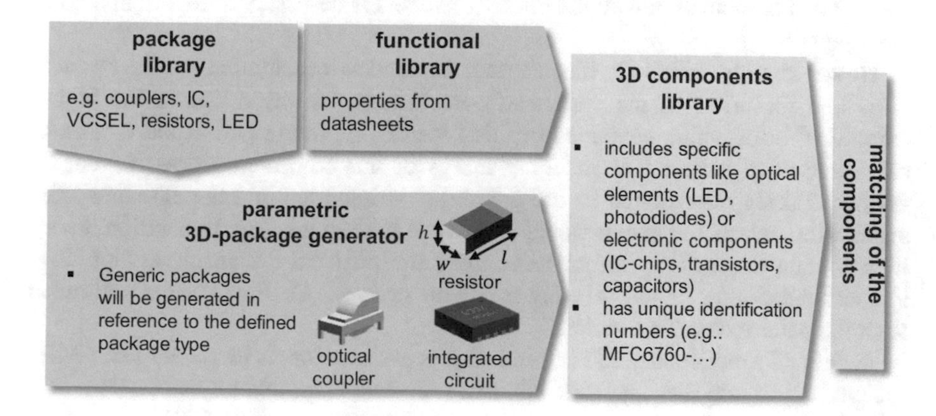

Fig. 2.11 Concept for a component library

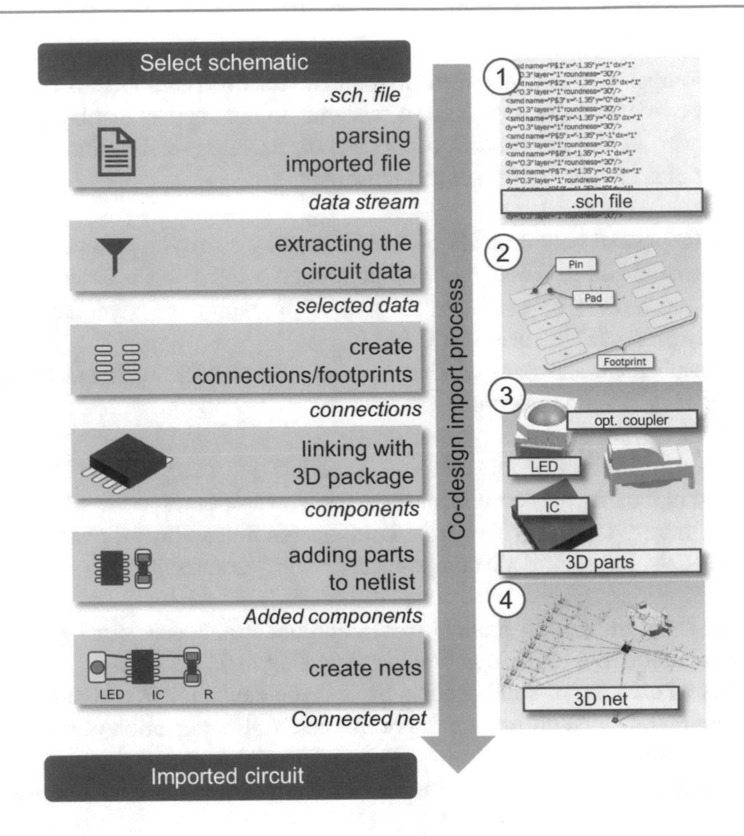

Fig. 2.12 Import of schematic in 3D

However, a component library alone is not sufficient to perform circuit synthesis. The information used to interconnect the individual components is also of decisive importance. For this reason, an import of the schematic file shown in Fig. 2.10 in combination with the library elements is necessary to represent a complete layout.

However, only schematic files contain detailed information about the connections and footprints of the individual components. However, due to the higher amount of information contained in this type of file, the import is more complicated and takes longer. The main advantage of this co-design process is that no 2D board design is required before the netlist is imported into the 3D-Opto-MID application, where the layout would have been revised anyway. In addition, board files contain unnecessary information about the physical design phase that is no longer needed after import. The entire import process (see Fig. 2.12) is optimized to fit the pattern described earlier.

A defined functionality can be used to process user inputs in the 3D-Opto-MID system and to add the import of other file formats from other systems. The data selection is done via a dialog box netlist import according to Fig. 2.13.

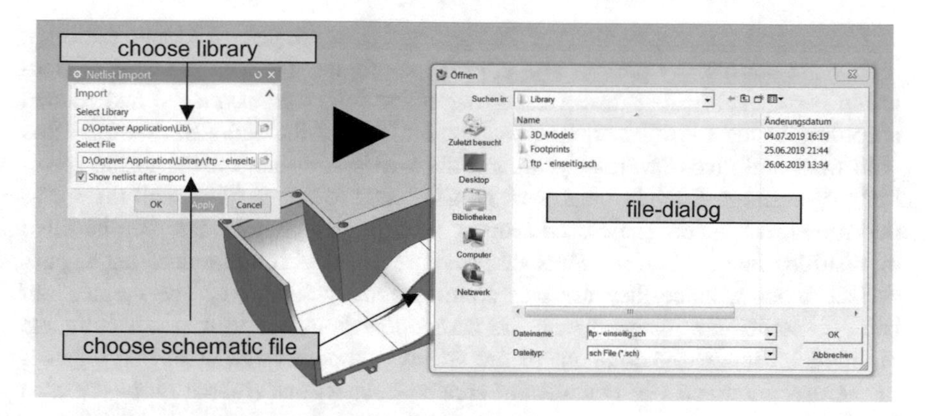

Fig. 2.13 File dialog for importing netlists

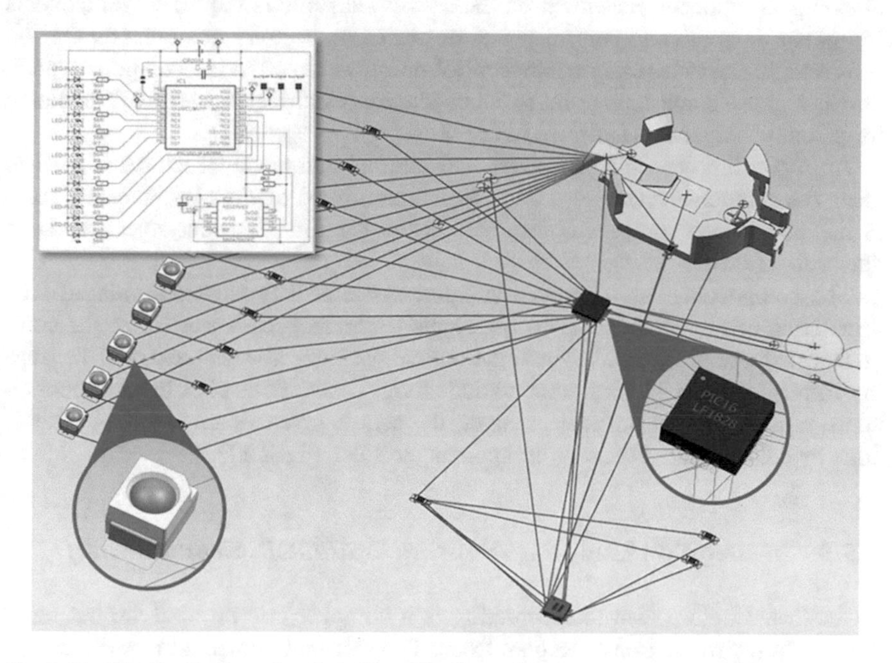

Fig. 2.14 3D visualization of netlist with rubberlines

The links between the elements (see Fig. 2.14) are modeled by a class and reference structure that requires an iterative search through already existing elements. Since only a fraction of the information in the file is used, not all attributes are analyzed. After the schematic file has been analyzed, the relevant data for the construction of the connections and networks is extracted from the imported file. Detailed information about schematic symbols mainly affects the circuit design and can be ignored. Additionally, some character values have to be converted into numeric types, such as roundness factors or angles in general.

In a second step, connections of the imported parts are generated automatically, if they are not already present in a physical subfolder. Thus, in the case of electronic parts, each pad is drawn using the shape information, placed and rotated according to the specified orientation. The outlines of the pad shapes are created with lines and arcs, first rotated around the z-axis and then moved to the specified x-y position. Straight lines must be shortened to create filets with the specified roundness factor. The closed loop of a single pad contour can then be filled in, resulting in a flat surface. Pads are not represented as solid models, but as pure surface models, since they are less computationally intensive. The connection/pad geometries are matched with the component housings to create a complete 3D model. Finally, the components are added to a netlist and displayed together in the user interface via dynamically created connections (rubberlines—see also Fig. 2.14).

As a final step in circuit synthesis and a transition to optical and electrical path planning, component placement on the circuit substrate is required. Components can either be applied to control points or placed at arbitrary positions on the foil substrate. In the latter case, a new control point is created at the component position as a future connection point to the conductor (electrical or optical). The dialog for placing components is shown in Fig. 2.15.

An algorithm consisting of active and internal command structures is used to place the components (see Fig. 2.16). When confirmed by the user, transient commands and objects (preview objects) are converted into persistent objects and can thus also be saved.

The components can be moved interactively after they have been placed. The orientation of the component always adapts to the respective normal of the component surfaces. Information about the logical wiring is always preserved, because the corresponding rubberlines are carried along. To avoid the placement of components, e.g., on curved surfaces or edges, the user is given visual feedback. This is done by coloring the respective component (see also Fig. 2.17).

2.5.3 Waveguide Modeling, Routing Optimization and Wiring

As indicated before, waveguide routing is a complex process, and various solutions exist for routing on B-Rep surfaces. The fact that numerical algorithms can be run on the polyhedra of a B-Rep model is crucial for the flow of 3D routing. The number of general routing parameters to be observed increases significantly. These are determined by the design, manufacturing and modeling constraints as well as the selected routing strategy or algorithm. The typification of the used surfaces is important here, especially because they have different internal representations. The used CAD system NX distinguishes between twelve types of solid surfaces, e.g., planar, cylindrical, conical, spherical and parametric surfaces. Of particular interest are surfaces that have boundary edges described as B-spline curves. However, the geometric properties of such surfaces can differ considerably.

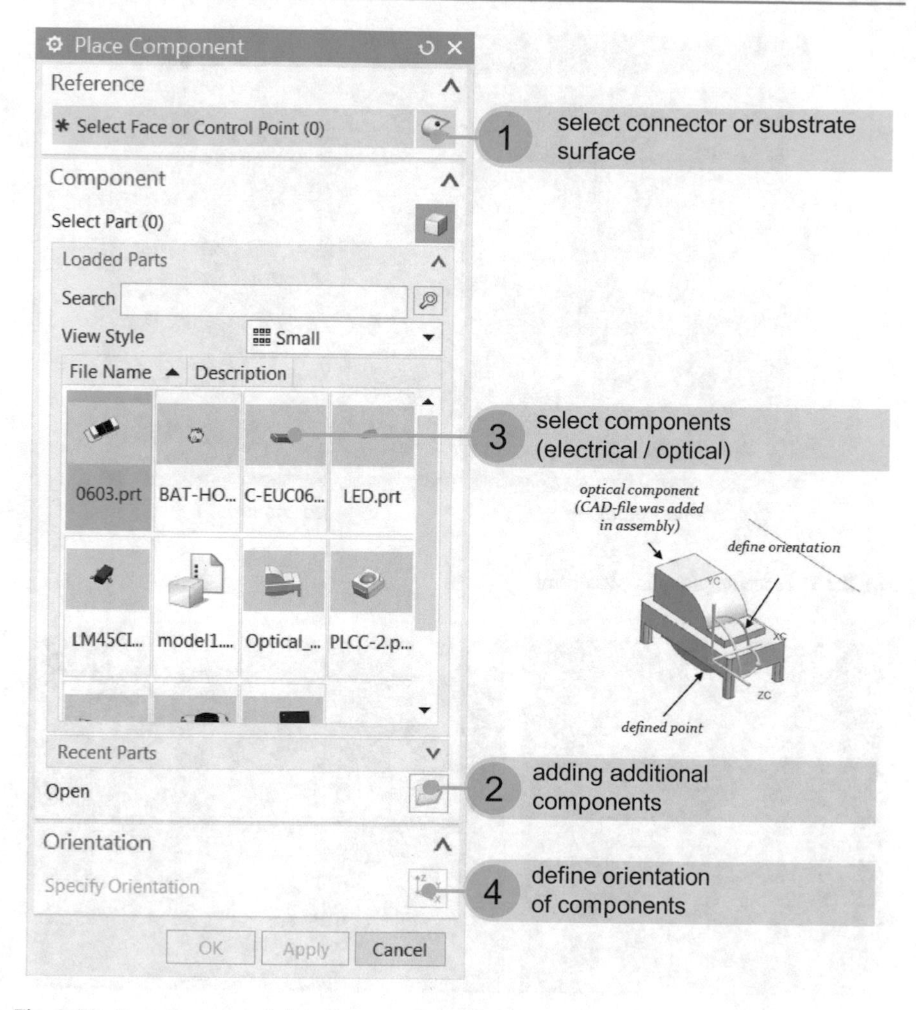

Fig. 2.15 Part placement dialog

This has special importance for the path planning of the ladder. Thus, individual routing algorithms have different strengths and weaknesses and often solve a routing problem only partially or in a specific context. Unlike conventional PCBs, most surfaces of 3D-MIDs are not rectangular in shape and must not have sharp edges in 3D-Opto-MID due to the transmission capability of optical signals. Therefore, most algorithms cannot be applied directly to a B-Rep surface and must be adapted to the respective surface contours. A surface enclosed by several different edge types, as well as oblique B-spline surfaces (surfaces with non-orthogonal U-V coordinates), makes it difficult to meet geometric boundary constraints. In addition, many surfaces have not only boundaries on the outside, but also one or more internal boundaries. This can result in gaps and holes. Figure 2.18 shows the surface types mentioned using MID design prototypes as an example.

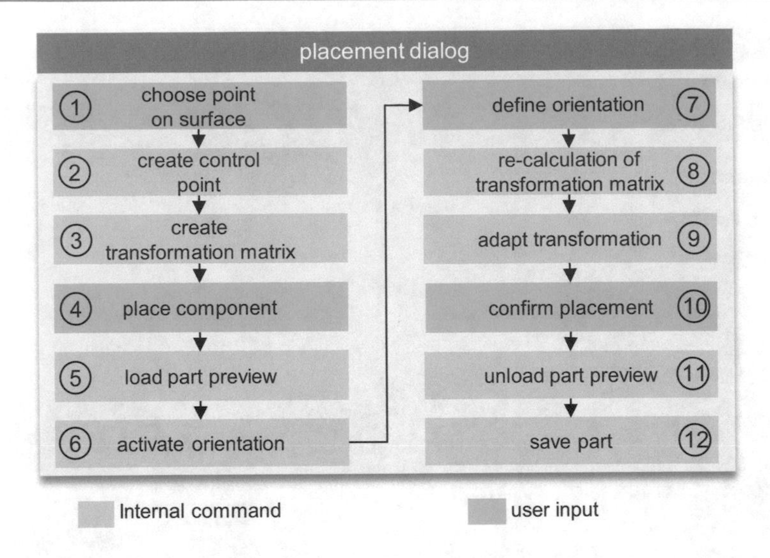

Fig. 2.16 Procedure for part placement

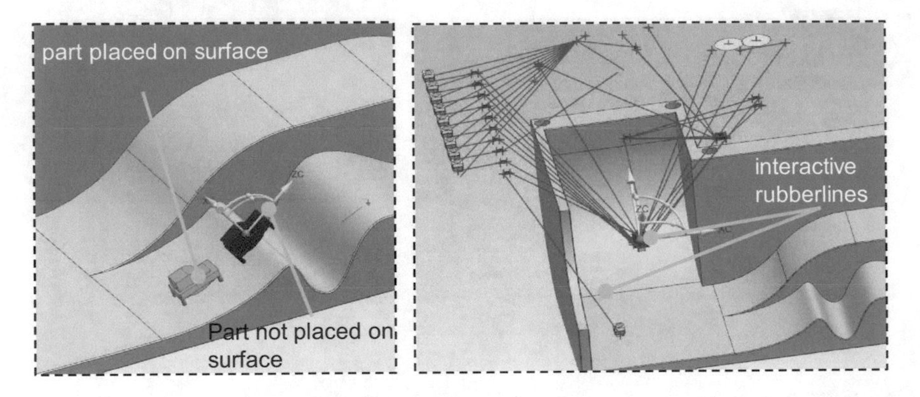

Fig. 2.17 Design rule checks for component placement

However, before an algorithm can be applied for such a purpose, the 3D routing problem for 3D-Opto-MIDs must be fully described due to the large number of input parameters. Questions that may arise in this framework are as follows:

1. To what extent is automated routing implementable?
2. Are obstacles present and how are they defined (e.g., type, shape, etc.)?
3. Which routing directions are possible?
4. Are rip-up, rerouting and shove-aside techniques required?
5. Are constraints imposed by the manufacturing process taken into account (manufacturability)?

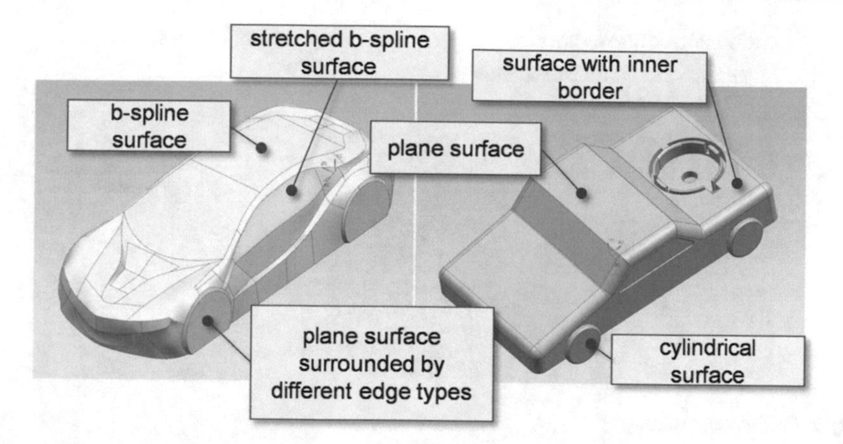

Fig. 2.18 Different surface types in 3D

The answer to these questions is highly dependent on the complexity of the planned layouts. Questions 2, 4 and 5 generally apply to both manual and automated routing. These can be handled comparatively easily by manual routing. However, for higher density layouts, the use of a (partially) automated procedure can be useful, which is why the basic question of an automated routing procedure as well as possible routing directions has to be answered.

In order to cover these questions, manual as well as partially automated approaches were pursued within the scope of the work. For the semi-automated routing a version of the Hadlock algorithm is used, which uses an A* heuristic adapted for 3D surfaces by a special gridding procedure. For manual routing, a specially developed algorithm for manual path planning will be used. First, the implementation in the 3D-Opto-MID system by means of manual path routing is explained.

2.5.3.1 Manual Routing

The planning of a manually created path is started via the route on surface dialog box (see Fig. 2.19). This consists of an object selection block, via which the necessary design operations can be carried out.

An internal selection filter restricts the selection to the optically defined surfaces (optical substrates). The selected surface is passed along with the selected point to a route on surface function, which calculates a preview of the segment between the current and the previous control point. If the user selects a surface (and no existing control point), the selection is overwritten by the current preview control point. In addition, a temporary marker (the orange dot in Fig. 2.19) is displayed at the selected position, marking the current endpoint of the segment path.

In addition to path planning, bends play a special role for optical waveguides. These are necessary so that light can be guided sensibly in the optical waveguide. Sharp edges pose a problem because the optical signal cannot be redirected in a

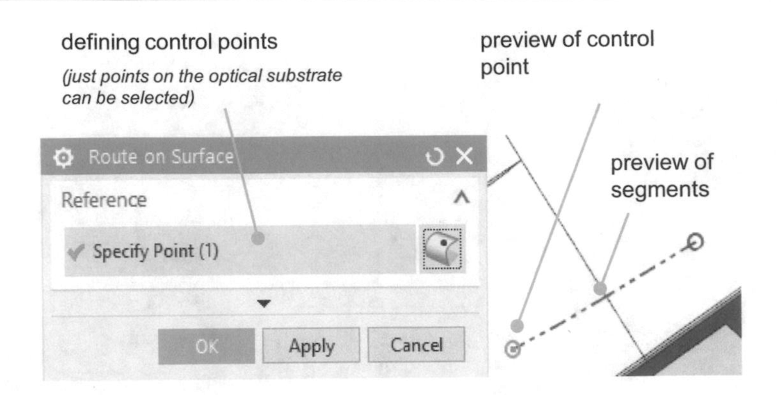

Fig. 2.19 Manual routing

meaningful way here. To ensure a sensible and uninterrupted process, a corresponding function is therefore implemented. In the implementation itself, the process looks as follows: First, the control point P with the sharp edge is selected. For this point, a bend (radius of curvature is defined in a dialog box), an arc segment and the position of two new control points P_1 and P_2 are calculated (see Fig. 2.20).

Figure 2.21 shows a corresponding path planning on an optical film substrate and placed components.

2.5.3.2 Automatic Routing

Another approach is partially automated routing. The overall routing process follows the auto-interactive approach of many PCB design tools. For semi-automated routing, the designer must be able to specify the desired source and destination points in order to lay a path between them. From the destination point, the routing can be continued in any direction by selecting another point.

The first approach is a so-called raster routing technique. The basic idea behind the presented raster routing technique is to reduce the 3D problem to a 2D problem. However, this is done by generating a grid over the 3D surface. This is a straightforward approach that has the advantage that once a grid has been applied to the surface, conventional routing algorithms can be used to search for paths. The grid is dynamically generated and adapts to different surface types. It is derived from local surface features and expands in both x- and y-directions. The position of the roughly equal grid points is approximated during the expansion of the grid. The finer the grid size, the better the approximation of the surface and the higher the quality of the routing path.

The technique does not aim to be applicable to all surfaces of a B-rep model. It includes most common problems, such as a series of connected continuous surfaces or complex surface structures like NURBS. A variety of raster-based algorithms can be used to find a path on the generated raster. Various methods can be used to generate the grid, such as the boundary element method (BEM) or the finite element method (FEM). In the following, however, a specially developed approach is presented, which is based on the existing B-Rep structure of the 3D

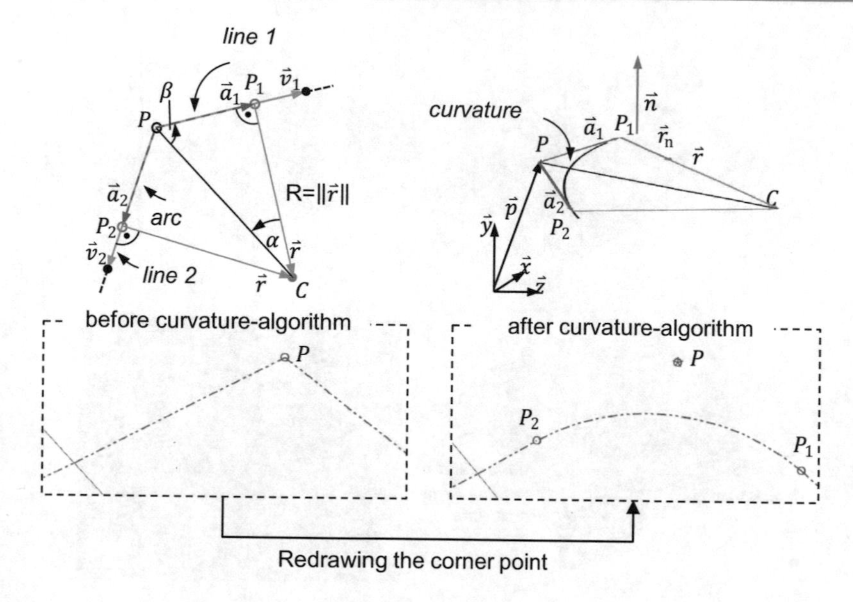

Fig. 2.20 Creation of bends

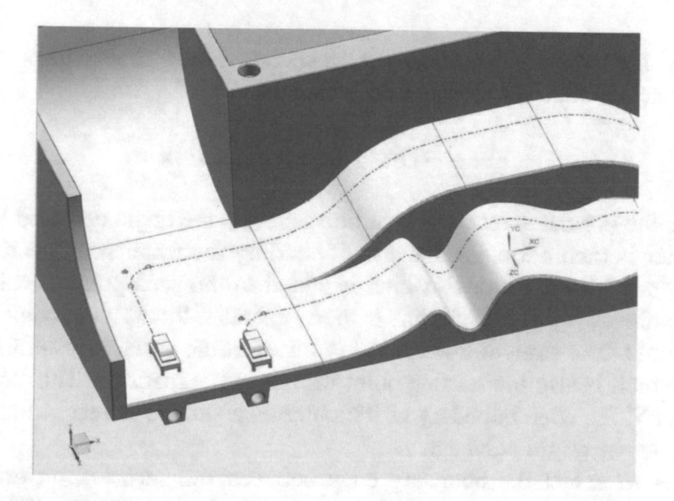

Fig. 2.21 Manual layout with curve segments

model and uses its advantages by reverting to the normal vectors of the defined surfaces. In this way, the original structure of the 3D model can be preserved and no conversions are necessary.

The following descriptions refer to the processes shown in Fig. 2.22. In the first step, a "seed point" and an expansion direction are defined by the user (1). The seed point represents the starting point for creating the grid. To find the next point in the grid, the grid point G, the direction vector $\vec{d}$ is projected into the plane

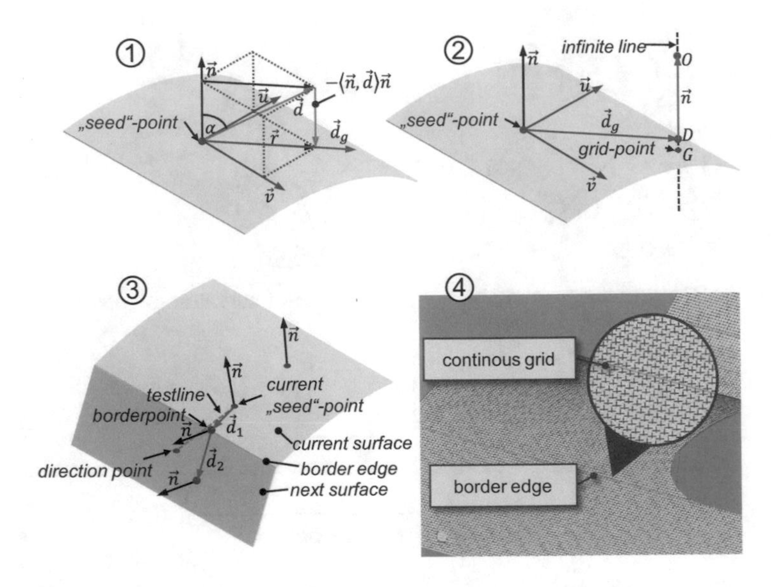

Fig. 2.22 Principle of the grid expansion algorithm

spanned by the local u-v coordinates at the specified seed point. The portion of the direction vector $\vec{r}$ that lies in this plane is calculated by

$$\vec{r} = \vec{d} - \cos(\alpha)\vec{n} = \vec{d} - \left\langle \vec{n}, \vec{d} \right\rangle \vec{n}$$

where $\vec{n}$ is the normal vector of the surface and α is the angle enclosed by $\vec{n}$ and $\vec{d}$. The vector is then normalized and multiplied by the raster size specified by the user. This provides a vector $\vec{d}_g$ which is added to the seed point to calculate the direction point D (2). A straight line is then generated through the point D and the offset point O. The straight line is used to intersect the surface, resulting in a grid point G, which is also the starting point for the next expansion. This procedure is repeated until the edge boundary of the outermost surface is reached or the algorithm reaches its origin point again.

In order to detect the boundary edge between the surfaces, an oriented test line is created between the current "seed point" and the evaluated direction point (3). The distance between this line and all edges of the current surface is evaluated. The boundary edge is the one closest to this line. The distance evaluation also provides the closest point on the edge, which is the closest "seed point", but which is not a grid point because it is not on the regular grid. The partial distance $|\vec{d}_1|$ is calculated, and in the next iteration, this value is subtracted from the grid size to obtain the partial distance $|\vec{d}_2|$. In this way, the grid remains evenly distributed for continuous surface transitions. If the transition between two surfaces is not continuous, the direction vector must be adjusted with respect to the local surface properties. The direction vector is rotated around the axis given by the vector product of the local normal vectors of the neighboring surfaces at the edge point. The result of this procedure is shown in (4).

The following applies to the method: The smaller the grid size, the lower the probability of local errors is. Although this method is particularly suitable for plane surfaces due to its high accuracy, it can also be used for 3D surfaces. The generation of the grids is dynamic: after an expansion in the main direction, the expansion function is recursively called to also expand the grid diagram in the opposite direction to the main direction. During the expansion process, memory is dynamically allocated so that any irregular surface shape fits into a rectangular grid.

By using the grid algorithm, the 3D problem is reduced to a 2D problem and maze routing algorithms can be used to find paths. In the example, Hadlock was implemented. Compared to other maze routing algorithms, the algorithm uses an A* heuristic to reduce time complexity and is particularly efficient when there is a low density of obstacles. By marking the created grid cells with the respective detour number, a shortest path can be found by backward search [28].

The optimal path depends on the desired layout and is not necessarily the shortest path. This can be explained by the fact that the implemented backtrace algorithm tries to minimize directional changes. Figure 2.23 shows one result of the algorithm.

The routing follows an auto-interactive approach. The change between red and yellow lines represents the calculated route section between the selected routing points.

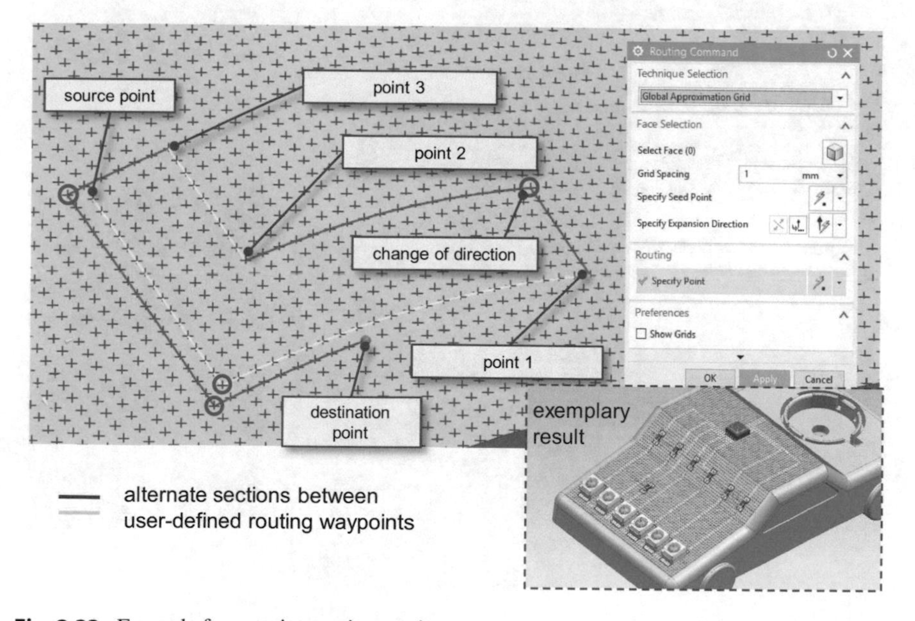

Fig. 2.23 Example for auto-interactive routing

In the prototypical implementation, new paths are created by simply selecting grid points. An automatically generated path is found between the last two selected points. If no path can be found, a warning is displayed to the user in a dialog box.

Volume Generation

Waveguides that are printed in one or more process steps usually have a parabolic or circular cross section. These conditions can be mapped by allowing the user of the 3D-Opto-MID application to choose between these shapes of the profile cross sections. In addition, individual cross sections can also be defined, but these will not be dealt with in the following. The associated conditioning lines result from the predefined geometry of the waveguide.

This definition is made in a separate dialog window (see Fig. 2.24) in which materials and roughness values for the optical simulation that will take place later are also entered. First, the manually or semiautomatically created paths are selected (1), a corresponding waveguide profile is selected (2), and the geometric dimensions of core and cladding are defined (3). In addition, the materials used can be defined (4), which are particularly necessary for a later calculation of the beam path. Since printed waveguides have imperfections, such as waviness or roughness, which result from the printing process, it is also intended to define waviness (correlation length) and roughness (RA value) (5).

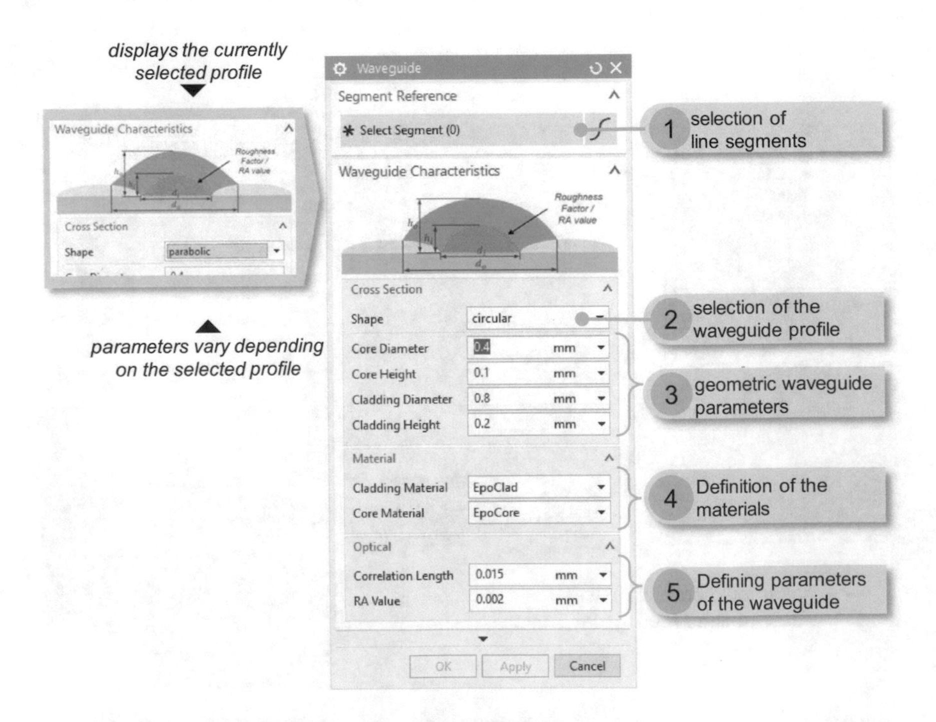

Fig. 2.24 Dialog for defining the waveguide parameters

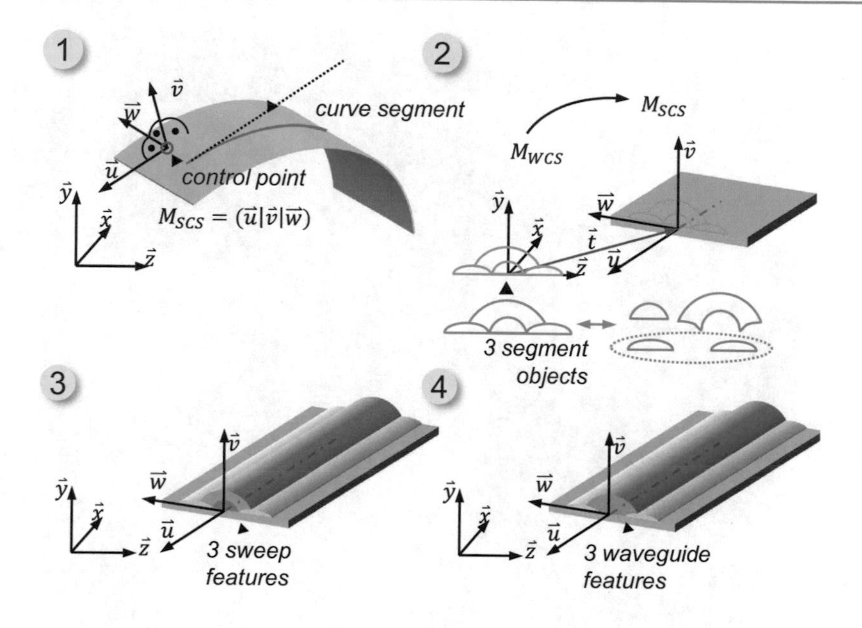

Fig. 2.25 Creation of a volume geometry for waveguides

The creation of corresponding 3D geometries of the user-defined waveguide function comprises four steps, which are shown in Fig. 2.25. In the first step, a transformation matrix is first calculated which transforms the intersection curves generated at the origin of the world coordinate system (WCS) into the respective segment coordinate system. The orthonormalized segment coordinate system (SCS) is spanned by three vectors perpendicular to each other. The first vector is the surface normal at the given control point, the second is the direction vector of the segment at its initial control point, and the third results from the normalized cross product of the first two vectors. The direction of a curve is a vector that is tangential to the curve at a given point.

In the second step, four independently closed profile sections are created, each representing a different component of the waveguide: the core, the cladding and the conditioning. The respective cross section depends on the type of waveguide chosen: circular or parabolic segment. The former consists of a circular, the latter of a parabolic core and cladding sections. In both cases, two elliptical conditioning sections are used. A sweep function is used to extrude the profile sections. Finally, the sweep features (extrusion) are converted into user-defined waveguide features. The user-defined waveguide parameters (core, cladding, conditioning) are now listed as separate elements in the component navigator and displayed in the 3D view (also investigated in [29], [30]).

An example of a layout with the appropriately placed components is shown in Fig. 2.26.

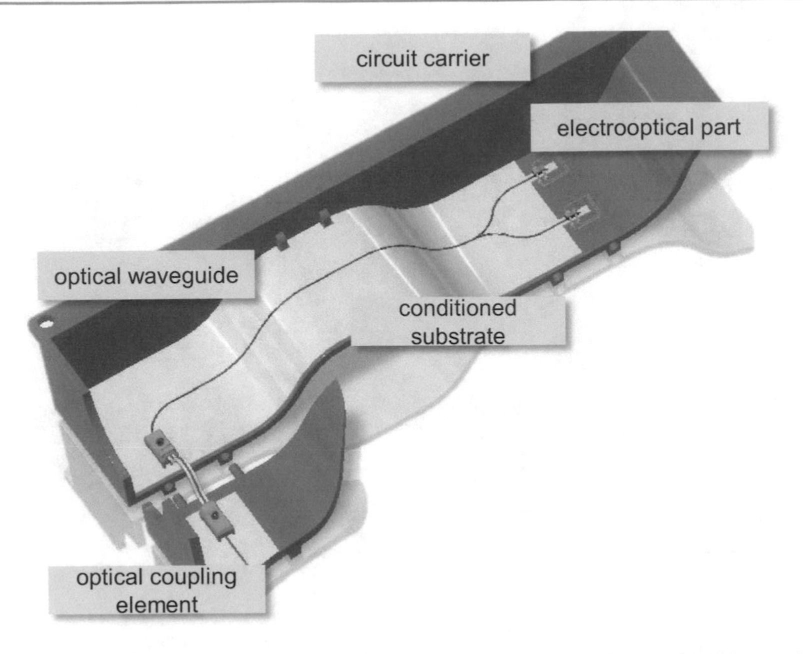

Fig. 2.26 Example of a layout with placed components, substrate and optical conductors in the 3D-Opto-MID tool

2.5.4 Design Rule Checks

Design and manufacturing rule checks (DRC or MRC) are inseparably linked to the steps of placing and laying optical components. These serve to detect and rectify errors in the layout as quickly as possible in order to guarantee functional efficiency. The laid, unbraided electrical/optical conductors form the basis of these checks. As soon as the initial layout has been completed, statements can be made about the individual elements by means of geometric tests. This information is then used to validate the optical layout. The functionality of the optical elements is checked with the help of the results identified in this process. If this is not sufficiently given, the spatial layout must be adjusted. The optical analysis must then be repeated. The steps described in the following chapter, in particular the optical simulation with subsequent detailed presentation and derivation of the results, represent a complex and time-consuming work step. For this reason, it is important to carry out design rule checks as an essential step in the preliminary phase in order to reduce the effort in the latter and to avoid unnecessary iterations. The early detection of design errors in the CAD environment enables direct error correction. It can thus be determined on the basis of geometric parameters in the layout design whether an optical simulation makes sense or whether adjustments must first be made. From the basics of optical signal transmission as well as the process of manufacturing polymer optical waveguides, a multitude of factors can be derived that influence the functionality of a 3D-Opto-MID. The following

Table 2.2 Physical and manufacturing influences on the quality of waveguides

Category	Factor	Dependent on parameter
physical	power loss: Light refraction dependent on angle of total internal reflection	• material combination cladding and core • curvature - microbends • curvature - curve radius • continuity (G_0 continuity) • kinks (G_1 continuity)
	power loss: attenuation	• absorption • scattering
manufacturing process	process-related	• material combination jacket / core (printing) • *curvature - curve radius (mechanical)*
	spatial requirements	• *distance between components* • *space on component surface*

overview (Table 2.2) lists some relevant factors that are particularly relevant for optical waveguides.

By checking them, conclusions can be drawn as to whether an optical system can function in principle. First, however, it is described how the integration of an initial validation of the layout can already be implemented during the spatial design process. Since in the application case of optical waveguides, no CAD system so far provides corresponding information for checking optical structures, and the parameters highlighted in italics in Table 2.2 remain as directly geometrically testable features (also examined in [31], [22]).

These parameters can be determined from purely geometric properties and thus also in the product model without background information on the material properties and manufacturing process. Both physical and manufacturing influences can be traced back to a large extent to geometric parameters. They thus form the basis for the following design rule checks:

- Steadiness check
- Radius check
- Distance check.

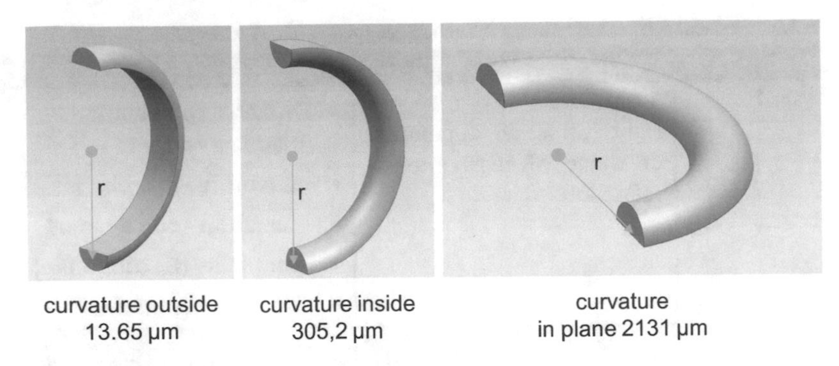

Fig. 2.27 Results for permissible radii of curvature according to [131, P13]

Curves of various types defined in the CAD product model, such as lines, splines or arcs, describe the course of installed optical waveguides. In order to fulfill the basic principles of signal transmission, these curves must be connected to each other in a certain way. If there are unwanted gaps between conductor segments, if the axes of both segments are not concentric to each other or if there are kinks, this can lead to power reductions or even total loss. To eliminate these errors in the layout, a continuity check of the conductors is necessary.

Furthermore, the radius of the curve is one of the factors that are optically of fundamental importance for the functionality of an optical waveguide. The occurrence of total reflection can only be ensured if the radius does not fall below a certain value, which is of great importance for the reduction of power loss along the optical waveguide. This value can vary depending on the arrangement of the geometry. For example, values were determined for an idealized printed waveguide according to [32], [33] for several different configurations (see Fig. 2.27).

A distance check is necessary because the machine kinematics play a decisive role in the aerosol jet printing of waveguides on complex spatial structures, and, for example, no manufacturability is given with undercuts.

Example of How to Perform Design Rule Checks

In order to subject laid waveguides to geometric checks as described, a design rule check tool was implemented in the user interface with which specific waveguide structures can be checked (Fig. 2.28).

The user can make an individual selection in the user interface to carry out checks. In the first block selection, the optical line segments of a waveguide laid in the model can be selected. If several waveguides are to be checked simultaneously, several segments can be selected in parallel. The following three blocks, continuity check, distance check and radius check, are used to set the geometry checks to be carried out.

In the first setting of each block, the layouter can determine whether the corresponding test is to be carried out. For the continuity analysis in the *continuity check* block, it is also possible to select which degree of continuity is to be

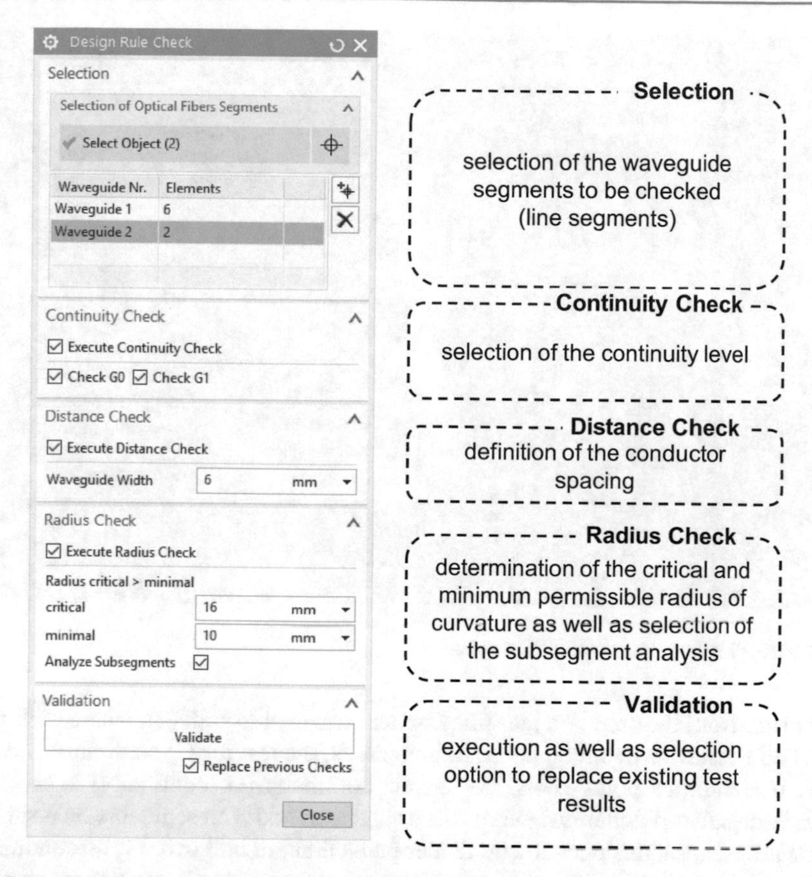

Fig. 2.28 User interface for running the design rule checks

considered. In the d*istance check* block, the width of the optical waveguide must be defined so that the space occupied by the guide can be considered. For the radius check (block: radius check), two radius values are to be specified: a critical radius value, the undershooting of which conditionally restricts the functionality of a waveguide, and a minimum permissible radius of curvature, the undershooting of which represents a total failure of the functionality of the guide. In addition, the option *analyze sub-segments* is offered. If this option is not selected, the entire curve of a conductor segment is marked as insufficient if the radius of curvature falls below the defined value. In the opposite case, the course of each curve is examined and only the sub-segments are highlighted. In the following, the respective results are explained in the user interface.

The result output of the *continuity check* in the working environment is shown in Fig. 2.29. The substrate foil has been colored red for better recognition in the transparent wireframe view. In the first example, it can be seen that there is no other segment at the end segment of a waveguide path that has a common start or end point with the end segment (1). In the verification algorithm, this case does

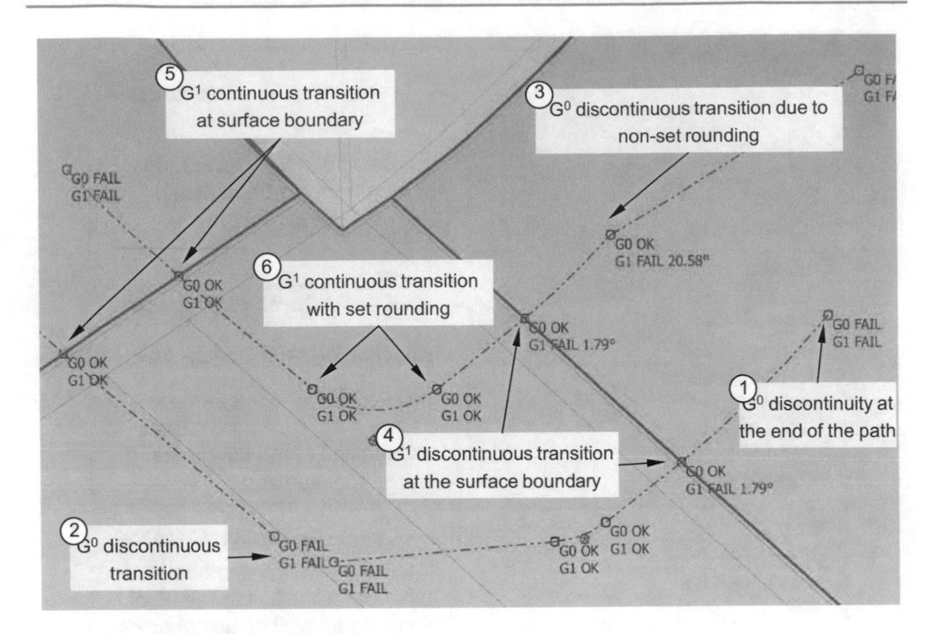

Fig. 2.29 Result of the continuity analysis

not differ from the case of a gap-filled optical waveguide path (2), where both path ends do not have a common point. In both cases, the test for G_0 continuity is negative. If a common point exists, the tangent continuity (G_1 continuity) is checked. This is negative if undefined curves occur between laid path segments on a surface (3). Furthermore, this occurs with ladder paths that are laid over G_1 discontinuous surface boundaries (4). G_1 continuous transitions occur at G_1 continuous surface boundaries (5) or by setting roundings between segment transitions (6).

The distance check is carried out by a separate module in which the width of the waveguide is first defined. Using available reference functions, the selected curve is offset parallel to the initial curve. This is used to create an edge area offset by a defined distance from the original curve and on the same surface. The offset curve is created on both sides of the optical path. Figure 2.30 shows different curves that are created as described above. In a subsequent step, it is determined whether offset curves exist that touch each other. If this is the case, the corresponding area is colored red. If corresponding areas are adjusted manually, segments of other ladder paths that were previously in conflict are highlighted in yellow. Thus, when adjusting the paths, the designer is shown in which area the old path was too close to a nearby conductor path and the new path to be routed requires an increased distance.

The radius check function is used to check the radii along the waveguide path. This is done by displaying functionally relevant value undershoots by color coding the path segments. Two values are defined: critical radii, where the function is impaired if the radius is undershot, and minimum permissible radii, where the function fails if the radius is undershot. The corresponding visualization is shown

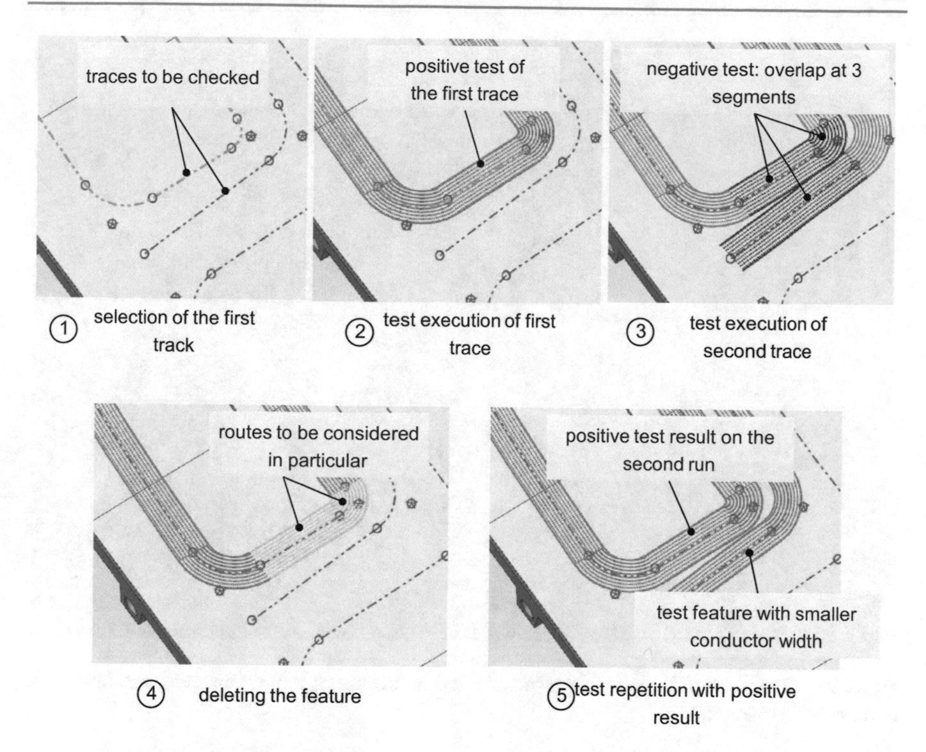

Fig. 2.30 Execution of the distance check

in Fig. 2.31. If one of the values is exceeded or not reached, the corresponding section of the laid waveguide path is shown either in red (if the minimum permissible radius is not reached) or in yellow (if the radius falls below the minimum permissible radius). All other segments remain in the original marking (green).

It is also defined whether the analysis of curves with a variable radius (splines, conic sections) only takes the minimum radius value of the curve into account or whether the entire curve is analyzed.

This may be necessary, for example, if only individual curve sections need to be checked. If the analysis is to be carried out on a curve with a variable radius, the different radii of the curve can be determined and the entire course of the curve can be colored according to the result (see Fig. 2.32).

2.5.5 Interface to Optical Simulation

For the design of 3D-Opto-MID, not only production-relevant descriptions, but also function-relevant properties are important. In addition to the design rule tests, optical simulation methods can be used to ensure these.

The optical simulation system RAYTRACE [30], developed by the Chair of Optics in the ODEM working group under the direction of Professor Lindlein,

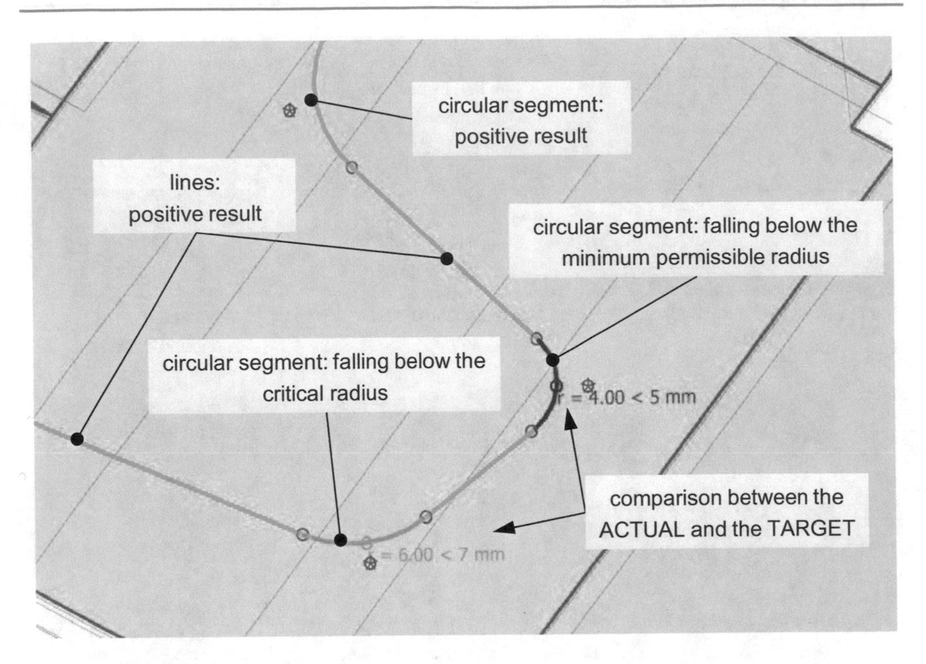

Fig. 2.31 Overview of the possible results for the radius check tool (exclusive consideration of minima for curvatures)

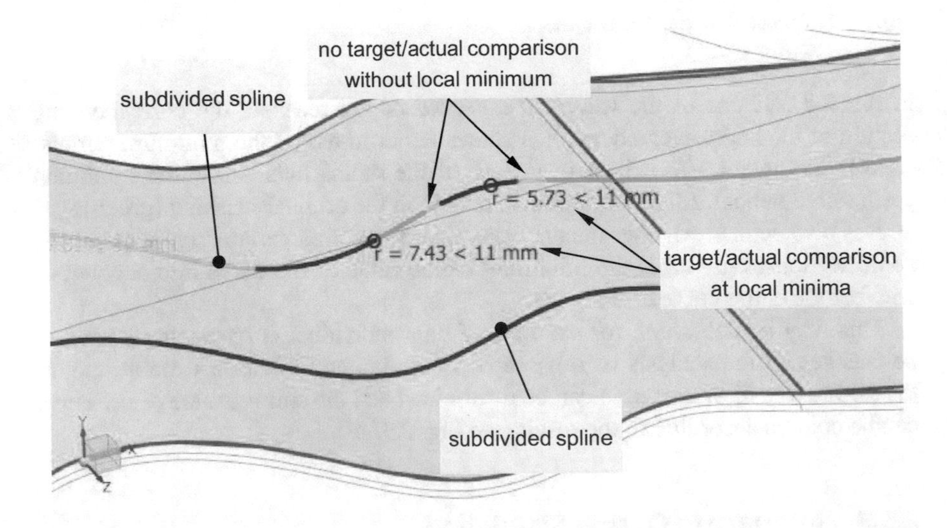

Fig. 2.32 Radius check tool (analysis of radii in the case of changing curves)

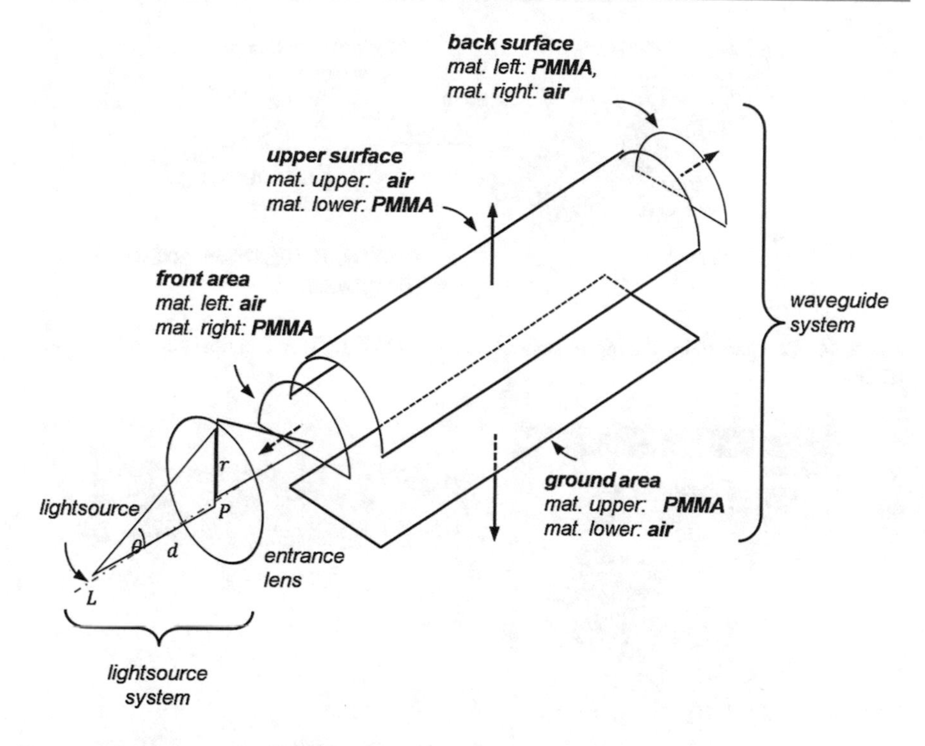

Fig. 2.33 Structure of a waveguide segment according to RAYTRACE topology

was used in this work. This was initially developed with the aim of simulating lens systems. In the broadest sense, a waveguide is a lens system with reflective surface properties. The different geometric surfaces representing a waveguide can be defined within the program. For this purpose, RAYTRACE provides so-called single surface elements that define the boundary between two different media. You can define the optical properties of the surface (refraction, absorption, scattering) and specify different materials. Each surface is distinguished between a left and a right side (or top and bottom), and different materials can be defined on each of the two sides. The normal vector of a surface always points to the right side of the surface.

A volume enclosed by these elements is designed as a waveguide. An example of a straight waveguide segment with defined surfaces is shown in Fig. 2.33. The flat (non-curved) front, back and bottom surfaces are easy to define, while the curved top surface must be generated by an explicit function. The surfaces can be rotated so that they are geometrically aligned with each other. Due to the arbitrary surface functions, more complex shapes such as curved, elliptical or spline waveguides are possible.

To calculate the attenuation and dispersion at the end of the waveguide, the back surface is set to be absorbing. The upper and lower surfaces have the scattering property set to simulate the scattering at the core-sheath interface.

From the previous descriptions, it can be seen that different representation approaches are used for the representation of geometry in the CAD system and

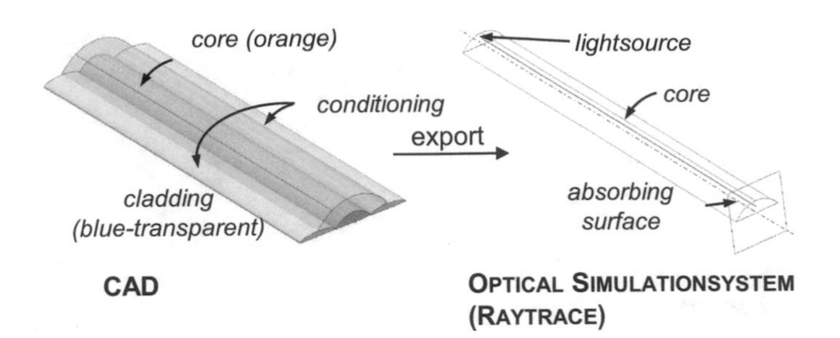

Fig. 2.34 Comparison of the representation forms of CAD (left) and optical simulation system (right)

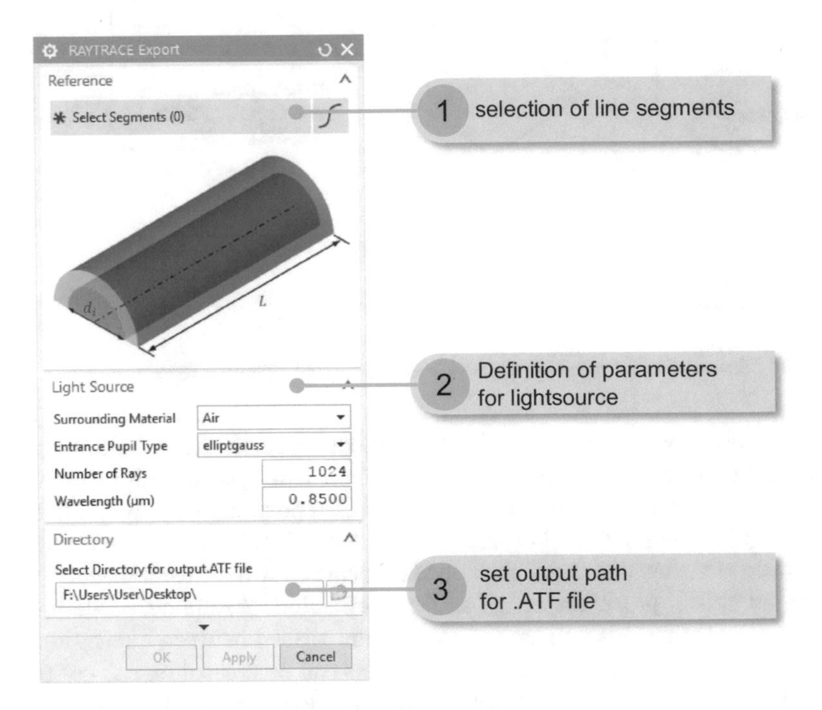

Fig. 2.35 RAYTRACE export dialog

in the RAYTRACE program. In order to be able to process the CAD data, it must be converted into a format that can be read by RAYTRACE. While CAD systems work with solid models, RAYTRACE uses a surface-based representation and calculation model (see Fig. 2.33 and 2.34) to establish the relationships between different objects. In addition, there is the requirement to assign key figures such as roughness, materials with optical properties or reflection properties to the CAD representation in order to obtain a holistic optically simulatable model.

Since the assembly does not contain any predefined light sources, this is defined by the user via a dialog box (Fig. 2.35) (1). The wavelength, the rays to be

Table 2.3 Specification of the exported segments

cross-section	circular segment	parabolic
core diameter	$0.4mm$	$0.4mm$
core height	$0.1mm$	$0.1mm$
length of waveguide	$46.8mm$	$46.8mm$
core material	EPOCORE ($= 1.58$)	EPOCORE
cladding material	EPOCLAD ($= 1.57$)	EPOCLAD
correlation length	$18\mu m$	$18\mu m$
roughness value (RA)	$0.2\mu m$	$0.2\mu m$
foil substrate material	PMMA	PMMA
entrance lens	elliptic	elliptic
wavelength	$850\mu m$	$850\mu m$

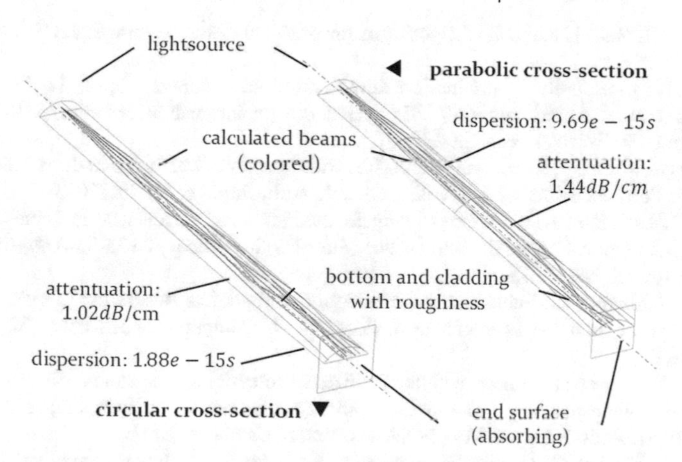

Fig. 2.36 Simulation of different waveguide types in RAYTRACE

calculated, the shape of the input lens and the ambient medium are defined. The selection of the waveguide segments to be simulated is determined by selecting the 3D view (2). Finally, an output path is defined to create the ATF file for later simulation.

For evaluation purposes, segments with parabolic and circular cross sections were exported for the non-sequential ray tracing simulation (see Table 2.3 and Fig. 2.36). In this example, EPOCORE and EPOCLAD from Microresist

Technologies are used as core and cladding material. These have a refractive index of n = 1.58 (core) and n = 1.57 (cladding). The light source is located halfway up the waveguide, and the light rays are distributed through the entrance lens. The numerical aperture is chosen so that all rays in the waveguide are reflected inwards. For more meaningful results on dispersion, the simulation should be carried out with more than 1,000,000 rays.

We sincerely thank the Deutsche Forschungsgemeinschaft for funding the research group OPTAVER FOR 1660.

References

1. Franke, J.: Three-dimensional molded interconnect devices (3D-MID): materials, manufacturing, assembly, and applications for injection molded circuit carriers. Hanser, München (2014)
2. Montrose, M.I.: EMC and the printed circuit board: design, theory, and layout made simple. IEEE Press; John Wiley & Sons, Inc, New York (1999) (IEEE Press series on electronics technology)
3. Bogaerts, W., Chrostowski, L.: Silicon photonics circuit design: methods, tools and challenges. Laser Photonics Rev. **12**(4), 1700237 (2018)
4. Jansen, D.: The electronic design automation handbook. Springer US, Boston, MA (2003)
5. Gerez, S.H.: Algorithms for VSLI design automation. Updated with corr. Wiley, Chichester (2005)
6. Alpert, C.J. (Ed.) Handbook of algorithms for physical design automation. CRC Press, Boca Raton (2009)
7. Gajski, K.: Guest editors' introduction: new VLSI tools. Computer **16**(12), 11–14 (1983)
8. Sherwani, N.A.: Algorithms for VLSI physical design automation, 3rd edn., p. 1999. Kluwer Academic Publishers, Dordrecht (2005)
9. Feldhusen, J., Grote, K.-H.: Pahl/Beitz Konstruktionslehre: Methoden und Anwendung erfolgreicher Produktentwicklung. 8, vollst. überarb. Aufl. Springer, Berlin (2013)
10. 2004. VDI-Richtlinie 2206 "Entwicklungsmethodik für mechatronische Systeme"
11. Optiwave: Optical SPICE, Screenshot aus Produktvideo. URL http://optiwave.com/. Zugegriffen: 9. Nov. 2015
12. Zhuo, Y., Alvarez, C., Feldmann, K.: Horizontal and vertical integration of product data for the design of moulded interconnect devices. Int. J. Comput. Integr. Manuf. **22**(11), 1024–1036 (2009)
13. Zhuo, Y.: Entwurf eines rechnergestützten integrierten Systems für Konstruktion und Fertigungsplanung räumlicher spritzgegossener Schaltungsträger (3D-MID). Fertigungstechnik Erlangen, Vol. 180. Meisenbach, Bamberg (2007)
14. Zeitler, J., Fischer, C., Goetze, B., Moghadas, S.H., Franke, J.: Integration of semi-automated routing algorithms for spatial circuit carriers into computer-aided design tools. In: VDE (Ed.) Proceedings of the 13th electronic circuits world convention. VDE Verlag GmbH, Berlin (2014)
15. Zeitler, J., Goetze, B., Fischer, C., Franke, J.: Novel approach for implementation of 3D-MID compatible functionalities into computer-aided design tools. In: Franke, J., Kuhn, T., Birkicht, A., Pojtinger, A. (Eds.) 11th international congress molded interconnect devices: scientific proceedings. Advanced materials research, Vol. 1038. Trans Tech Publications Ltd, Pfaffikon (2014)
16. Krebs, T.: NEXTRA. http://www.mecadtron.com/. Zugegriffen:18. Febr. 2018

17. Krebs, T., Franke, J.: Konstruktionswerkzeuge für elektronisch/mechanisch integrierte Produkte: ECAD- und MCAD-Funktionen in einem dreidimensionalen Entwicklungssystem integriert. Elektronik **2005**(18), 60–66 (2005)
18. Brüderlin, B., Roller, D.: Geometric constraint solving and applications. Springer, Berlin (2012)
19. Anderl, R. (ed.): STEP: Standard for the exchange of product model data; eine Einführung in die Entwicklung, Implementierung und industrielle Nutzung der Normenreihe ISO 10303 (STEP). Teubner, Stuttgart (2000)
20. AS 3643.1-1989. Computer graphics – Initial graphics exchange specification (IGES) for digital exchange of product definition data – General (1988)
21. Bierhoff, T., Schrage, J.: Rechnergestützter Entwurf und Simulation von optischen Verbindungen in Leiterplatten (2008)
22. Zeitler, J., Reichle, A., Franke, J., Loosen, F., Backhaus, C., Norbert, L.: Computer-aided design and simulation of spatial opto-mechatronic interconnect devices. 26th CIRP Design Conference, pp. 727–732 (2016)
23. Franke, J., Zeitler, J., Reitberger, T.: A novel engineering process for spatial opto-mechatronic applications. CIRP Ann. Manuf. Technol. **65**(1), 153–156 (2016). https://doi.org/10.1016/j.cirp.2016.04.091
24. Vajna, S., Weber, C., Bley, H., Zeman, K., Hehenberger P.: CAx für Ingenieure. Springer, Berlin (2009)
25. Loosen, F., Carsten, B., Zeitler, J., Hoffmann, G.-A., Reitberger, T., Lorenz, L., Lindlein, N., Franke, J., Overmeyer, L., Suttmann, O., Wolter, K.-J., Bock, K., Backhaus, C.: Approach for the production chain of printed polymer optical waveguides-an overview. Appl Opt **56**(31), 8607–8617 (2017). https://www.osapublishing.org/ao/abstract.cfm?uri=ao-56-31-8607
26. N.N.: EAGLE: Leiterplatten – Layouts leicht gemacht. https://www.autodesk.de/products/eagle/overview. Zugegriffen: 3. Sept. 2019
27. Octopart.com: Octopart API. Zugegriffen: 15. Juli 2019
28. Pecht, M.: Placement and routing of electronic modules. Electrical engineering and electronics, Vol. 82. Dekker, New York (1993)
29. Loosen, F., Backhaus, C., Lindlein, N., Zeitler, J., Franke, J.: Implementation of a scattering method for rough surfaces in a raytracing software linked with a CAD (Computer-Aided Design) toolbox. In: Optical Society of America (Ed.) Frontiers in Optics 2016 (2016)
30. Loosen, F., Backhaus, C., Lindlein, N., Zeitler, J., Franke, J.: Concepts for the design and optimization process of printed polymer-based optical waveguides (scattering processes). In: Abbe School of Photonics (Ed.) DoKDoK 2015 Proceedings (2015)
31. Loosen, F., Backhaus, C., Lindlein, N., Zeitler, J., Franke, J.: Design and simulation rules for printed optical waveguides with implemented scattering methods in CAD and raytracing software. In: DGao (Ed.) 117th DGaO Proceedings (2016)
32. Backhaus, C., Dötzer, F., Hoffmann, G.-A., Lorenz, L., Overmeyer, L., Bock, K., Lindlein, N.: New concept of a polymer optical ray splitter simulated by Raytracing with a new Bisection Algorithm. In: Optical Society of America, 2019 (OSA Technical Digest) (Eds.) Frontiers in Optics 2019. Washington, DC United States (2019)
33. Backhaus, C., Lindlein, N., Zeitler, J.T., Franke, J.: Beeinflussung der optischen Eigenschaften von Polymer Optischen Wellenleitern durch das Druckpfad-Design. In: DGao (Hrsg.): DGaO Proceedings

3 Three-Dimensional Simulations of Optical Multimode Waveguides

Carsten Backhaus, Florian Loosen and Norbert Lindlein

The aim of this chapter is the development of algorithms to simulate optical multimode waveguides with an uncommon cross section and a three-dimensional path function. While current commercial applications struggle to replicate the experimental setup, new simulation techniques within known simulation methods such as raytracing and wave propagation method (WPM) have to be developed. Therefore, this chapter answers the question:

How to simulate optical multimode waveguides on three-dimensional objects with an uncommon cross-section by either geometrical optics or physical optics?

3.1 Demands on the Simulation

In order to simulate optical multimode waveguides integrated in optical bus systems, numerous requirements have to be met:

- Three-dimensional simulation
- Fast simulation algorithm
- Relevant experimental parameters as input
- Relevant experimental parameters as output
- Selecting the adequate simulation method.

C. Backhaus (✉) · F. Loosen · N. Lindlein
Institute for Optics, Information and Photonics, Friedrich-Alexander-Universität Erlangen-Nürnberg, Erlangen, Germany
e-mail: carsten.backhaus@fau.de

F. Loosen
e-mail: florian.loosen@fau.de

N. Lindlein
e-mail: norbert.lindlein@fau.de

J. Franke et al. (eds.), *Optical Polymer Waveguides*,
https://doi.org/10.1007/978-3-030-92854-4_3

In the application of multimode waveguides in the device communication, typically three-dimensional structural elements are used, and the path of the waveguide is no longer planar but three-dimensional. Consequently, a simulation tool must be able to replicate the physical system and has to include **three-dimensional simulations**. Commercially available simulation tools by market leaders such as Zemax [1], Synopsys [2], VirtualLab Fusion [3] and Lambda Research Corp. [4] do not present acceptable workflows to simulate several centimeter-long waveguides with a non-symmetrical cross section on highly three-dimensional structural elements. Workarounds for a correct geometrical representation can be achieved by using Boolean operators and subtracting a cube from a cylinder and deform the mathematical object to meet the three-dimensional structure of the surface. However, this leads often to inconsistent models, which is followed by an enormous increase in computation power. To this point, an efficient and easy-to-use application is of urgent need.

In today's fast-moving world, time is more valued than ever, and there is no exception for the computation time of optical simulations. The demand of **fast simulation algorithms** for geometrical and physical optics does grow even further when three-dimensional structures are of interest. In recent years, the advancement of graphic processing units (GPU) has enabled the use of massive parallel computations [5] and is the future for real-time simulations [6]. With its multithreaded architecture, GPUs can be used to perform fast Fourier transformations (FFT) efficiently [7], which is the center operation in standard physical optics simulations with algorithms such as the wave propagation method (WPM) [8]. In addition, in geometrical optics, an even further step is taken: Graphic cards have been designed explicitly to solve raytracing problems [9]. This being state of the art, new simulation applications have to use parallel computation methods to fully use its computation potential. While the methods itself, raytracing and FFTs on GPU, are a well-researched field, its application is usually used for cinematic rendering (few ray reflections) or geometrically simple objects (e.g., rotation symmetric). For a non-symmetrical waveguide of several centimeter length on three-dimensional objects, both methods have to be optimized on geometrical descriptions to make use of its enhanced performance on the computer's GPU. Special mathematical models have to be derived to meet the criteria of mimicking in sufficient detail the experimental setup as well as a mathematical model which is optimized for parallel computing.

The optical simulation accuracy of optical multimode waveguides certainly depends on the correct assumptions of **relevant experimental parameters as input,** especially in a highly three-dimensional environment. The applications goal is to give an engineer a tool to test different experimental setups, without having to perform them, but rather has a simulation result. Therefore, the input parameters for the simulation should give as much freedom as possible while being not too overwhelming or being irrelevant. In any case, geometrical parameters to describe the non-symmetrical waveguide have to be defined (Fig. 3.1). Further, waveguide imperfections introduced by production techniques (waviness of the waveguide) and optical behavior of used materials (roughness of polymers) have to be considered in the simulation. Also, parameters to replicate the nature of light present in the experimental setup (e.g., a laser) have to be set in the simulation.

The quality of an optical simulation application is measured by its practicality in the daily use and accompaniment for an engineer in the setup process of the experimental arrangement. The result of the optical simulation should serve as real added value for the engineers in the understanding of the experimental setup and therefore has to include **relevant experimental parameters as output**. In order to characterize the quality of a polymer optical waveguide (POW), the attenuation and the intensity distribution in the waveguide is measured. Consequently, output parameters of the simulation have to include the intensity distribution $P(\vec{r})$ with dependency in space and attenuation α:

$$\alpha = \frac{10}{L} \cdot \log\left(\frac{P_0}{P_L}\right), \tag{3.1}$$

with length L of the waveguide, P_0 initial power and P_L power at the end of the waveguide. In addition, the signal dispersion is calculated, which is of special interest in multimode waveguides due to its effect on the maximum data transfer rate.

Finally, **selecting the adequate simulation method** does play a key role in the optical simulation of POWs. Due to the multimode nature of the waveguides, several different approaches of simulation methods are of interest but exclude quantum optical methods and are therefore situated in the classical optics. The field of classical optics can again be divided in geometrical and physical optics, latter having numerous different simulation approaches (Fig. 3.2).

The most common simulation method for single-mode waveguides is the beam propagation method (BPM) [10]. In the case of incoherent spatial modes, multimode waveguides can also be simulated with this approach by superimposing the intensities of the propagated single modes, although the computing time increases proportionally to the number of spatial modes. In addition, to simulate a temporal signal, temporal modes of different wavelengths must be coherently superimposed. For highly multimode waveguides, the beam propagation approach would therefore not be practicable, since the computing times would be enormous.

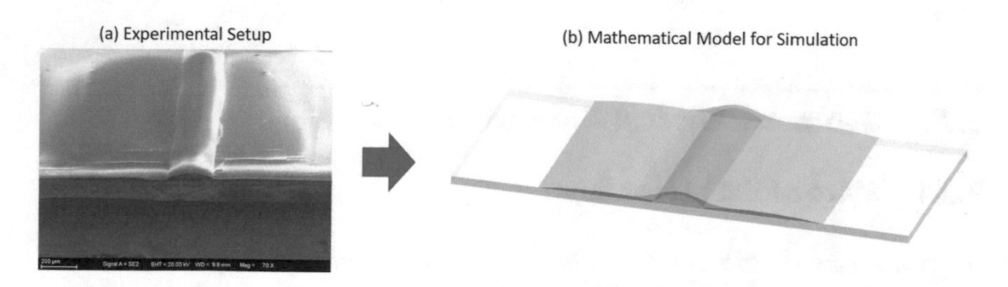

Fig. 3.1 Creating a mathematical surface model from the experimental setup is crucial for the simulation

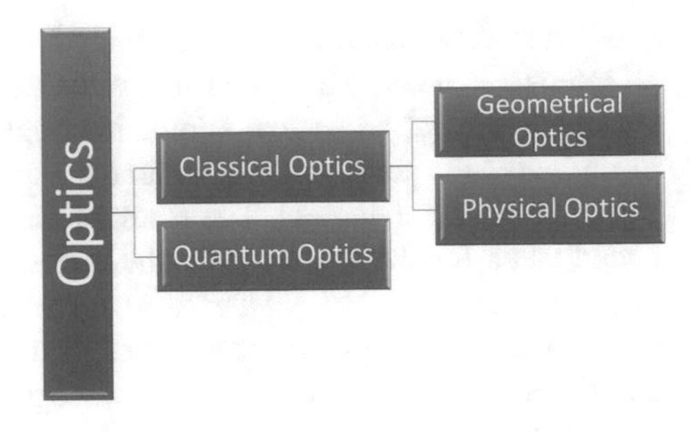

Fig. 3.2 Overview of fields in optics

Another approach to simulate waveguides, the "finite difference time domain method" (FDTD) [11] performs a direct numerical integration of Maxwell's equations in the time domain and has the advantage that short pulses can be simulated directly and not only by coherent superposition of modes of different wavelengths. However, here is the problem that the spatial sampling of the field must be done in a grid of a fraction of a wavelength, so that a waveguide with, for example, 100 μm width, 20 μm height and several millimeters length cannot be simulated. In multimode waveguides, the FDTD method also has to calculate each of the incoherent spatial modes separately and sum up the intensities at the end so that the computational effort is even higher to the BPM approach.

With increasing number of modes, however, a simulation approach based on normal geometric optical ray tracing becomes more accurate [12] and is many times faster than an approach based on BPM or FDTD. As a consequence, when the simulation of a POW is needed, the ray tracing algorithm is best suited to give fast and detailed results.

However, there may occur cases where geometrical optical methods are no longer sufficient, such as when coupling two optical fibers. In need of a fast physical optical simulation method, the wave propagation method is deployed. Originally developed by Brenner and Singer [8] for the simulation of micro lenses, the algorithm enables a non-paraxial simulation in contrast to the paraxial simulation of a simple BPM. In addition, it was also shown that the WPM is significantly faster than the non-paraxial Padé-based BPM with the same or even higher accuracy [13]. Therefore, if physical optical simulations are required for example for the simulation of an asymmetric bus coupler with POWs [14], we will deploy the WPM.

3.2 Geometric Optical Simulation

3.2.1 State of the art for Raytracing

As being part of the classical optical field, the basic theory of raytracing is descended by Maxwell's equations [15, 16]. The central assumption in geometrical optics must be made: $\lambda \rightarrow 0$ and from this $k_0 \rightarrow \infty$ (with λ being the wavelength and k_0 the modulus of the wave vector). In practice this means: If the geometry of the system is large compared to the wavelength, geometric optics can be used. If one now inserts the equations with this condition into each other, one obtains the so-called eikonal equation with the optical path length L and the refractive index n:

$$\left(\vec{\nabla}L(\vec{r})\right)^2 = n^2(\vec{r}) \tag{3.2}$$

It is the basis for the assumption of raytracing. A ray is defined in such a way that its trajectory is always perpendicular to the wavefronts, so that $\vec{\nabla}L(\vec{r})$ indicates the direction. For homogeneous materials, the refractive index does not change $\vec{\nabla}n(\vec{r}) = 0$, so the following simple differential equation applies:

$$\frac{d^2\vec{r}}{ds^2} = 0 \tag{3.3}$$

with s being the arc length of the ray. The solution of this equation is the well-known ray equation:

$$\vec{r}(s) = \vec{P} + s \cdot \vec{Q} \tag{3.4}$$

with $\vec{P}$ the starting point of the ray and $\vec{Q}$ the direction vector of the ray. Then $\vec{r}(s)$ is the position of a point along the ray at arc length s relative to the starting point.

Now that the physical basis has been laid, the raytracing algorithm is to be introduced. Spencer and Murty [17] proposed the following procedure:

1. Find the point of intersection of the ray with a surface, giving the new starting point of the ray.
2. Find the change in direction (reflection, refraction, diffraction, etc.), giving the new direction vector of the ray.
3. Calculate the new ray equation.
4. Repeat steps 1–3 until the ray hits either no surface or a surface that is marked as absorbing (e.g., the detector).

The realization of step 1 and step 2 however does bear difficulties. Finding the point of intersection can be simplified, if it is known, which surface the ray hits first, second and so forth. In the so-called sequential raytracing, the order of the surfaces a ray will hit is known and also each surface is only met once. A more complex calculation has to be performed, when a ray can hit a surface or several surfaces more than once. The so-called non-sequential raytracing [18] typically uses an approach, where the points of intersection of the ray with all surfaces present in the simulation system are calculated and that point of intersection with the smallest positive distance to the starting point is the correct one.

Regarding step 2, finding the change in direction is the center of all raytracing algorithms but also is most likely where algorithms differ. Optical transmission, absorption, diffraction and scatter models are implemented in various ways, and depending on the main use of an application, the efficient calculation is in many cases only available for one specific field (e.g., transmission). Including all algorithms, the number of applications available for optical design includes more than 200 companies [19].

In the working group optical design and measurement (ODEM) at the Chair of Optics of the Institute of Optics, Information and Photonics of the Friedrich-Alexander-University of Erlangen-Nuremberg the development of the program package RAYTRACE started more than 30 years ago, starting with the diploma thesis of the group leader Norbert Lindlein [20–23]. This program allows raytracing on quite arbitrary refracting and reflecting surfaces (plane surfaces, spheres, aspheres, freeform surfaces), which can be arranged arbitrarily in three-dimensional space. From the calculated ray data like ray direction, optical path length, ray power, etc., various information can be taken. The RAYTRACE program package can perform sequential raytracing as well as non-sequential raytracing. For an optical waveguide, where the number of reflections within the waveguide is not known at the beginning or varies depending on the angle of the ray to the axis of the waveguide, i.e., depending on the mode, only non-sequential ray tracing can be used.

3.2.2 Mathematical Properties for the Description of a POW

Several centimeter-long waveguides with a non-symmetrical cross section on complex three-dimensional structural elements need a new algorithm for geometric optical simulations, since it cannot be represented by already implemented simple geometrical models. To take a step further, one could ask for an algorithm that describes with a continuous mathematical model a waveguide consisting of an arbitrary chosen cross section. In order to describe the algorithm, several mathematical descriptions and properties have to be discussed before diving into the algorithm itself. In the following subchapters, the possibility of defining a local coordinate system, a mathematical description of a cross section and an intersection of a ray with a plane is briefly described.

Local Coordinate System

Before any further thoughts, the defining parameters of the waveguide have to be evaluated. Besides the cross-sectional definition, the path of the waveguide on top of the three-dimensional element has to be converted into a mathematical function. For that, one can define a general three-dimensional parametric function $\vec{f}(u)$ to represent the path of the waveguide [24, 25]. Since this description has no information about the surface normal, one also has to define a function $\vec{n}(u)$, which represents the surface normal at any given point $\vec{f}(u)$. It is important to notice that the parameter u is in both functions the same parameter.

This general approach has to be more specified in the way, that one can demand:

$$\vec{f}(u) \exists \vec{f'}(u) \forall u \in \mathbb{R}. \tag{3.5}$$

This demand has two benefits: First, one can find two functions $\vec{f}_1$ and $\vec{f}_2$ where $\vec{f}'_1(u_1) = \vec{f}'_2(u_2)$ and $\vec{f}_1(u_1) = \vec{f}_2(u_2)$ applies. This allows to stitch functions after one another and gives high flexibility for the path function design. The path function $\vec{f}(u)$ can now be described as a composite function with subintervals $\vec{f}_i(u)$. This function definition holds for a cubic spline function, which is used in the simulation model.

Second, it allows to define a local coordinate system at any given point $\vec{f}(u)$ with the local z-axis $\vec{n}_z$:

$$\vec{n}_z(u) = \frac{\vec{f'}(u)}{\left|\vec{f'}(u)\right|}. \tag{3.6}$$

The local coordinate system is further defined by the local y-axis $\vec{n}_y$ (depending on the surface normal)

$$\vec{n}_y(u) = \frac{\vec{n}(u)}{\left|\vec{n}(u)\right|} \tag{3.7}$$

and the local x-axis $\vec{n}_x$ as the cross-product of both (Fig. 3.3)

$$\vec{n}_x(u) = \vec{n}_y(u) \times \vec{n}_z(u). \tag{3.8}$$

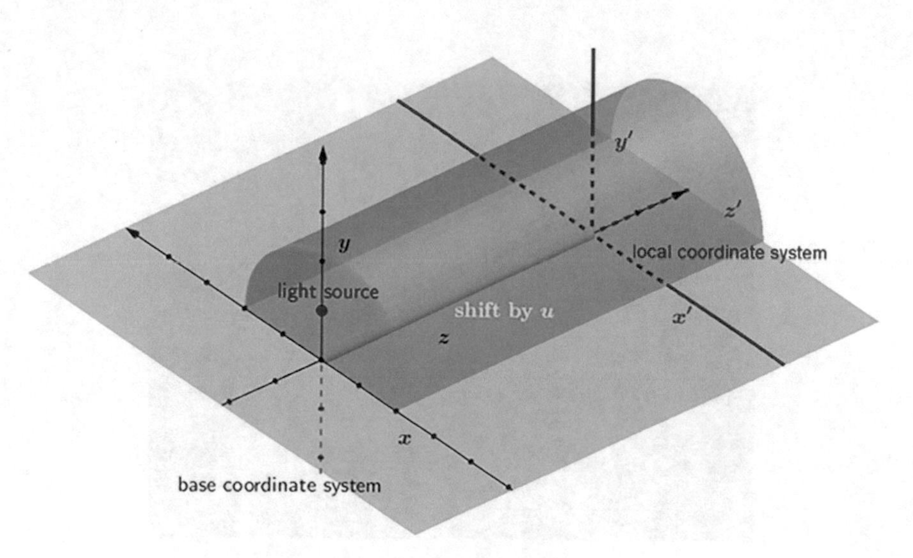

Fig. 3.3 For any given point u on the print path, a local coordinate system can be defined by the surface normal (y'), the tangent of the path function (z') and the cross-product of both (x')

Cross section of a Waveguide

The purpose of defining a local coordinate system at any point $\vec{f}(u)$ is evident: The cross-sectional function $g(x)$ defining the geometry of the waveguide is always represented in the local n_x-n_y-plane. The function $g(x)$ itself has to define a closed path (e.g., a circle, three lines with three intersections, two functions with two intersections, etc.). For the simulation of a POW, one has to determine the cross section and parametrize it. In Fig. 3.4, a REM-picture of a cut POW displays the cross section and allows to define the parameters height h and width b.

With these two parameters a mathematical replicate can be defined. It is obvious to choose two functions: One for the straight surface $g_2(x)$ between substrate and the waveguide and one for the curved surface $g_1(x)$ between air and waveguide. A quite good fit to the curved surface is to choose a circle segment. This results in (see Fig. 3.5):

$$g_1(x) = \sqrt{\left(\frac{4h^2 + b^2}{8h}\right)^2 - x^2} + h - \frac{4h^2 + b^2}{8h} \qquad \forall x \in [-b/2, b/2] \quad (3.9)$$

$$g_2(x) = 0 \qquad \forall x \in [-b/2, b/2]. \quad (3.10)$$

In the following chapters, this definition of $g(x)$ will be used. But keep in mind, that a general definition of $g(x)$ is also valid, and therefore, a waveguide with an arbitrary cross section can be simulated.

By knowing the path function $\vec{f}(u)$, the possibility to define a local coordinate system by knowing the surface normal $\vec{n}(u)$ and having the definition of the cross section $g(x)$, a waveguide is defined at any given point in the system.

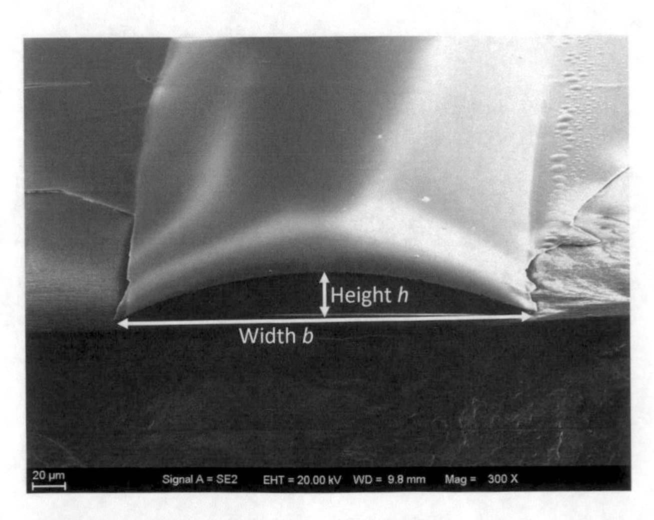

Fig. 3.4 Picture taken by a scanning electron microscope (SEM) of a polymer optical waveguide at the trimmed edge displaying the D-shaped cross section

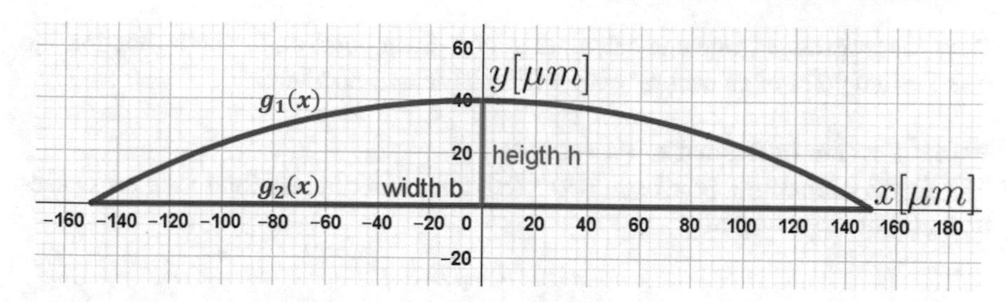

Fig. 3.5 Mathematical reconstruction of the cross section of a POW by the usage of the two parameters height h and width b

Intersection of Ray and Plane

To calculate the intersection point of a ray and a plane, one has to define both elements. In Eq. 3.4, a ray is already defined and outstanding definition of a plane in the normal form is given in the following:

$$E(\vec{x}) = (\vec{x} - \vec{p}) \cdot \vec{n} = 0, \tag{3.11}$$

with $\vec{p}$ being a point on the plane and $\vec{n}$ the surface normal. To find the intersection, $\vec{x}$ is substituted by the ray Eq. (3.4):

$$E(s) = \left((\vec{P} + s \cdot \vec{Q}) - \vec{p}\right) \cdot \vec{n} = 0 \tag{3.12}$$

The reason for this introduction of intersection calculation is that one can calculate the intersection of a ray with a local n_x-n_y-plane. In Eq. 3.12, the point $\vec{p}$ is given by the point on the path function $\vec{f}\,(u)$ and the surface normal $\vec{n}$ by the tangent function $\vec{f'}\,(u)$ at that point. Substituting this in Eq. 3.13 and solving for s

$$\left(\left(\vec{P} + s \cdot \vec{Q}\right) - \vec{f}\,(u)\right) \cdot \vec{f'}\,(u) = 0 \tag{3.13}$$

$$\Rightarrow s = \frac{\left(\vec{f}\,(u) - \vec{P}\right) \cdot \vec{f'}\,(u)}{\vec{Q} \cdot \vec{f'}\,(u)}, \tag{3.14}$$

allows the calculation of the intersection point $\vec{I_{gl}}$ in the global coordinate system by evaluating Eq. 3.4 with the calculated s. The intersection point $\vec{I_{loc}}(x, y)$ in the local coordinate system Eq. 3.7 and 3.8 can be calculated with the help of the Hessian normal form:

$$x(u) = \left(\vec{I_{gl}} - \vec{f}\,(u)\right) \cdot \vec{n}_x(u) \tag{3.15}$$

$$y(u) = \left(\vec{I_{gl}} - \vec{f}\,(u)\right) \cdot \vec{n}_y(u). \tag{3.16}$$

With these formulas, one can calculate at any given point on the path function the intersection point of an arbitrary ray with the local n_x-n_y-plane.

Points Position in Regard to Waveguide
The found intersection point $I_{loc}(x(u), y(u))$ can be set in relation to the cross section $g(x)$ of the waveguide:

$$g(x) = \begin{cases} g_1(x) \\ g_2(x) \end{cases}, \tag{3.17}$$

which can be projected onto the plane (see Fig. 3.6). One can define three areas, the point can be found in:

1. Outside the waveguide: $y(u) > g_1(x(u)) \vee y(u) < g_2(x(u))$
2. Inside the waveguide: $y(u) < g_1(x(u)) \wedge y(u) > g_2(x(u))$
3. On the surface of the waveguide.

The surface is in fact an area and not just a line, since we assume points that have a distance smaller than a defined tolerance ε to be on the surface. The surface area is defined by the following equation:

$$-\varepsilon < y(u) - g(x(u)) < \varepsilon \tag{3.18}$$

3.2.3 Algorithm for Arbitrary Cross sections

After the basic mathematical properties have been described in the previous chapter, we now can introduce the algorithm which will be able to simulate several centimeter-long waveguides with a non-symmetrical cross section on complex three-dimensional structural elements. Figure 3.7 gives an overview of the general implementation of the algorithm.

Input Parameters
To start the algorithm, the basic parameters have to be defined. First, the cross section has to be specified, which is done by stating the height h and the width b of the waveguide (Fig. 3.8).

Fig. 3.6 In a plane perpendicular to the print path, three regions can be defined: Outside of the waveguide (green), inside the waveguide (red) and the surface of the waveguide (blue)

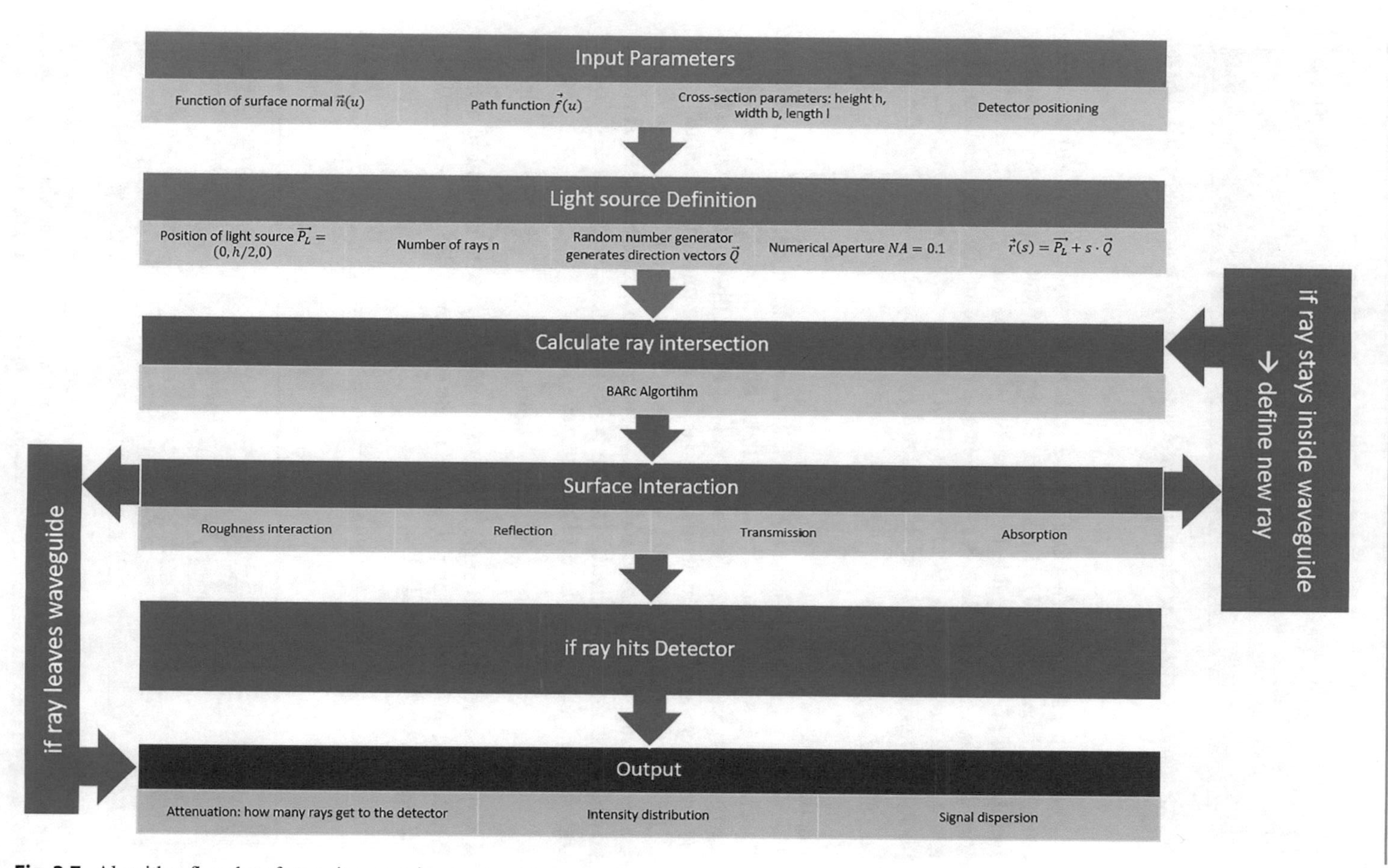

Fig. 3.7 Algorithm flowchart for tracing a ray in a waveguide

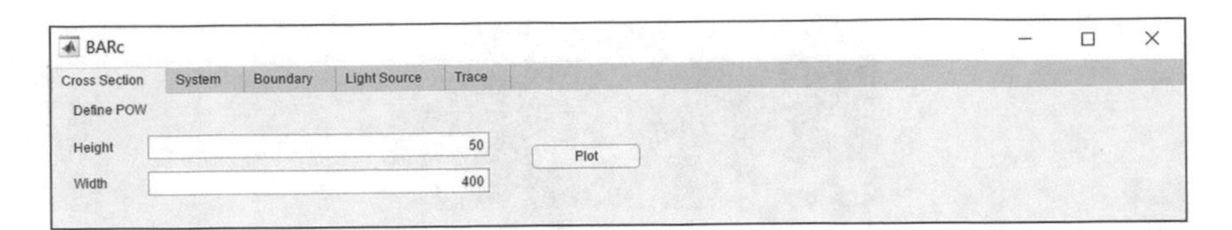

Fig. 3.8 Input mask for the definition of the waveguides cross section by defining the height and width

Further, a parametric function for the path of the waveguide $\vec{f}(u)$ has to be entered in dependence of u. Under "Set Range of u," one defines the beginning and the end of the waveguide by evaluating $\vec{f}(Start)$ and $\vec{f}(End)$. In addition, one has to specify (see Sect. 3.2.2) the first derivative of the path function $\vec{f}'(u)$ and the function for the surface normal $\vec{n}(u)$. With these definitions, a detector plane is automatically set, which is defined by the point $\vec{f}(End)$ and the plane normal $\vec{f'}(End)$ (Fig. 3.9).

Light source definition

The next step in the algorithm is the definition of the light source. By default, the light source is a point source ("width of light source = 0," see Fig. 3.10) starting in the center of the front face of the waveguide at $P(x = 0|y = height/2|z = 0)$. One can manually vary this entrance point by adding a three-dimensional offset.

In reality of course, the light source does not sit directly in the front face of the waveguide, but is focused on it or is butt coupled [26] by a fiber. Therefore, an indirect definition of the light source is implemented. First, typically the light source is defined by its numerical aperture (NA_{ls}):

$$NA_{ls} = n_i \cdot \sin(\theta_{max}), \tag{3.19}$$

which defines a maximum angle θ_{max} for incoming rays with the optical axis (see Fig. 3.11) in the present material in front of the waveguide defined by its refractive index n_i. By applying Snell's law

$$n_i \cdot \sin(\theta_{max}) = n_{POW} \cdot \sin(\theta_t) \tag{3.20}$$

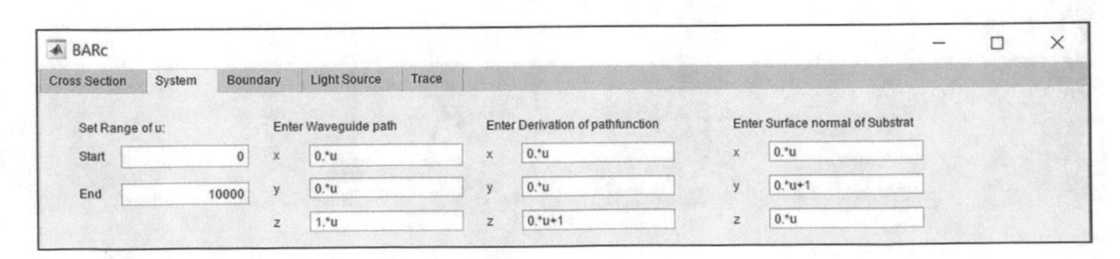

Fig. 3.9 Input mask for the definition of the print path and the surface normal, both depending on the parameter u, which is limited by "start" and "end." Since the application is written in MATLAB, the functions are written for elementwise multiplication of matrices with ".*"

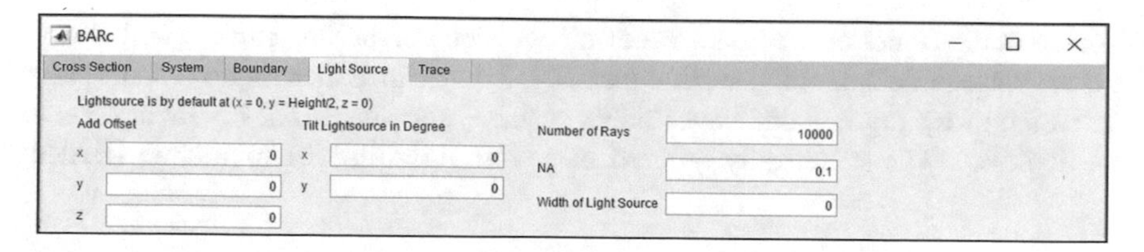

Fig. 3.10 Definition of the light source for the simulation

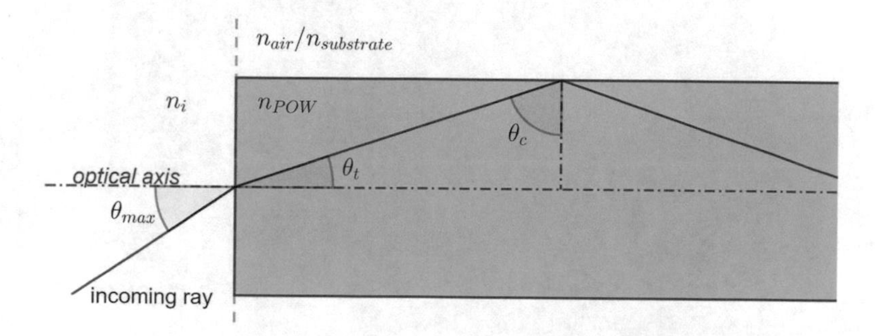

Fig. 3.11 Overview of the angles needed for the definition of the initial numerical aperture

and using the refractive index n_{POW} of the POW, a maximum angle θ_t can be calculated. For the indirect definition of the light source at the point $P(x = 0|y = height/2|z = 0)$, the NA is now entered as:

$$NA = n_{POW} \cdot \sin(\theta_t). \tag{3.21}$$

With the parameter NA, one has defined a cone in three-dimensional space, in which rays can travel. The next step is to define N ("=number of rays") rays, each having its starting point in $\overrightarrow{P}$ and a direction vector $\overrightarrow{Q}$ so that they propagate only in the three-dimensional cone. In order to get good statistical representation, the rays are uniformly distributed over the whole cone.

The orientation of the cone can be adjusted. By default, the center of the cone is along thc optical axis, which is in these representations the z-axis. In order to simulate a tilt of the light source, for example, the butt-coupling fiber is misadjusted, the cone can be tilted by defining two angles for the x- and y-axis rotation.

In addition, when using a butt-coupling fiber, the entrance point of the light into the waveguide cannot be considered as a single point but has to be assumed as an extended point (circular area). The diameter of this circle can be adjusted by entering its value in "width of light source." The starting point P for each ray is then randomly generated within this circle.

Calculate Ray Intersection by BARc Algorithm

After generating N initial rays $r_i(s)$ with $i \in \{1, \ldots, N\}$, the algorithm has to find the intersection of each ray with the first surface it hits. When starting with a light source in the center of the front face of the waveguide, the first surface will always be the waveguide's defining surface (see Fig. 3.12). Since there is no default plane in which every ray hits the waveguide's surface, non-sequential raytracing has to be deployed. The specifically tailored bisection algorithm for raytracing (BARc)

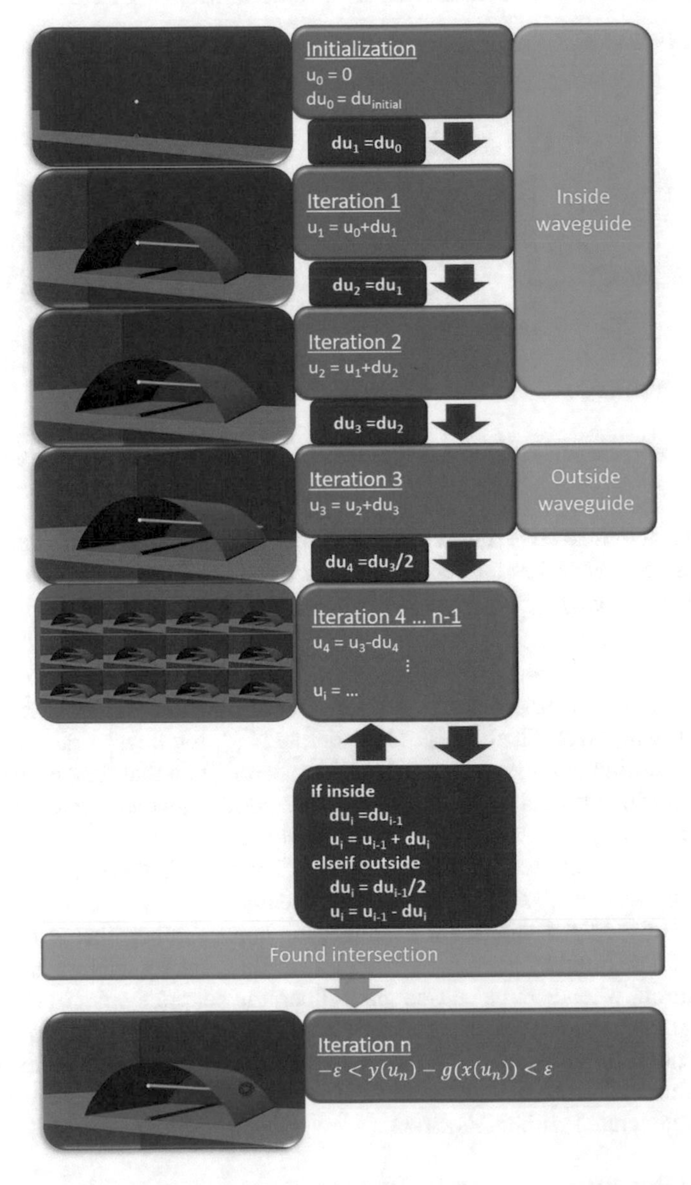

Fig. 3.12 Overview of the BARc algorithm with schematic display for a straight waveguide and one ray (green). The light source is depicted in the schematic in yellow and the path function in red

uses a non-sequential raytracing technique combined with the bisection method [27] to simulate light propagation in POWs.

For simplicity, we will show the algorithm only for one ray $r(s)$ but keep in mind that the algorithm does all following steps simultaneously for all rays. First, one has to define an initial distance du_1 to calculate a new point $\vec{P}_{new}$ on the path function $\vec{f}(u)$. When u_0 is defined by the plane in which the light source lies, we can formulate the following equation for the new point with $u_1 = u_0 + du_1$:

$$\vec{P}_{new} = \vec{f}(u_1) \tag{3.22}$$

As shown in Sect. 3.2.2, one can define a plane through $\vec{P}_{new}$ and calculate an intersection point I_{loc} of the ray and the plane. The next step of the algorithm is decided by the position of the intersection point I_{loc} in relation to the waveguide (see Fig. 3.6). Depending on whether the point is inside the waveguide, outside the waveguide or on the surface, different actions have to follow. Here, assume the intersection point I_{loc} is still inside the waveguide (as shown in Fig. 3.12). For this result, propagate the plane along the path function for the same distance ($du_2 = du_1$) as before:

$$\vec{P}_{new} = \vec{f}(u_2 = u_1 + du_2) \tag{3.23}$$

For an intersection point I_{loc} in the (i−1)-th iteration, which is inside the waveguide, a general formulation of the next u_i can be written:

$$du_i = du_{i-1} \tag{3.24}$$

$$u_i = u_{i-1} + du_i \tag{3.25}$$

With this formulation, one propagates the ray indirectly along its path by propagating along the path function. However, at some point, the local positioning check will state that the ray is no longer inside the waveguide but outside the waveguide (in the example in Fig. 3.12 in iteration 3). Therefore, the ray has to be propagated backward in the next step, since the intersection of ray and surface was over jumped. In order to propagate backward along the path function, the new point has to be calculated as follows:

$$\vec{P}_{new} = \vec{f}(u_4 = u_3 - du_4) \tag{3.26}$$

with $du_4 = du_3/2$.

A general formulation can be derived:

$$du_i = du_{i-1}/2 \tag{3.27}$$

$$u_i = u_{i-1} - du_i \tag{3.28}$$

With this formalism, the algorithm iterates until Eq. (3.18) is satisfied, and the intersection point $I_{gl}(x, y, z)$ is determined.

Surface Interaction

After finding the intersection point of ray and surface, the influence of the surface on the ray's direction has to be defined. In general, three interactions are possible: absorption, reflection and transmission (see Fig. 3.13). Absorption, defined as the transformation of electromagnetic radiation into another type of energy, leads to a cancelation of the ray. However, this interaction depends on the absorptance of the second medium, which is extremely small for optical materials such as the polymer varnishes used for the print of the waveguides, and therefore, absorption is neglected in the simulations. In contrast, when using optical materials, the properties of reflection and transmission are essential. In order to define an ongoing ray after the intersection point, one has to apply Snell's law. If the refractive index of the first material is higher than the refractive index of the second material ($n_1 > n_2$), a critical angle α_c can be calculated

$$\alpha_c = \sin^{-1}\left(\frac{n_2}{n_1}\right) \tag{3.29}$$

at which the ray is no longer transmitted but reflected with the same angle as the incident angle α_{in}. In the algorithm, the angle α_{local} between the local surface normal at point I_{gl} and the ray is calculated and compared to the critical angle α_c. In the case of $\alpha_{\text{local}} \leq \alpha_c$, the ray would be transmitted. Since we are not interested in the behavior of the ray outside the waveguide, the ray is deleted and is count as a lost ray, since it does not reach the detector at the end of the waveguide. In the case of $\alpha_{\text{local}} > \alpha_c$, the ray is reflected at the surface, and a new ray starting in point I_{gl} is accordingly calculated. For rays that do not exit the waveguide, the steps "define new ray," "calculate ray intersection" and "surface interaction" are looped through in a circle, as shown in Fig. 3.7. This loop is only broken, if the ray hits the detector plane at the end of the waveguide.

Output

When simulating the propagation of light in the waveguide, the key parameter is the attenuation, since it defines the quality of the waveguide. The definition of attenuation D in terms of raytracing is given by

$$D = \frac{10}{l} \cdot \log\left(\frac{P_0}{P_l}\right), \tag{3.30}$$

where P_0 is the initial number of rays and P_l the number of rays which are still in the waveguide and hit the detector after distance l.

Another output of the simulation besides the attenuation D is the intensity distribution at the end of the waveguide. In order to get this distribution, for each ray, the intersection point with the end face of the waveguide is calculated and subsequently registered in a two-dimensional pixelated picture. The number of rays per pixel determines the intensity value, and the size of the pixels determines the resolution of the intensity distribution plot. In Fig. 3.14, one can see the influence of the pixel size. For a higher resolution also, more rays have to be used.

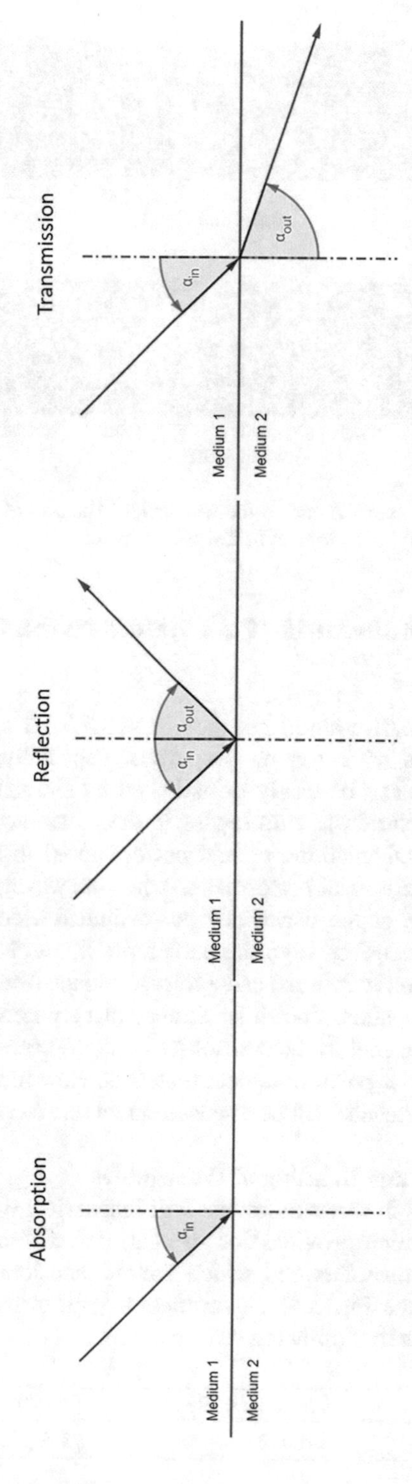

Fig. 3.13 Three possible surface interactions are shown: absorption, reflection and transmission

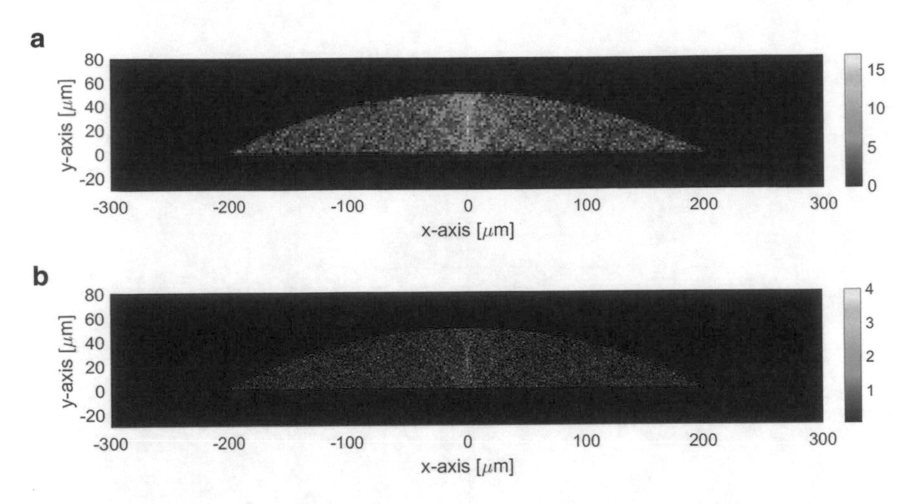

Fig. 3.14 Intensity distribution at the end of the waveguide. The size of the pixels determines the resolution of the picture: **a**) 56 × 376pixel **b**) 221 × 1501 pixel

3.2.4 Impact of Manufacturing Parameters on the Optical Properties

After implementing the BARc algorithm (see Sect. 3.2.3), it is possible to simulate different waveguides with varying properties. Especially, influences of the manufacturing procedure can be easily investigated by simulations. One obvious influence by the manufacturing is with regard to the cross section: How does the D-shaped cross section influence the optical performance? In addition, how does the cross section effect the optical performance when the waveguide is bended?

Further, the production of the waveguide has a limited accuracy in the surface profile. A deviation of the surface from the perfect profile can be subdivided in two categories: a macrodeviation (waviness) and a microdeviation (roughness). Both influences are subject of investigation, since it is assumed that a waviness is introduced by a non-constant flow of material in the aerosol-jet printing process (see chap. 5) and a roughness by the usage of polymer optical materials. How much each effect does influence the optical performance will be discussed in this chapter (Fig. 3.15).

D-shaped Cross section and Bending of Waveguides

Before discussing defects in waveguides, we will first review what attenuation values the simulation algorithm provides for straight, defect-free POWs. In current productions, POWs are manufactured which have dimensions of 300 μm width and 30–40 μm height (see Fig. 3.4). A simulation with different heights of the POW was performed with the following settings:

Parameter	Length l	Number of Rays	Width b	Height h
Value	10 cm	20000	300 μm	10–60 μm

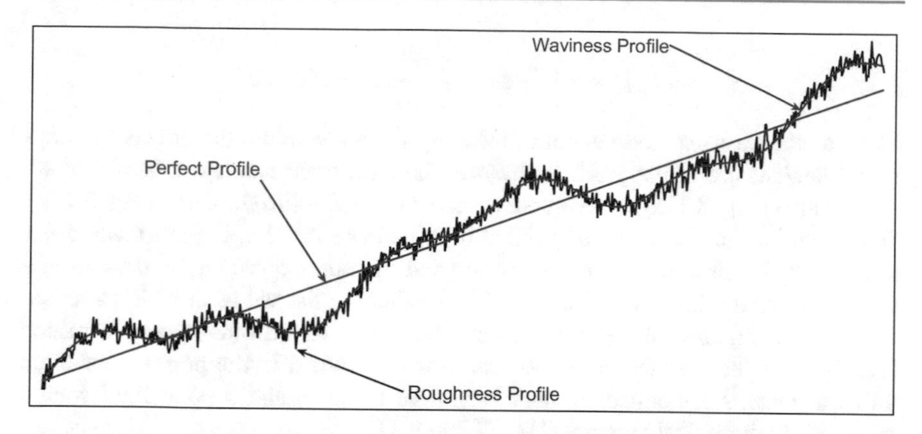

Fig. 3.15 Example of a detected surface profile (black). The mean of the surface is the waviness profile (red), which differs from the perfect profile (blue). The difference of the detected surface and the waviness profile is the roughness profile. [25]

With the simulation results from Fig. 3.16, it can be seen that the damping is exactly zero for all height settings. Therefore, the D-shaped cross section does not contribute by default any attenuation and the printing process resulting in such a cross section has no disadvantage.

This result is consistent with expectations because rays are counted as losses in the algorithm only if they fall below the critical angle of total internal reflection at one of the interfaces. This is not the case for a minimum angle of incidence $\theta_{min} = 86.4122°$ at a numerical aperture NA$=$0.1, when the largest critical angle is $\theta_{\mathrm{crit,PMMA}} = 68.4368°$. The minimum angle of incidence for a core with refractive index $n_{cor} = 1.598$ results from:

$$\mathrm{NA} = n_i \cdot \sin(\theta_{max}) = n_{cor} \cdot \sin(90^o - \theta_{min}) \tag{3.31}$$

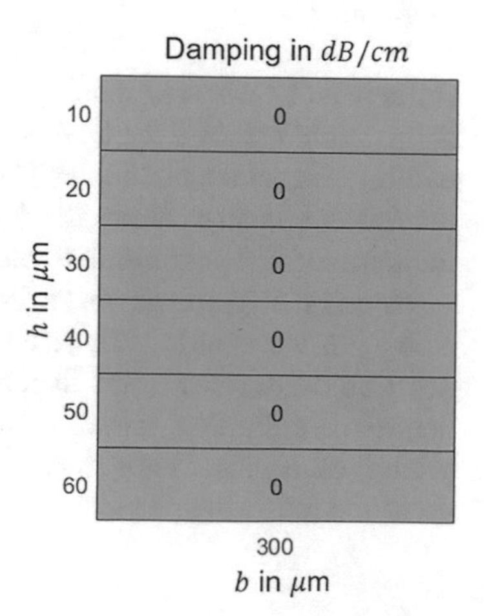

Fig. 3.16 Damping of a perfectly produced straight polymer optical waveguide

$$\Rightarrow \theta_{min} = 90^\circ - \sin^{-1}\left(\frac{NA}{n_{cor}}\right) = 86.4122^\circ.$$

Since a straight waveguide is not influencing the attenuation, the effects of bends in waveguides are investigated. Therefore, three different bends can be classified: Outer curve (Fig. 3.17a), inner curve (Fig. 3.17b) and left/right curve (Fig. 3.17c). While a left/right curve can be printed on a planar substrate, the inner curve has to be printed inside of a bended substrate and the outer curve on the outside of a bended substrate. In order to find a critical value for the radius at which attenuation does occur, simulations with varying radii were for all three cases performed [28, 29]. For the simulation, a 180° curve was assumed in the presence of three different materials, defined by their refractive index n_i and a point light source with a given numerical aperture (NA) (Table 3.1).

The result of the simulation shows that an outer curve does have the smallest critical value (13.65 μm) and should therefore be the preferred curve when designing the printing path. The critical value for the here simulated inner curve was 305.2 μm and for the left/right curve 2131 μm. Since rays are always guided on the outside of the waveguide at a curve, the difference in the behavior can be found in the difference of the material on the outside of the curve. While it is the cladding (here air) for the outer and left/right curve, it is the substrate for the inner curve, resulting in a much smaller difference between the refractive indices of the two materials and therefore more likely for rays to exit the waveguide (see surface interaction Fig. 3.13). The high critical radius for a left/right curve however can be explained by the very small corners of the D-shaped cross section of the waveguide. Since again the rays are guided on the outside of the waveguide, the angle between the ray and the surface exceeds the critical angle and therefore light couples out. From these simulations, two design rules can be formulated:

1. Radii smaller than the critical radius should not be used in the design process.
2. Outer curve should be preferred over inner curve, inner curve over left/right curve.

Influence of Imperfections

If the geometry of the POW is changed along the waveguide path in height and width, different attenuation values are to be expected, since the angles of incidence of rays can change along the length of the POW. Here, we will investigate the influence of inclusions and droplets (see Fig. 3.18) on the attenuation behavior.

In order to simulate these imperfections, a method to model such a change along the waveguide path is necessary. Therefore, one has to find functions to describe the surface limits in regard to the elongation direction of the waveguide defined as z. Further, these investigations are done on straight waveguides that are printed on a planar substrate. Finally, to make a reasonable choice for the functions, the following properties of imperfections are defined:

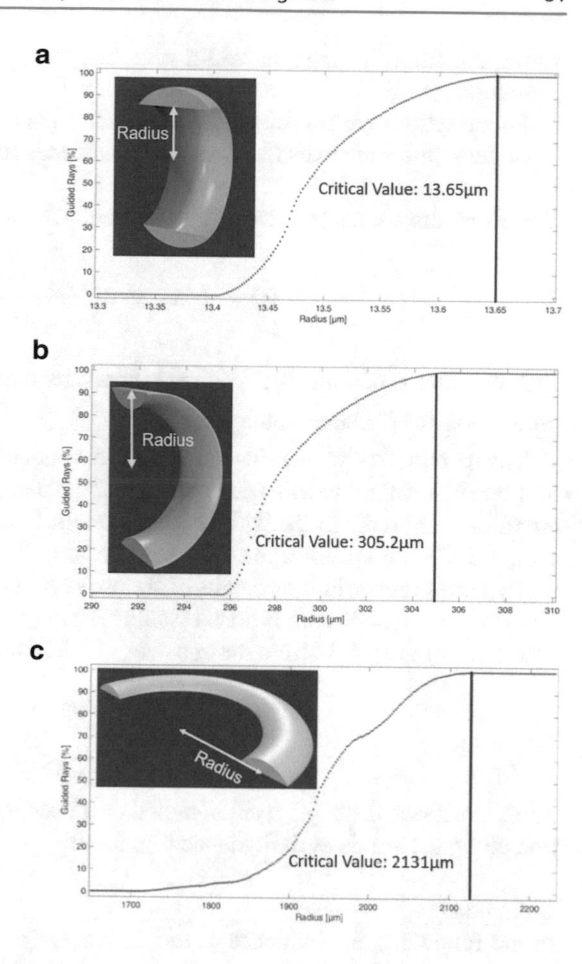

Fig. 3.17 By varying the radius for **a**) an outside curve, **b**) inside curve and **c**) left/right curve, the critical radius is defined as the smallest radius at which 100% of the rays are guided

Table 3.1 Parameters for the simulation of bended waveguides

Parameter	Height	Width	Core n_{cor}	Cladding n_{clad}	Substrate n_{sub}	NA
Value	40	300	1.598	1	1.49	0.1

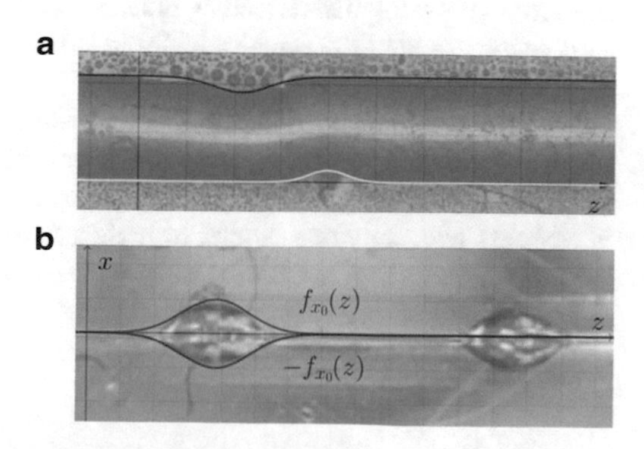

Fig. 3.18 Imperfections of waveguides do occur during the printing process. **a**) Inclusions are the result of not enough applied material. **b**) Droplets are the result of too much material. Both imperfections can be modeled by an enveloping function

1. Imperfections change in width and height in a (mathematical) continuous matter.
2. Imperfections are considered to be symmetrical around the print path.
3. Imperfections are extended over short distances (usually a few hundred μm).

Therefore, the choice falls on a Gaussian function of the form

$$f_{x_0}(z) = a \cdot x_0 \cdot \exp\left(-\frac{(z-\mu)^2}{2 \cdot \sigma^2}\right). \tag{3.32}$$

This function is continuous, symmetric about its maximum in z = μ and converges rapidly to zero for larger multiples of σ, i.e., $f_{x_0}(z) \overset{z=\mu\pm(k\cdot\sigma)}{\rightarrow} 0, k \gg 1$.

The parameters of the function are the z-position μ at which the imperfection has its center, the half width σ of the imperfection along the z-axis, a relative amplitude *a* and the initial width b' resp. height h' of the waveguide x_0. The factor $a \cdot x_0$ allows the specification of a relative amplitude *a*, independent of the currently considered width or height of the waveguide. The modulation of the POW along the waveguide path is done by subtracting (inclusion) or adding (droplet) the function from Eq. 3.32 from the constant width b' or height h':

$$b(z) = b' \pm f_{x_0=b'}(z)$$

$$\mathrm{h}(\mathrm{z}) = h' \pm \mathrm{f}_{\mathrm{x}_0=h'}(\mathrm{z})$$

Thus, functions for b and h are obtained as a function of z, changing the cross section along the z-axis as desired (see Fig. 3.19).

Inclusion

In the following, the influence of the dimension of the imperfection is systematically analyzed. First, an inclusion is the subject of investigation. The two geometric parameters, namely amplitude *a* and the half width σ, are varied separately, and their influence on the attenuation is determined separately.

To investigate the amplitude, a waveguide with an inclusion of fixed extension ($\sigma = 50\,\mu\text{m}$) is simulated and different relative amplitudes are run through. While the parameters for the refractive indices and the initial height and width of the waveguide are kept the same as in Table 3.1, the settings of Table 3.2 were chosen for the simulation.

Before we can analyze the attenuation result of the simulation, a remark regarding its values: In consistency with all other simulations, the attenuation is defined as dB/cm. This length dependency however can be misleading, since a longer waveguide with only one droplet or inclusion and no other defects would decrease the attenuation value in dB/cm in its own. In order to compare the influence of the

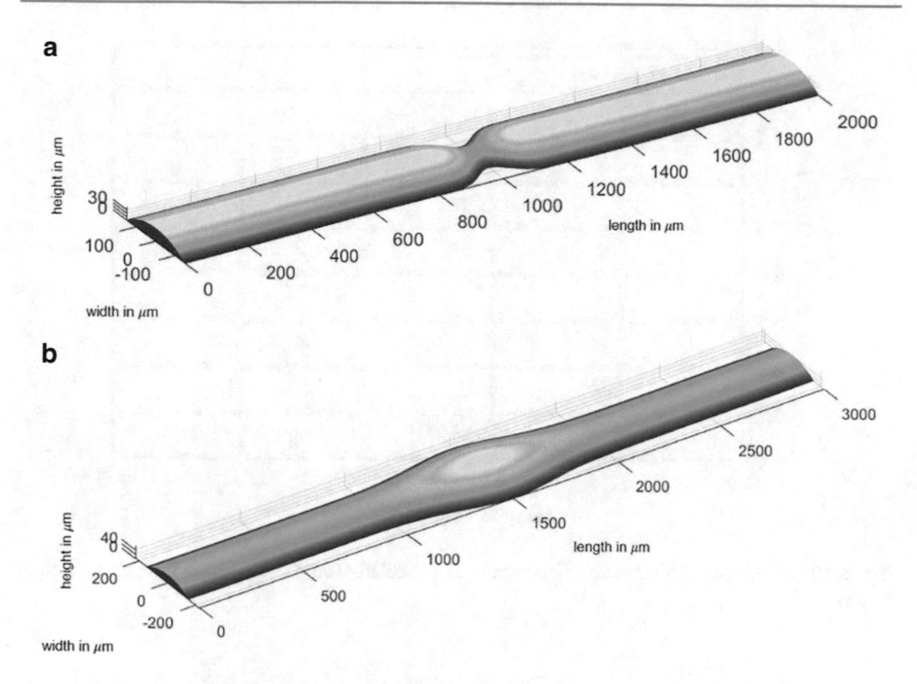

Fig. 3.19 Resulting model for **a**) an inclusion and **b**) a droplet

Table 3.2 Simulation parameters

Parameter	#Rays	Length l	Half width σ	Position μ
Value	20000	2 cm	50 μm	l/2

droplets and inclusions, the length is always set to the same length, so that it does not influence the comparability. This holds for both cases, inclusions and droplets.

Figure 3.20 shows that the attenuation is zero for amplitudes $a < 0.1$ and increases from $a \geq 0.1$ for larger amplitudes. It also shows that the damping does not increase linearly. The increase in attenuation can be derived by considering the "hit area of inclusion." Rays are calculated as loss when the angle of incidence of the ray on a boundary surface falls below the critical angle of total reflection. This is possible at the front side of an inclusion, i.e., in the $z \leq \mu$ region. Considering a ray in the plane composed of its direction vector and the unit vector of the z-coordinate $\vec{e_z}$, the angles are obtained from Fig. 3.21. The vector $\vec{r}$ is obtained by projecting the direction vector of the light ray $\vec{Q_1}$ onto the x-y-plane.

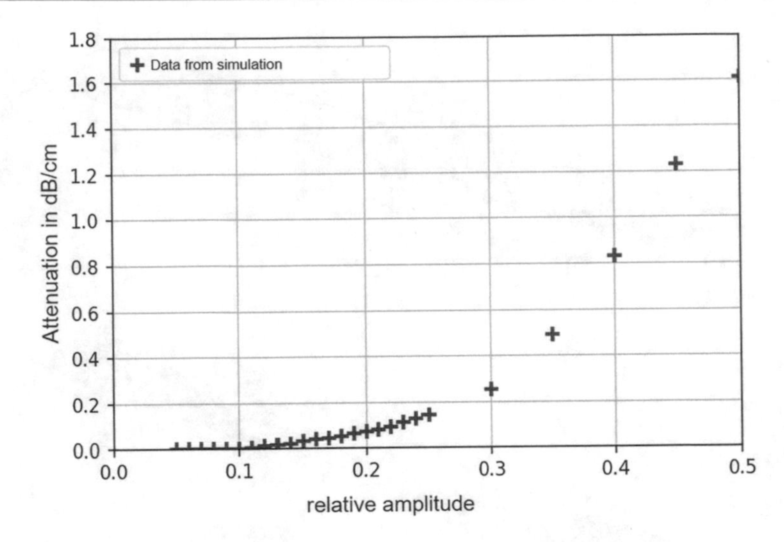

Fig. 3.20 Graph of the attenuation depending on the relative amplitude for a given half width of 50 μm

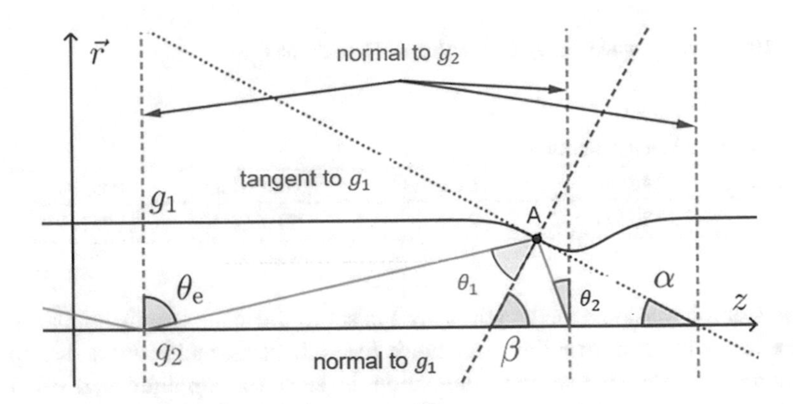

Fig. 3.21 Angle dependencies are shown for an incident ray reflecting on the inclusion

Here, it can be clearly seen that above a certain amplitude there is a possibility that rays fall below the critical angle of total internal reflection either directly at the interface between air and POW, but only for very large amplitudes a, or at the interface between substrate and POW, after being reflected at the top surface. The hit area of the inclusion, for which the criterion $\theta_r < \theta_{crit}$ applies, increases with increasing amplitude. Obviously, the larger the hit area, the higher the probability that rays of light will hit it and thus count as lost rays. This means that the larger the hit area of the inclusion, the higher the attenuation.

The more exact curve of the attenuation in Fig. 3.20 can be explained qualitatively as follows:

A minimum value of the amplitude must be reached so that a hit surface greater than zero is created at which $\theta_1 < \theta_{crit}$ or $\theta_2 < \theta_{crit,2}$ becomes possible. As long as this minimum amplitude is not exceeded, the damping is zero.

The hit area increases from $a \geq 0.1$ and so does the attenuation. The increase is not linear because the hit area does not grow linearly. A more detailed consideration of the area is quite complex, since it depends on the geometry of the cross section and the minimum angle of incidence at a numerical aperture of here 0.1.

For relative amplitudes greater than 0.5, the damping is expected to continue to increase, until it reaches the value $D = \infty$ at $a = 1$ (for this value, the waveguide would have height and width of zero, so that no light can pass), since the hit area increases until no more rays can pass the inclusion.

In a next step, the length of the inclusion is varied and analyzed on its influence on the attenuation. Here, the amplitude is fixed in three different settings, and the following settings were selected for the simulation (see Fig. 3.22). The full set of parameters for the simulation is given in the following:

Parameter	#Rays	Length l	Rel. Amplitude a	Position μ
Value	20000	2 cm	$a \in \{0.1, 0.3, 0.5\}$	l/2

From Fig. 3.22, it can be seen that the damping for each of the three selected amplitudes decreases as the expansion increases. The maximum of the attenuation is always at thc lowest expansion, e.g., at $a = 0.3$ the maximum is at $\sigma = 25$ μm with an attenuation of D = 1.056 dB/cm. In all three cases, after a certain expansion, losses no longer occur. A decrease in damping with increasing expansion can be explained as follows: Decisive for the angle of reflection at the boundary surface of the inclusion is the orientation or inclination of the normal vector at the point of intersection of the ray with the boundary surface. If the ray path is again considered in the $\vec{r}$-z-plane, the sketches in Fig. 3.23 result.

Thus, the angle θ_r depends on the inclination β of the normal to the graph of the boundary surface g_1 at point B with respect to the z-axis, which is given by the relation $\beta = 90° - \alpha$ from the slope of the tangent to the graph of g_1 at point B. The larger α, the smaller β and the smaller the angle of reflection θ_r becomes.

One can again define a hit area. It denotes the area of inclusion $\mu - c_1 \cdot \sigma \leq z \leq \mu - c_2 \cdot \sigma$ with $c_1, c_2 \in \mathbb{R}$ where $\theta_r < \theta_{crit}$ holds.

Attenuation in dB/cm

relative Amplitude a \ Expansion σ of Enclosure	25	50	75	100	125	150	175	200	225	250	275	300	325	350	375	400
0.1	0.01908	0.001086	0	0	0	0	0	0	0	0	0	0	0	0	0	0
0.3	1.056	0.25	0.1844	0.1177	0.0552	0.003912	0	0	0	0	0	0	0	0	0	0
0.5	2.506	1.594	0.7251	0.4605	0.2301	0.0315	0	0	0	0	0	0	0	0	0	0

Fig. 3.22 Result of the simulation of an inclusion/enclosure with dependency on the expansion and relative amplitude

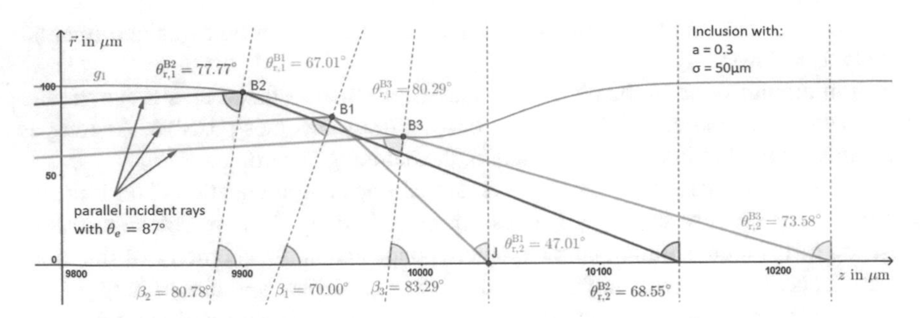

Fig. 3.23 Ray path displayed for three different but parallel rays incident with an angle of 87° to the z-axis

Conclusion: The attenuation is larger the higher the amplitude and the smaller the width of the inclusion is.

Droplet

The same parameter variation can be done for the droplets. Again, we first vary the amplitude and investigate its influence on the attenuation. The parameters of the simulation are:

Parameter	#Rays	Half Width σ	Length l	Position μ
Value	20000	500 μm	2 cm	1/2

Looking at the progression of the attenuation as a function of relative amplitude in Fig. 3.24, a flat increase can be observed for $a<2$. After that, the attenuation increases much faster from 0.38 dB/cm at $a=2$ to 1.06 dB/cm at $a=2.5$, and for $a>2.5$, the further increase is again flatter and comparable to the first section $a<2$. A particularly important observation from Fig. 3.24 is that the attenuation keeps increasing, even for relative amplitudes of $a>3$. Even if such high values are never reached in the optical waveguides produced, this is definitely of interest from a theoretical perspective.

The values can be explained if the slope of the interface of the droplet toward the cladding (g_1) is considered. Analogous to inclusions, the slope must be large enough so that the reflected ray makes an angle $\theta_{r,2} \leq \theta_{crit}$ to the normal to the interface g_2 between the PMMA substrate and the POW. Compare Fig. 3.25, which shows various rays incident on the interface g_2 of a droplet in the POW.

From Fig. 3.25a, it can be seen that the position of the reflection in the droplet has an influence on $\theta_{r,2}$, and in Fig. 3.25b, that the relative amplitude influences the reflected angle. The larger the amplitude, the smaller the angle $\theta_{r,2}$. Similarly, for higher amplitudes, the hit area of the droplet becomes larger, increasing the probability of rays hitting the same and then being counted as losses. This then results in a higher attenuation, as it can be observed in Fig. 3.24.

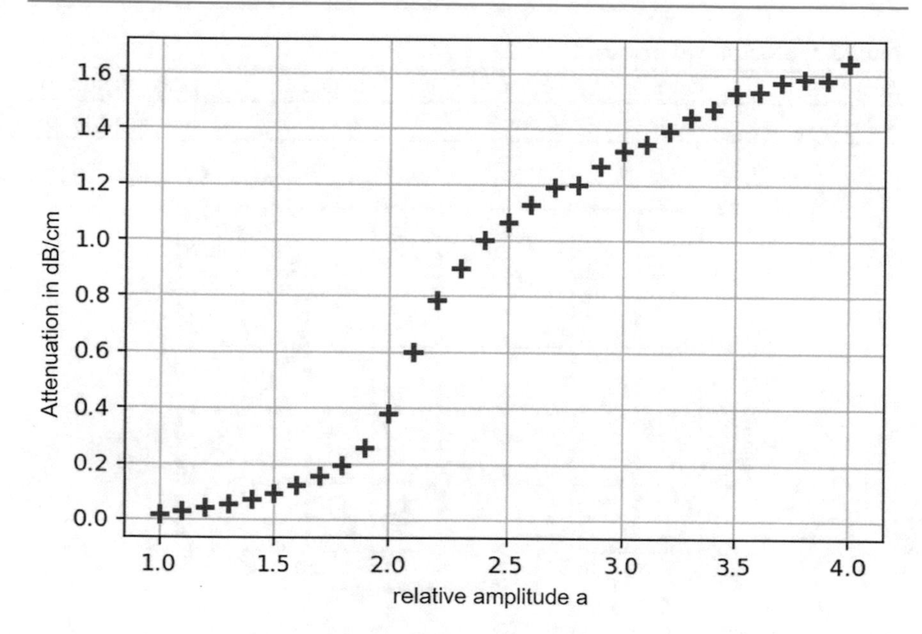

Fig. 3.24 Attenuation in dependency of the relative amplitude a for a droplet

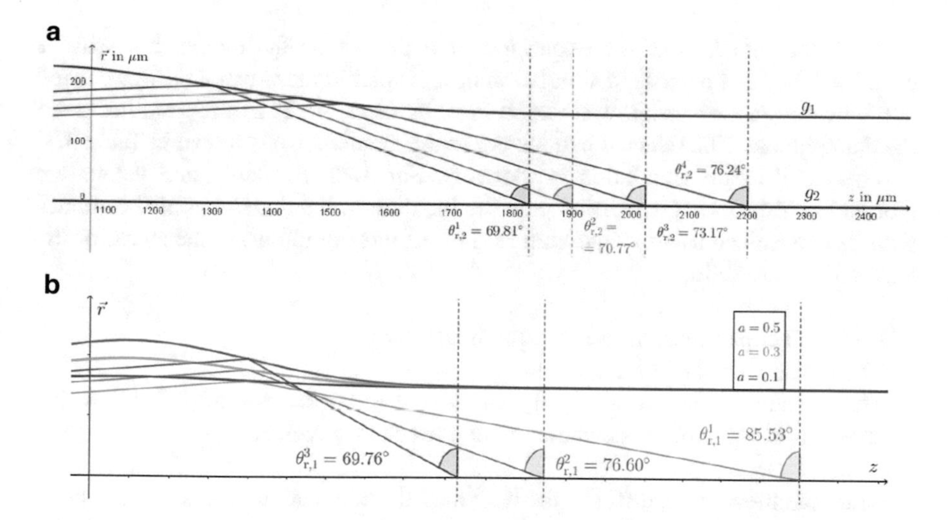

Fig. 3.25 **a**) Position of reflection point influences the angle with the substrate. **b**) Graph shows the influence of different relative amplitudes

Table 3.3 Simulation parameters

Parameter	#Rays	Length l	Position μ	Rel. Amplitude a
Value	20000	2 cm	l/2	$a \in \{0.5,1.0,1.5\}$

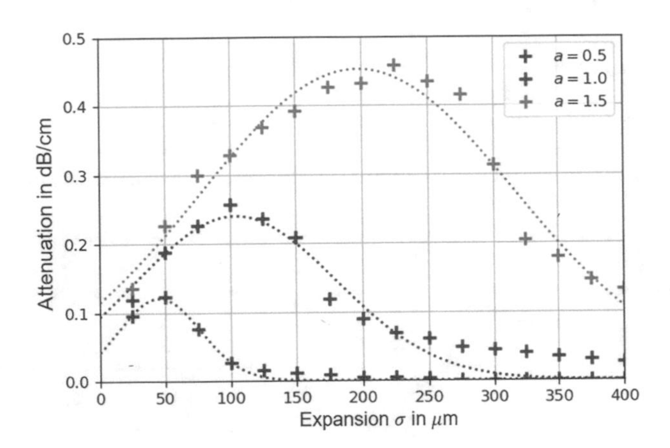

Fig. 3.26 Graph shows the dependency of the attenuation on the expansion σ for three different relative amplitudes a

Next, the influence of the expansion of a droplet on the attenuation will be examined. For this purpose, a simulation of a droplet in the waveguide was carried out, whereby the expansion was varied in steps of $\Delta\sigma = 25\,\mu\text{m}$ for each of three fixed amplitudes. The selected parameters of the simulation are shown in Table 3.3.

The result of the simulation is shown in Fig. 3.26. The simulated values are plotted as a function of expansion, and the location of the maximum of the attenuation is determined using a Gaussian fit. For all three amplitudes, the shape of the attenuation is as follows:

1. For small expansions, the attenuation increases.
2. The attenuation reaches a maximum.
3. The attenuation decreases rapidly directly after the maximum. After that, the attenuation continues to decrease but at a much slower rate.

The measurement series differ in the height of the attenuation values and the position of the maxima. Thus, the higher the amplitude, the greater the attenuation for the same extension. The position of the maximum of the attenuation shifts more and more toward higher expansion for larger amplitudes.

The observed curve of the attenuation can be explained by the superposition of two phenomena. Firstly, the development of the slope of the interface g_1 and secondly, the development of the droplet hit area (see Fig. 3.27). This is discussed and analyzed in detail in [30].

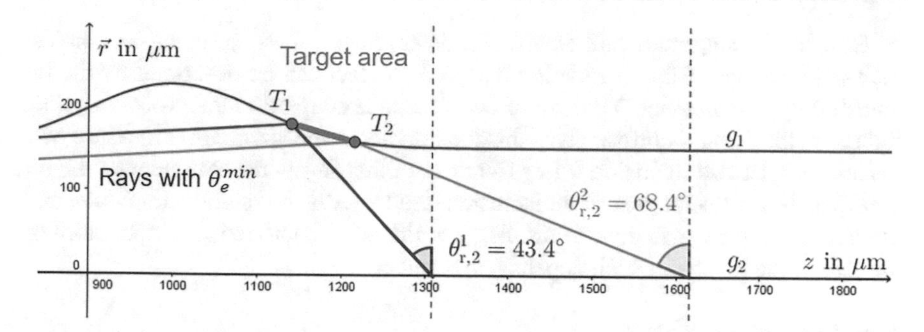

Fig. 3.27 If rays hit with a minimal incident angle of θ_e^{min} the target area, they leave the waveguide toward the substrate

Conclusion: For droplets, a maximum of attenuation exists for a certain extension sigma, whereby the position of this maximum increases with the relative amplitude.

Comparison of Inclusions with Droplets

Following the discussion of the results, a summary comparison will be added to show the important differences and similarities between the influence of inclusions and droplets on attenuation. Basically, two parameters determine the attenuation behavior of droplets and inclusions, the relative amplitude a and the half width σ of the expansion. A quantitative comparison of the attenuation ($\overline{D} = D \cdot l$) of droplet and inclusion is given for different parameters a and σ in the table below:

Relative amplitude a	Half width σ [μm]	Inclusion [dB]	Droplet [dB]
0.1	50	0.002	0
0.5	25	5.012	0.190
0.5	50	3.188	0.244
0.5	100	0.921	0.051

That is, for the same parameters, the influence of inclusions on the attenuation behavior is at least one order of magnitude higher than the influence of droplets. The difference is particularly strong for small expansions of the defects, as can be seen from the difference $D_{inclusion} \approx 26 \cdot D_{droplet}$. The reason for the high differences becomes clear when the factors leading to losses are compared:

A boundary condition for the occurrence of losses at defects in the POW is provided by the *maximum slope* of the interface between the POW and the surrounding medium of the defect (here always air). Reflection at g_1 at the defect results in such a small angle of incidence at the interface g_2 that total reflection is no longer possible. This angle of incidence depends on a and σ and is identical for inclusions and droplets.

However, if amplitude and extent of a defect have values, so that $\frac{\sigma}{a} = c$ holds, then the behavior of the attenuation due to the defect can be described by the hit area and its development. The size of the hit area is coupled to the probability that light rays hit it and therefore gives information about the attenuation measure of a defect. Droplet and inclusion differ from each other in the development of the hit area. The larger the hit area of an inclusion, the larger is the relative amplitude and the smaller is the expansion. In droplets, on the other hand, exists a maximum of the size of the hit area, which depends on a and σ.

Roughness of Surfaces

So far, only perfect surfaces are considered in the simulations. Since the printing process uses polymer materials, it can be assumed that the surfaces do have a certain roughness. To take the roughness into account in the simulation, a mathematical model has to be found to represent the physical behavior.

In general, a rough surface will lead to a diffuse scattering of the light. Figure 3.28 shows the principle sketch of a horizontal section through the waveguide. In this, a ray with an input angle lying in the angular range between zero and the critical angle θ_{kr} propagates through the waveguide [31]. Each time the ray hits the interface between the core and cladding material of the waveguide, the ray is scattered into the complete solid angle by the roughness of the surface. Therefore, from one traced ray, many rays with intensity weighting are generated according to the scattering spectrum and propagate further through the system (reflected rays) or couple out of the waveguide (transmitted rays).

Figure 3.29 illustrates the angular relationship between the angle of incidence of an incident ray and the angular combination of a scattered ray in reflection in hemispherical representation. Here, θ_0 describes the angle of incidence of the incident ray in the local plane of incidence (x-z plane). Accordingly, the scattered rays can be determined by the hemisphere. With the combination of φ (azimuth angle) and θ(polar angle), all scattered rays are exactly assigned. Also indicated in the figure, it is possible to subdivide the hemisphere into equally sized area elements (spatial angle elements).

In order to characterize a rough surface, parameters of the surface have to be defined. The characteristic properties of the interface between core and cladding material (surface properties) are described by the correlation lengths of the surface and the root mean square (RMS) of the height distribution. The correlation lengths

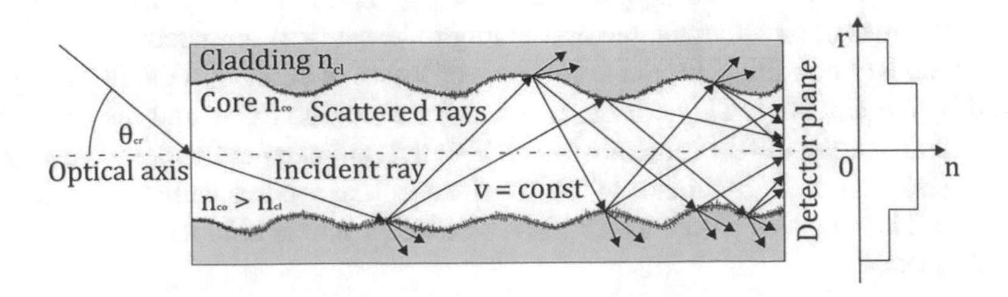

Fig. 3.28 Sketch of a ray traced in a waveguide with a rough surface. [32]

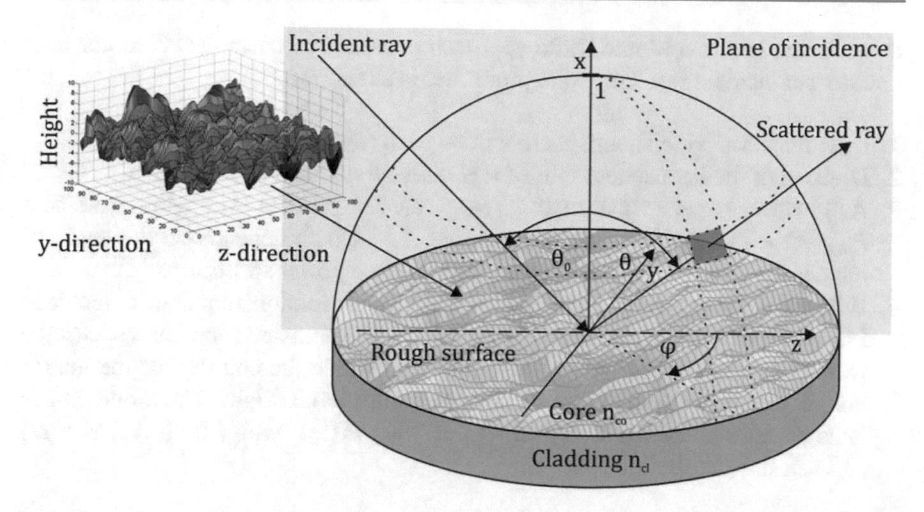

Fig. 3.29 Hemisphere description of the ray interaction with the surface between core material with refractive index n_{co} and cladding material with refractive index n_{cl} with local y and z directions of the surface depending on the incident ray [32]

can be determined via the autocorrelation function (ACF), which describes the correlation of a function with itself. The surface is shifted with itself by a defined value $(u_y|u_z)$, and the expected value defined in Eq. (3.33) is obtained.

$$C(u_y, u_z) = \langle x(y, z) \cdot x(y + u_y, z + u_z) \rangle \tag{3.33}$$

The correlation length is directional and is defined as $C(\vec{\omega}) = C(\vec{0})/e$, where $\vec{\omega}$ is the direction vector from one point to another point in the plane of the surface.

Further, the RMS value, which gives the root mean square of the height distribution, can be given as:

$$\sigma = \sqrt{\frac{1}{n} \sum_{i=1}^{n} (x_i - \langle x \rangle)^2}$$

Here, x_i describes a certain value of the numerical values used for the averaging, n the number of these and $\langle x \rangle$ the mean value.

Monte Carlo Algorithm

The scattering spectrum necessary for Monte Carlo raytracing can be obtained from a 1st order perturbation theory approach [32–35]. The generated scattering spectrum is applied for each ray interacting with the statistically described rough surface. An exponential autocorrelation model is used for the description of the rough surface, and the height distribution of the surface is normal distributed. The main problem of diffuse scattering is the sheer number of rays generated, as with every reflection rays get added to the count (see Fig. 3.28). In order to overcome this limiting factor, a Monte Carlo approach is used. The basic idea is to consider one ray per incident ray. The exact procedure is as follows:

1. Determination which hemisphere will be used (reflection or transmission).
2. Division of the hemisphere into M × N discrete spatial angle elements.
3. A power is assigned to each spatial angle element $P_{m,n}(\theta_0) = P_0 \cdot S^{(s,p)}(\theta, \phi, \theta_0) \forall m \in M, \forall n \in N$, which depends on the angle of incidence θ_0, the initial power P_0 and the discrete power spectrum $S^{(s,p)}$.
4. With the goal of forming a one-dimensional distribution function, a vector is first generated from the matrix $P_{m,n} \to P_i$. The transfer function $i = S(m, n)$ forms a column vector with M x N entries in which the columns of the matrix are put one beneath the other in the order from left to right. Then, one cumulates the entries of the vector $K_j(\theta_0) = \sum_{i=1}^{j} P_i(\theta_0)$ with $j \in \{1, \ldots, M \times N\}$ and gets the distribution function:

$$K(j) = \frac{K_j(\theta_0)}{K_{M \times N}(\theta_0)}$$

5. With the help of this distribution function the classical Monte Carlo algorithm can be established. For this purpose, it is necessary to generate a random number $r \in [0,1]$ from the uniform distribution.
6. The random number r is inserted into the distribution function: $K(j) \leq r < K(j + 1)$
   From the position j, the solid angle element and thus the angle combination (φ, θ) can be inferred, so that the direction of the continuing propagation is determined.
7. Depending on the decision made in step 1 (reflection/transmission), the ray carries only the power of transmission or reflection. Therefore, the lost power must be subtracted from the total power of the ray: $P^r_{m,n} = P_0 - \sum_{m,n} P^t_{m,n}(\theta_0)$ or $P^t_{m,n} = P_0 - \sum_{m,n} P^r_{m,n}(\theta_0)$

If a sufficiently large number of rays is used, the power spectrum is obtained as a good approximation.

Characterization of the used Polymers

After setting up a mathematical model of a rough surface, the parameters *coherence length* and the *RMS* have to be determined of the polymers, which are used in the printing process. The film substrate on which the conditioning lines as well as the core are printed is made of polymethyl methacrylate (PMMA). For the printing process, the UV-curing print varnish J+S 390119 from the company Jänecke+Schneemann Druckfarben GmbH is used as the core material. To accurately determine the surface parameters, the materials were measured using the Profilm3D optical profilometer at the Filmetrics, Inc. application laboratory (Fig. 3.30; Table 3.4).

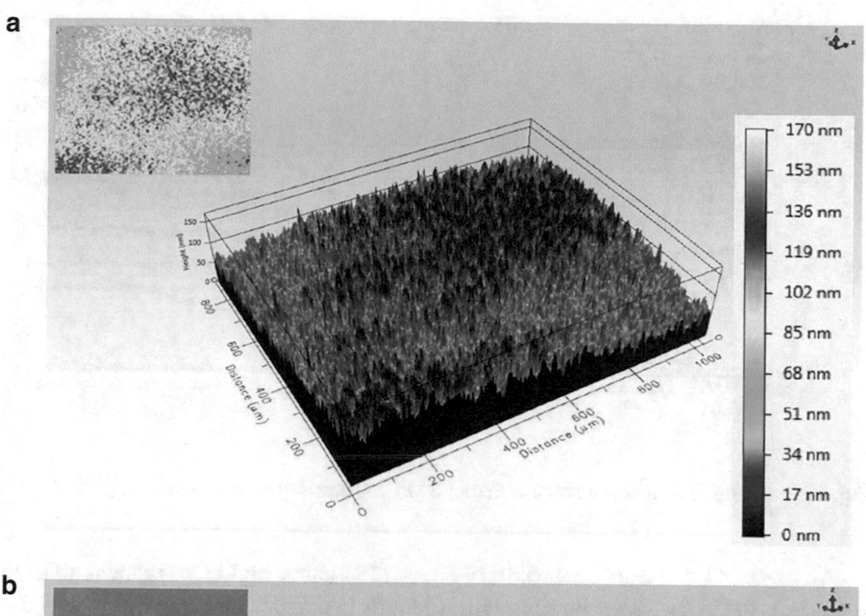

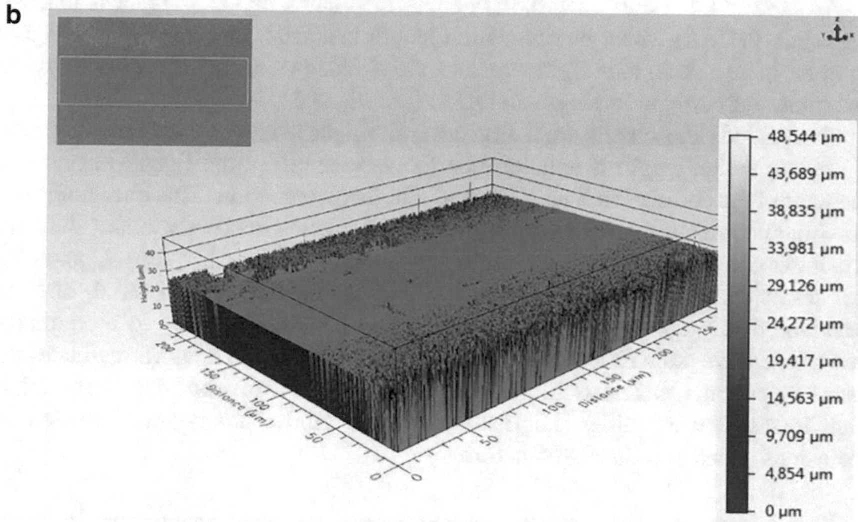

Fig. 3.30 Surfaces captured by the Profilm3D profilometer of (**a**) the substrate PMMA and (**b**) the core material J+S 390119

Table 3.4 Measured parameter for the materials

Parameters	Correlation length	RMS
J+S 390119	14.51 µm	300 nm
PMMA	73.32 µm	16 nm

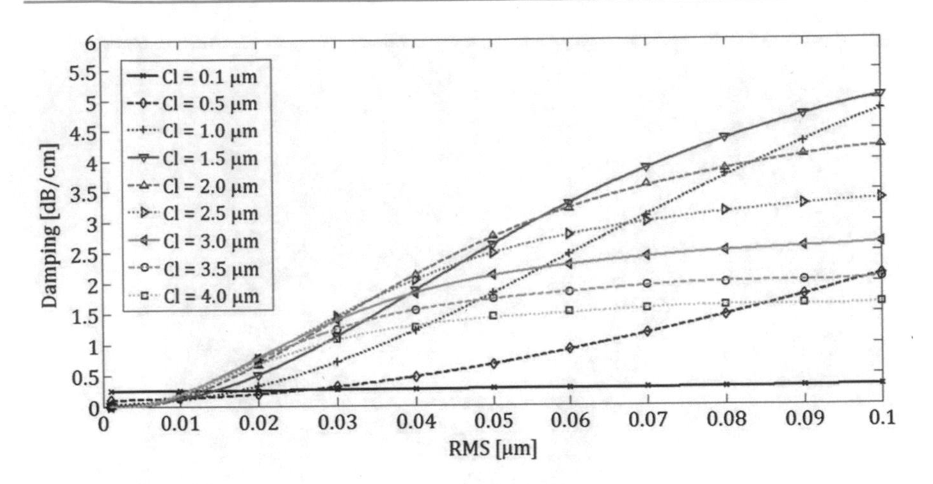

Fig. 3.31 Damping in relation to the different RMS and correlation length values. [36]

As expected, the core material (J + S 390119) has a higher roughness than the substrate (PMMA), since its correlation length is smaller and the RMS is higher. In order to be able to classify the results, simulations were carried out with different values for correlation length and RMS (see Fig. 3.31).

To conclude the results from Fig. 3.31, the light averages over the roughness if the correlation length is smaller than the wavelength of the light. However, an increase of the correlation length leads to a higher attenuation. This only holds to a maximum correlation length after which the attenuation decreases again, since the structures causing roughness are more spread across the surface [36]. Comparing this result with the measured roughness of substrate and core material, one can conclude that roughness does not contribute in a significant way to the attenuation of the waveguide (attenuation < 0.01 dB/cm). This result however validates the usage of polymer materials in 3D-printed optical structures but does not explain high losses in waveguides. High losses in waveguides are typically caused by inclusions which was discussed before.

3.3 Physical Optical Simulation

3.3.1 State of the art for Wave Propagation Method (WPM)

After using raytracing for the simulation of waveguides in Sect. 3.2, we now want to introduce a physical optical simulation method, namely the wave propagation method (WPM). The use of a wave-optical propagation method is necessary, when either interference effects are of importance or the geometric dimensions are in the order of the wavelength. Latter does apply, when the central assumption for raytracing ($\lambda \to 0$) no longer holds and a different way of simulating light propagating in a material has to be defined.

Again, the Maxwell equations [15, 16] are used as the initial description of propagating light. From that, the wave equation for the electric field E of light can be calculated by rearranging and substituting Maxwell's equations into each other and results for a monochromatic wave in the so-called scalar Helmholtz equation describing propagation of light in a (at least piecewise) homogeneous material with refractive index n (k_0 is the modulus of the wave vector in vacuum):

$$\left(\nabla^2 + k_0^2 n^2(x, y, z)\right)E = 0 \tag{3.34}$$

Scalar plane waves represent a solution of this equation and are given by:

$$E(\vec{r}) = E_0 \cdot e^{i\vec{k}\cdot\vec{r}} \tag{3.35}$$

with wave vector $\vec{k} = \left(\frac{2\pi}{\lambda}\right)\vec{e}$, angular frequency ω and complex amplitude E_0. Here, the vector $\vec{e}$ is a unit vector in the direction of propagation of the plane wave.

Another solution of the scalar Helmholtz equation is also given by a sum of plane waves with different propagation directions. This means that a continuous spectrum of plane waves is also a solution. In order to be able to calculate such a solution, it is necessary to integrate over two components of the wave vector. Since only a forward propagation of the light is considered, these two components are defined in x-direction and y-direction. If one introduces now the spatial frequencies

$$\vec{\upsilon} = \frac{1}{2\pi}\vec{k} = \begin{pmatrix} \upsilon_x \\ \upsilon_y \\ \upsilon_z \end{pmatrix} \tag{3.36}$$

one receives the so-called complex amplitude $u(x, y, z)$ as superposition of plane waves:

$$u(x, y, z) = \iint_{-\infty}^{\infty} \tilde{u}(\upsilon_x, \upsilon_y)\exp\left(2\pi i\left(\upsilon_x x + \upsilon_y y + \upsilon_z z\right)\right)d\upsilon_x d\upsilon_y. \tag{3.37}$$

Therefore, the complex amplitude in a layer perpendicular to the z-axis at the point z=0 can be written from Eq. (3.37) by the following equations:

$$u(x, y, 0) = \iint_{-\infty}^{\infty} \tilde{u}(\upsilon_x, \upsilon_y)\exp\left(2\pi i\left(\upsilon_x x + \upsilon_y y\right)\right)d\upsilon_x d\upsilon_y \tag{3.38}$$

$$\tilde{u}(\upsilon_x, \upsilon_y) = \iint_{-\infty}^{\infty} u(x, y, 0)\exp\left(-2\pi i\left(\upsilon_x x + \upsilon_y y\right)\right)dxdy, \tag{3.39}$$

where $\tilde{u}$ is the Fourier transformation of the complex amplitude u. Since for υ_z, the following holds:

$$\upsilon_z = \sqrt{\frac{n^2}{\lambda^2} - \upsilon_x^2 - \upsilon_y^2} = \frac{1}{\lambda}\sqrt{n^2 - \lambda^2(\upsilon_x^2 + \upsilon_y^2)}, \tag{3.40}$$

one can calculate the complex amplitude for a plane at $z_0>0$:

$$u(x,y,z_0)=\iint_{-\infty}^{\infty}\tilde{u}(\upsilon_x,\upsilon_y)\exp\left(2\pi i\frac{z_0}{\lambda}\sqrt{n^2-\lambda^2(\upsilon_x^2+\upsilon_y^2)}\right)\exp(2\pi i(\upsilon_x x+\upsilon_y y))d\upsilon_x d\upsilon_y. \tag{3.41}$$

Discretization of Simulation Field

In general, a computer cannot calculate with continuous functions but always evaluates functions on discrete values (comp. Fig. 3.32). Comparable with pixels on a screen, the continuous appearance is caused by the fact that the distance between the values or pixels is chosen very small. In order to use Eq. (3.41) in a simulation, the field distribution must be discretized or sampled.

Since the field is discretized, the Fourier transformation in (3.41) is also no longer continuous but discrete. In general, the Fourier transformation can be used to decompose arrays into their k-space components. The discrete Fourier transform assumes that the input array repeats periodically. Thus, the discrete Fourier transform corresponds to a Fourier transform on $[0,2\pi]$, i.e., a circle. The entries of the array are evenly distributed on the interval. Thus, each entry is assigned an angle from $[0,2\pi]$. The formula for calculating a discrete Fourier transform converts an N-element array in spatial space to an N-element array in k-space:

$$\mathcal{F}[x(n)](k)=\sum_{n=1}^{N}x(n)\cdot\exp\left(-2\pi i\cdot(k-1)\cdot\frac{n-1}{N}\right). \tag{3.42}$$

In 1965, Cooley and Tukey [37] introduced the well-known "fast Fourier transformation" algorithm (FFT), which made the calculation very efficient for array structures of $N=2^m$ on binary computers. Therefore, in order to have efficient simulations, the number of sample points should be selected to be $N=2^m$ with $m\in\mathbb{N}$.

Light Source Definition

For the initial layer of the WPM, a light source has to be defined. A realistic example of a light wave is described by a Gaussian beam (Fig. 3.33).

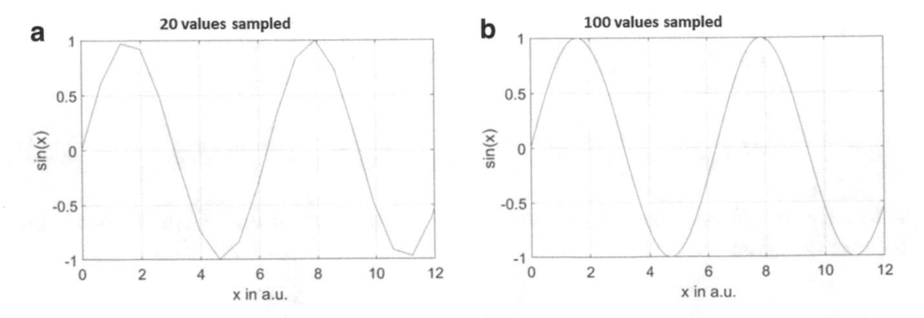

Fig. 3.32 **a**) a sine function with only 20 values sampled. The individual values can be easily distinguished. **b**) Sine function with 100 sampled values, the function appears continuous

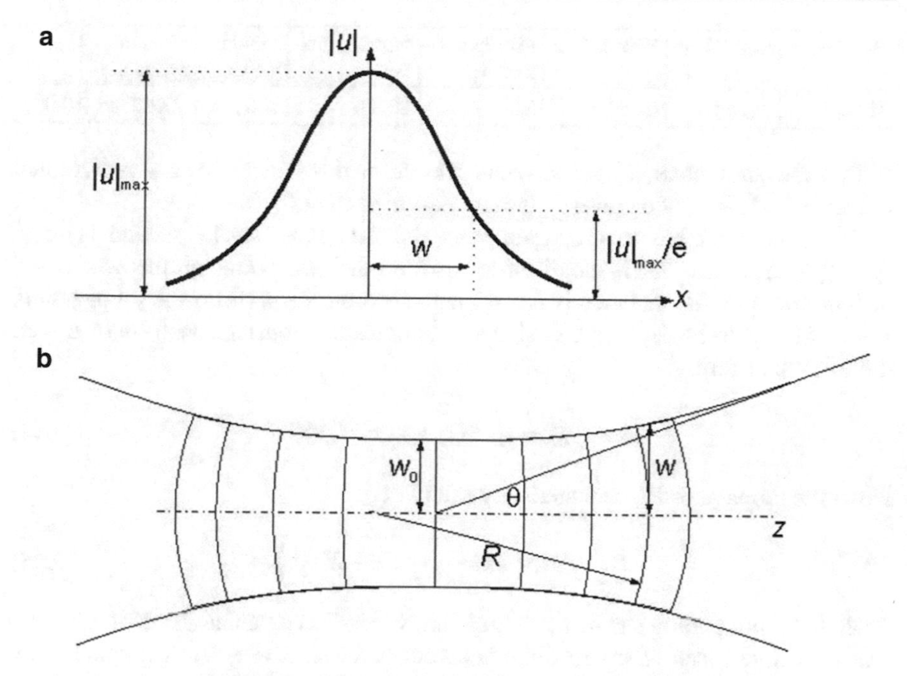

Fig. 3.33 **a**) Amplitude of a Gaussian beam in a plane perpendicular to the propagation direction. **b**) Propagation of a Gaussian beam along the z-axis

$$E(r,z) = E_0 \cdot \frac{w_0}{w(z)} \cdot \exp\left(-\frac{r^2}{w(z)^2} - ik\frac{r^2}{2R(z)} - i(kz - \phi(z)) \right). \quad (3.43)$$

Here, $w(z) = w_0\sqrt{1 + \frac{z^2}{z_R^2}}$ is the width of the Gaussian beam until the field decreases to 1/e of the maximum amplitude. $R(z) = z\left(1 + \frac{z_R^2}{z^2}\right)$ describes the radius of curvature of the wave front and $\phi(z) = \arctan\left(\frac{z}{z_R}\right)$ the phase shift at the wave front. For compact notation, the Rayleigh length was introduced with $z_R = \frac{\pi w_0^2}{\lambda}$. Further, the beam waist w_0 and the far field angle θ are connected to each other via $w_0 = \frac{\lambda}{\theta\pi}$, where θ can be defined by the NA of the light source with $\theta = \sin^{-1}\left(\frac{\text{NA}}{n}\right)$.

Algorithm Explanation at the Example of Free Space Propagation of Gaussian Beam

In order to explain the algorithm in detail, we simulate light propagating in free space from a Gaussian light source in two dimensions. Therefore, we have to set parameters for the light source as well as the material specific properties:

Parameter	Refractive Index	Wavelength	#Gridpoints x-axis	#Gridpoints z-axis	Field in x-axis	Field in z-axis	NA
Value	1 (air)	0.85 μm	256	20000	500 μm	20000 μm	0.005

With these parameters, a 1×256 vector E is defined by Eq. (3.43) and represented in Fig. 3.34. It is the first layer in the simulation at $z_0 = 0$.

Now, the light has to propagate from the first layer to the second layer at $z_2 = z_0 + \Delta z$. Here, Δz is the distance between the grid points on the z-axis, and in this example, the distance is $\Delta z = 1\ \mu\text{m}$. In order to perform the propagation, Eq. (3.41) has to be deployed. First, one calculates the Fourier transform of E with the FFT-algorithm:

$$\tilde{E}(\upsilon_x, z_0) = \int E(x, z_0) \cdot e^{-2\pi i(\upsilon_x x)} dx \tag{3.44}$$

Next, the propagator P_m is numerically calculated.

$$P_m = \exp\left(2\pi i \frac{\Delta z}{\lambda} \sqrt{n^2 - \lambda^2 \upsilon_x^2}\right). \tag{3.45}$$

From the input parameters of the simulation, $k_0 = \frac{2\pi}{\lambda}$ is calculated, and the respective refractive index of the medium is taken; here, air $n = 1$. υ_x then denotes the entries in an array of the υ-space defined as follows:

For a given width of the electric field $E_x([-x_{max}, x_{max}])$, the υ-space corresponds to a width of $\left[-\upsilon_{x_{max}}, \upsilon_{x_{max}}\right]$ and a distance between entries of $\Delta\upsilon_x$. Since only a limited lateral range is sampled in the spatial domain of the field, the υ -space must also be limited accordingly. Just as $2x_{max} = N_x \cdot \Delta x$ is valid in spatial domain, $2\upsilon_{x_{max}} = N_{\upsilon_x} \cdot \Delta\upsilon_x$ is valid in υ-space where $\Delta\upsilon_x = \frac{1}{2x_{max}}$. This results in

$$\upsilon_{x_{max}} = \frac{N_x}{4 \cdot x_{max}} \text{ with } N_x = N_{\upsilon_x} \tag{3.46}$$

Thus, the propagator is an array of N_x entries that is multiplied by the spectrum. This multiplication in υ-space can be done element wise with the vector of the spectrum $\tilde{E}(\upsilon_x, z_0)$ and the vector of the propagator P_m and results in the vector

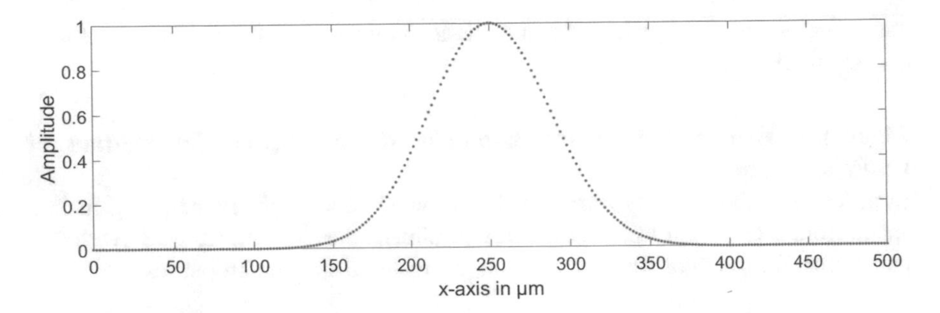

Fig. 3.34 Initial distribution of light in the first slice of the simulation field

donated as E_p. The final calculation of Eq. (3.41) is the inverse Fourier transformation of E_p, which again is easily performed with the FFT-algorithm. The result is the field distribution $E(x, z_2)$ of the second layer at $z_2 = z_0 + \Delta z$.

To further propagate the light, the next layer (layer number three) has to be calculated with the same scheme from the second layer. Equation (3.44) is now stated as follows:

$$\tilde{E}(\upsilon_x, z_2) = \int E(x, z_2) \cdot e^{-2\pi i \upsilon_x x} dx \tag{3.47}$$

This formalism is repeated until the last layer is reached, here at $z_{N_z} = 20000$ µm. The intensity distribution over the two dimensions can now be plotted (see Fig. 3.35).

The here shown formalism can also be used for three-dimensional simulations. Each individual layer is then no longer a vector but a matrix. In Fig. 3.36, a three-dimensional simulation with the same parameters as for the two-dimensional simulation in Fig. 3.35 is shown.

So far, only free space propagation was considered. However, with the goal of simulating a POW, up to three materials with different refractive indices are present (core, cladding, substrate). But, in Eq. (3.41), no dependency of the refractive index on its spatial property is considered. In order to use the WPM to simulate a POW, a new formulation of the WPM has to be introduced. In general, for M different refractive indices, the following formalism is considered [38]:

1. Make the Fourier transformation from the spatial domain to the Fourier domain.
2. Copy the field for every refractive index n_j with $j \in (1, \ldots, M)$ to get M fields $\tilde{u}_{0,j}$.
3. Multiply each of the fields with the propagator factor in the respective medium using the respective refractive index.

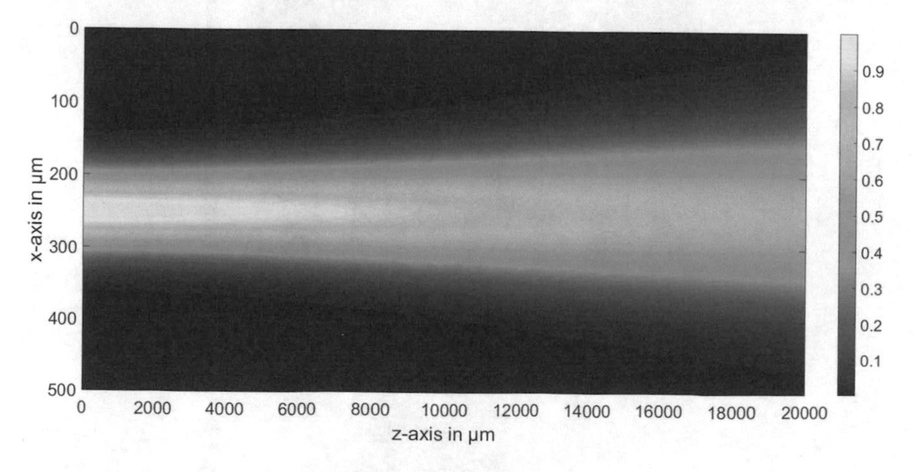

Fig. 3.35 Simulation of a Gaussian beam in two dimensions

Fig. 3.36 Simulation of a Gaussian beam in three dimensions

4. Make the Fourier back transformation of all M fields.
5. Depending on the refractive index of each pixel, take the complex amplitude out of one of the fields.

Of course, whereas for the free space propagation, the axial step size Δz has been arbitrary, i.e., it can also be quite large, for the WPM in a waveguide, the axial step size Δz has to be quite small. Otherwise, the plane waves with large angles would propagate from one pixel to the next from one medium to a different medium, which is not allowed. So, it is clear that the WPM converges only to the exact result for axial step sizes Δz in the order of a wavelength or even smaller.

3.3.2 Impact of Periodicity of Discrete Fourier Transform

Problem Description

The periodicity of the discrete Fourier transform does however bring a disadvantage with it. Light, which exits the field of simulation, is again introduced on the opposite side of the field. A simple example is shown in Fig. 3.37, where a Gaussian beam, propagating under a certain angle with the z-axis, is propagated out of the field at $z = 5800\,\mu\text{m}$. At the same z-position, one can observe the re-introduction of the beam at $x = 0\,\mu\text{m}$, which cannot be explained by physical behavior but the periodicity of the discrete Fourier transform.

Moreover, if a waveguide is simulated (see Fig. 3.38), it can lead to instabilities in the simulation. Especially if a very small sampling rate for the x-axis (and y-axis in 3D) is used, the instabilities occur and do make a valid simulation impossible.

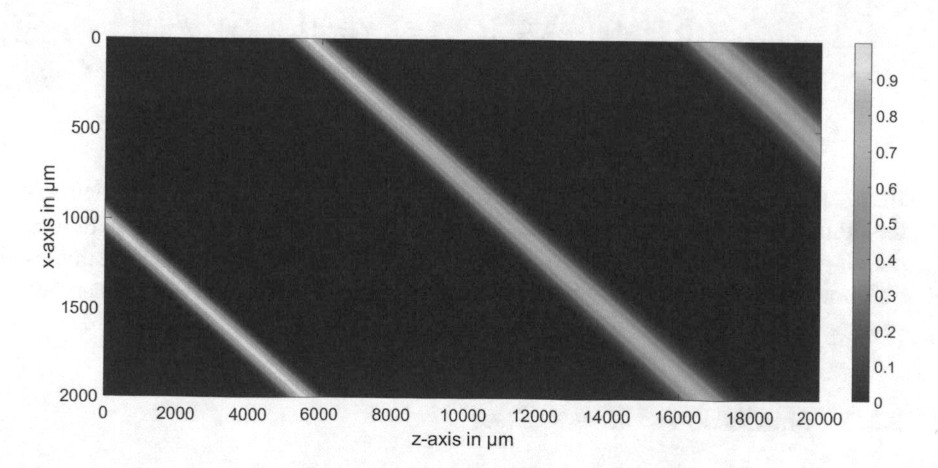

Fig. 3.37 Simulating a Gaussian beam with a propagation direction different to the z-axis leads to "re-entering" of the light on the other side of the simulation caused by the periodicity of the fast Fourier transformation

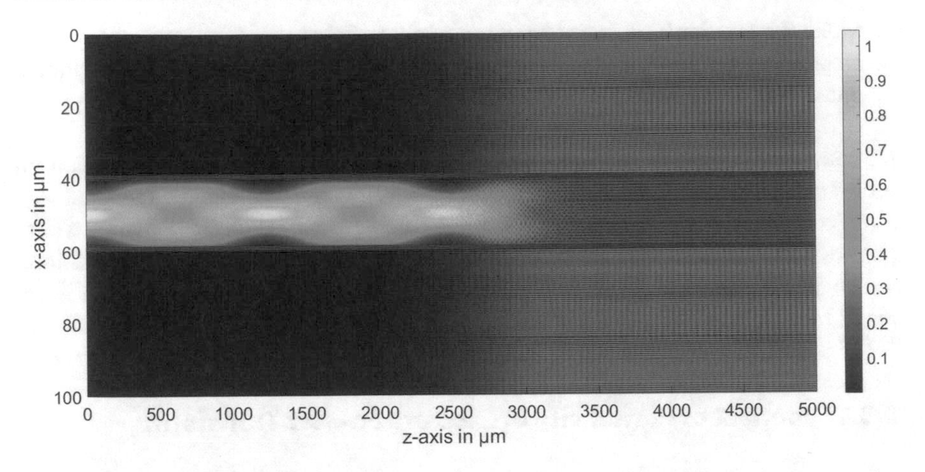

Fig. 3.38 Occurring instabilities during the simulation of a straight waveguide

Beam Waist	Wavelength	Core	Cladding	Nx	Nz	Waveguide Width
5 µm	0.5 µm	1.5	1.4	1024	15000	20 µm

Solution to Problem

In order to avoid light coming back to the field of view in an un-physical manner, one can prevent that intensities are present at the edges of the field by using absorbing boundary conditions as it is also made in several beam propagation methods. Therefore, a function can be multiplied onto the intensity in each slice that lets the intensity decay at the field edges. We introduce the following function:

$$t(x_i) = \begin{cases} \exp\left(-\frac{(i-p)^2}{2\sigma^2}\right), & \forall i \in \{1, \ldots, p\} \\ 1, & \forall i \in \{p+1, \ldots, N_x - p\} \\ \exp\left(-\frac{(i-(N_x-p+1))^2}{2\sigma^2}\right), & \forall i \in \{N_x - p + 1, \ldots, N_x\} \end{cases}$$

Here, x_i is the x-coordinate of the i-th element of the discrete vector E, p denotes the number of pixels used to decay the intensity and σ is the standard deviation of the Gaussian function. With this function, a beam propagating out of the simulation field is correctly simulated (see Fig. 3.39). The usable field however becomes smaller by $2p$ pixels and has to be adjusted in order to simulate the same field.

3.3.3 Simulation of Waveguides

The WPM was originally developed to simulate microlenses in order to obtain a non-paraxial simulation. Both the suitability and the subsequent simulative application of the WPM for step-index waveguides have never been studied. Therefore,

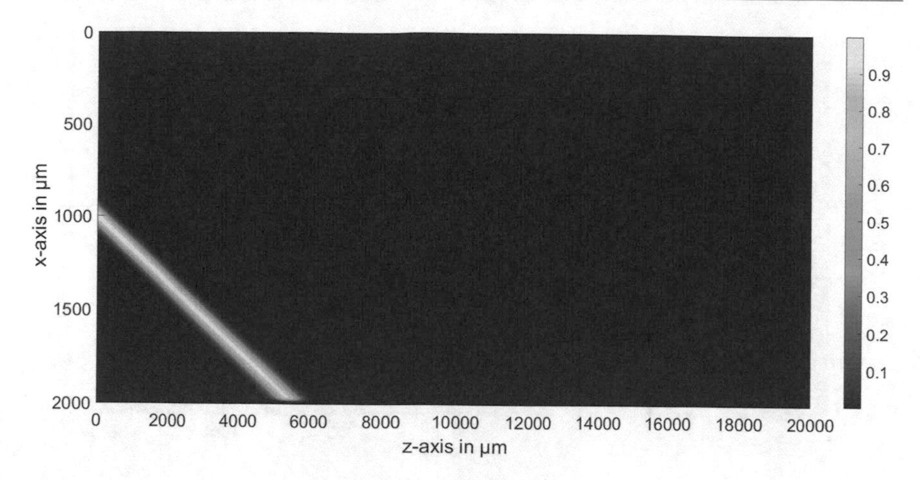

Fig. 3.39 Same simulation as in Fig. 3.37 but with the function $t(x_i)$ and its parameters ($\sigma = 20$, $p = 10$)

two 2D simulations of a waveguide are performed to validate its usage for waveguide simulations. For both simulations, the parameters are set as follows:

Parameter	Beam Waist	Wavelength	Core	Cladding	Nx	Nz	Waveguide Width	Field length
Value	10 µm	0.5 µm	1.5	1.4	1024	36000	80 µm	1200 µm

For the first simulation, the angle of the beam in regard to the surface normal is set to 76°. One can calculate the critical angle $\alpha_c = \sin^{-1}\left(\frac{1.4}{1.5}\right) = 68.96°$ for total internal reflection, and the expected behavior of reflection is shown in the simulation (Fig. 3.40).

For the second simulation (Fig. 3.41), the beam is incident with an angle of 68.2° to the surface normal and is therefore beneath the critical angle α_c of total internal reflection. The result is a beam that gets divided in a transmitted and a reflected beam following the equations of Fresnel (like for TE polarization). These examples show that the WPM is also suitable for waveguides and can automatically perform the splitting into totally reflected and transmitted beam.

After proving the suitability of the WPM as simulation algorithm for waveguides, a few remarks regarding the parameters of the simulation will be made. To start with, the sample size of x-axis and z-axis can freely be chosen by the user. However, the size does impact the simulation in several ways: Firstly, the simulation does take longer the more pixels are used and the goal of the simulation setup should include to minimize the number of pixels. On the other hand, the more pixels are used, the more detail is given by the simulation. In order to calculate an optimized pixel size, the following equation can be given:

$$\varphi \cdot \Delta z \leq c_1 \cdot \Delta x$$

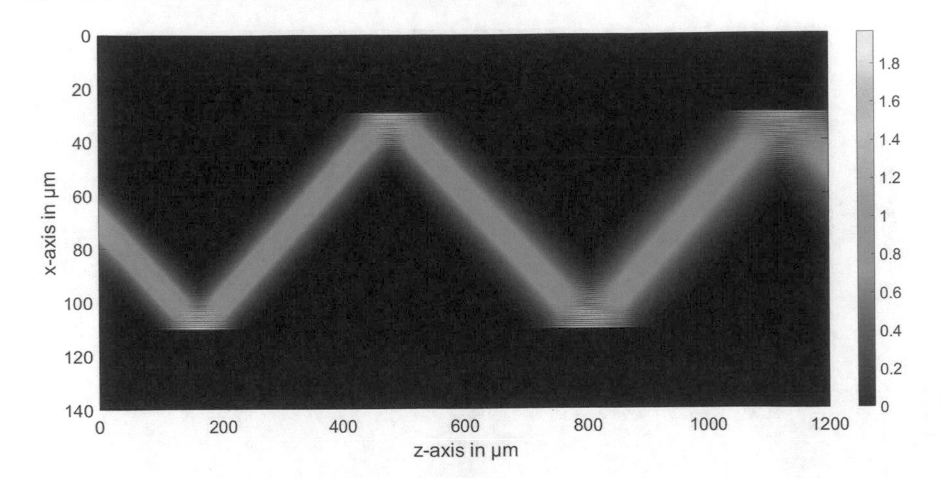

Fig. 3.40 Two-dimensional simulation of a beam guided inside a waveguide

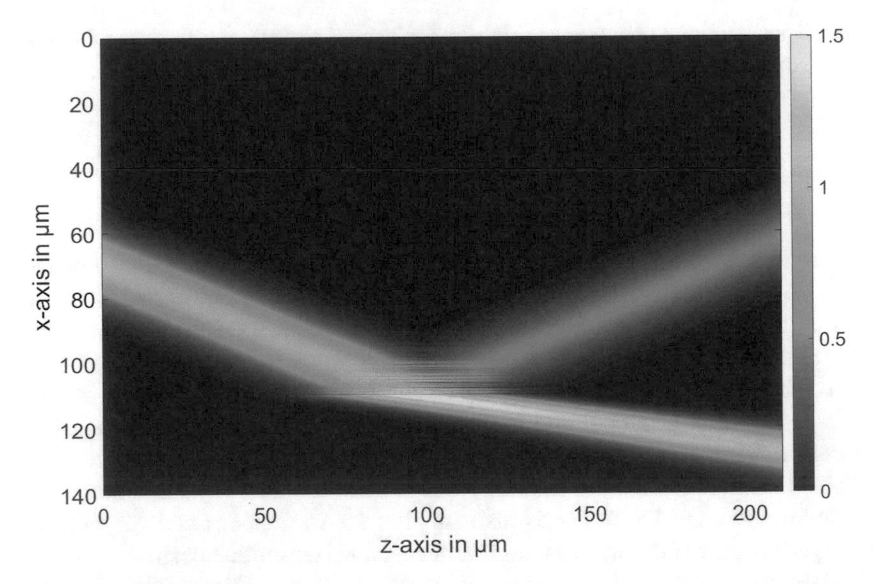

Fig. 3.41 Same as Fig. 3.40 but 21.8° as incidence angle relative to the z-axis, i.e., 68.2° incidence angle relative to the surface normal

Here, φ denotes the maximal angle of the wave (in radians), Δz the step size along the z-axis and Δx the step size along the x-axis. The constant c_1 is typically chosen to be smaller than 0.5. This ensures that when the waves propagate in the z-direction at the boundary between the two refractive indices of the core and cladding (or other materials), they hardly "run into" the neighboring pixel and the WPM becomes accurate enough.

Another optimization condition is given with

$$\Delta n \cdot \Delta z < c_2 \cdot \lambda,$$

with Δn the maximal refractive index difference in the system and λ the wavelength. The constant c_2 should be smaller than 0.1. The smaller Δn, the larger steps can be taken. This however is logical, because in the limiting case of $\Delta n \to 0$, i.e., homogeneous medium, Δz can also become very large (infinite).

As the light source, a Gaussian beam is chosen as described in Sect. 3.3.1 and sampled along the x-axis. Its parameters are given by the beam waist and the wavelength. In the simulation field, the light source is introduced on the left side. Further, the materials of the core and the cladding have to be set with their respective refractive index and its distribution in the simulation field. The easiest simulation is a straight waveguide, in which the core is given in the center of the simulation field as a rectangular field surrounded on the top and on the bottom by the cladding. For the simulation, the user has to enter a waveguide width. To give an example for a simulation, the following parameters are entered:

Parameter	Beam Waist	Wavelength	Core	Cladding	Nx	Nz	Waveguide Width b
Value	5 μm	0.5 μm	1.5	1.4	1024	15000	20 μm

The result of the simulation is given in Fig. 3.42. The simulation runs steady and shows that the light is guided in the waveguide with a repetitive pattern. This pattern is similar to a pattern due to the self-imaging of periodic objects illuminated coherently which was first studied in 1836 by Henry Fox Talbot and is therefore called "Talbot effect." It describes the behavior of light after a periodic diffraction grating or other periodic object and defines a distance z_T after which the intensity has the same pattern as the intensity in the starting plane. Here, the periodic

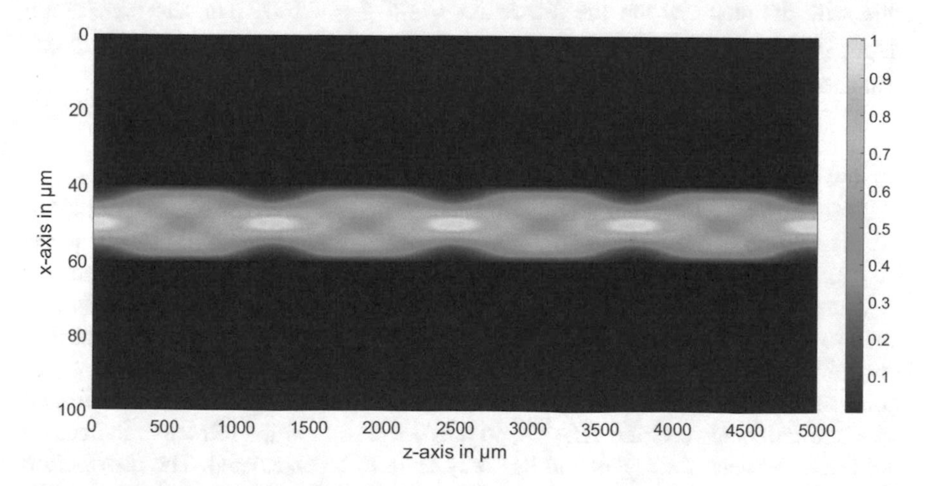

Fig. 3.42 Two-dimensional simulation of a fully illuminated waveguide

diffraction grating has to be substituted by the waveguide, but the general effect is still the same. The calculated distance

$$z_T = \frac{n_{core} \cdot b^2}{\lambda} = \frac{1.5 \cdot (20\,\mu\text{m})^2}{0.5\,\mu\text{m}} = 1200\,\mu\text{m} \tag{3.48}$$

after which the initial pattern repeats agrees only fairly well with the one simulated which is about 1250 μm (Remark: In the normal Talbot effect z_T would be twice as large, and at the distance calculated in the equation above, the intensity pattern would look the same, but laterally shifted by half a period.).

One more remark can be made with regard to the "Goos-Hänchen effect," which describes the lateral shift of light when totally internal reflected. This shift is automatically simulated with the WPM algorithm, but only if the resolution of the simulation is high enough. Obviously, this effect might be of high importance when dealing with waveguides since they solely depend on total internal reflection. Therefore, the effect caused by the finite penetration depth of the wave acts as if the edge of the waveguide is shifted/enlarged with respect to the surface normal by the penetration depth *d*. The penetration depth of the evanescent wave (1/e-drop of the amplitude) amounts to

$$d = \frac{\lambda}{2\pi n_{core}\sqrt{\sin(\theta)^2 - \left(\frac{n_{clad}}{n_{core}}\right)^2}}. \tag{3.49}$$

Here, θ is the angle of incidence of the beam in the core at the core/shell interface. Thus, $\theta = 90°$ denotes a beam parallel to the waveguide axis. Hence, for a Gaussian beam with the parameters used in Fig. 3.42, the angle of incidence of the edge ray is $\theta = 90° - \sin^{-1}\left(\frac{\lambda}{n_{core}\pi w_0}\right) = 88.8°$. For the simulation shown in Fig. 3.42, the penetration depth can be calculated to $d = 0.15\,\mu\text{m}$. If one compares this with the step size on the x-axis $\Delta x = \frac{100\,\mu\text{m}}{1024} = 0.0977\,\mu\text{m}$, the penetration depth is represented by roughly two pixels in this simulation. The axial Goos-Hänchen shift can now be calculated with

$$2 \cdot d \cdot \tan(\theta) = 14\,\mu\text{m}.$$

So, the Goos-Hänchen effect can be important for the case of grazing incidence.

Moreover, by taking into account the penetration depth *d*, the effective width of the waveguide is $b = 20\,\mu\text{m} + 2d = 20.3\,\mu\text{m}$. So, the effective Talbot length using Eq. (3.48) is $z_T = 1236\,\mu\text{m}$ which then agrees very well with the simulated one.

These remarks given above also hold for three-dimensional simulations and have to be extended by the y-direction. Therefore, one has to define the distribution of the refractive indices in each layer. The description of the printed waveguide is the same as in Sect. 3.2.2. Here, we simulate a straight waveguide printed on a substrate and surrounded by air, so the distribution of the refractive indices do not differ between the layers and has only once to be calculated. The distribution for the following example is shown in Fig. 3.43.

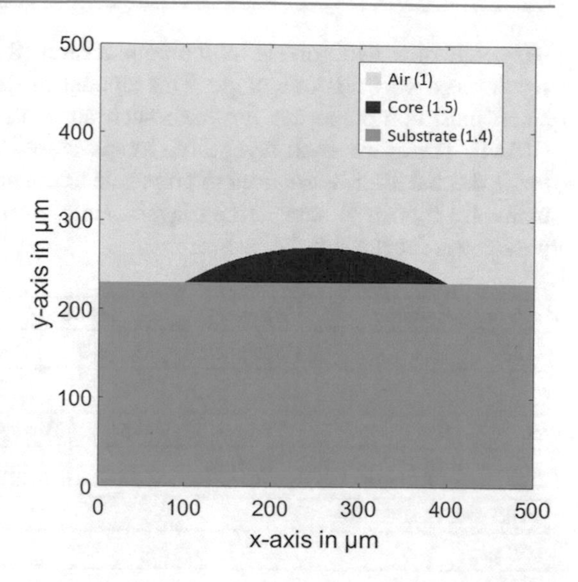

Fig. 3.43 Refractive index distribution for one slice

Since every layer consists now of a $N_x \times N_y$ matrix, the computation time of the simulation increases. In addition, the simulation field can no longer be stored in one now three-dimensional matrix because the size of the matrix becomes very large. In the shown example in Fig. 3.44, the matrix would consist of

Fig. 3.44 Evolution of the intensity distribution over the length of 3 mm

1024 × 1024 × 5900 pixels in double precision (8 Bytes per pixel) and therefore would need 46.1 GB of storage. This amount of data has to be carefully handled since most computers do not have such an amount of random access memory (RAM). Therefore, each layer is no longer stored until the end of the simulation but is deleted after it was used to propagate to the next layer. The user however can manually choose to save certain layers. Here, every 100th layer was stored. The parameters for the simulation are:

NA of light source	Wavelength	Core	Cladding (Air)	Substrate
0.25	0.85 μm	1.5	1	1.4

N_x	N_y	N_z	Waveguide width	Waveguide height	Waveguide length
1024	1024	5900	300 μm	40 μm	2950 μm

Field width x-direction	Field width y-direction
350 μm	87.5 μm

The light source is coupled into the center of the waveguide's front face and is here focused on a spot with a beam waist of 1.07 μm. Propagating now along the z-axis for the first three shown layers, one observes a free space propagation of the Gaussian beam. This changes when light hits the surfaces of the waveguide at about z = 150 μm and is reflected. Due to the reflection, intensities interfere with still outward propagating light and therefore leads to destructive and constructive interference, resulting in the center of the waveguide in an almost perfect interference pattern. After about 500 μm propagation distance also, the very corners of the waveguide are reached by the light, and so, light is also reflected there. They also lead to a clearly visible interference but does not have a distinct pattern. From z = 1000 μm and onward, the two interference patterns (left/right and up/down) also do interfere and form several different interference patterns as the light is propagated along the z-axis. From about z = 2000 μm, the nearly lamellar interference pattern does start to vanish and results in a fully illuminated waveguide that only shows interference phenomena similar to speckle.

But, how does this intensity distribution correspond to the intensity distribution which was calculated using ray tracing, i.e., an incoherent superposition of light? There, a stable intensity distribution with high intensity along a central vertical line results for large z-values (see for example Fig. 3.14). For that, the intensity distribution calculated by the WPM is averaged over many slices along the z-axis in the range of z = 3 mm to z = 6 mm, which is shown in Fig. 3.45. Then, the speckle-like structure disappears and the averaged intensity shows a vertical line with high intensity like in the ray tracing result. Of course, there are additional structures in Fig. 3.45 like a spot at the original position of the light source due to constructive interference at this point, and an arc-shaped horizontal structure generated also due to interference. This averaged intensity pattern remains stable if the light is further

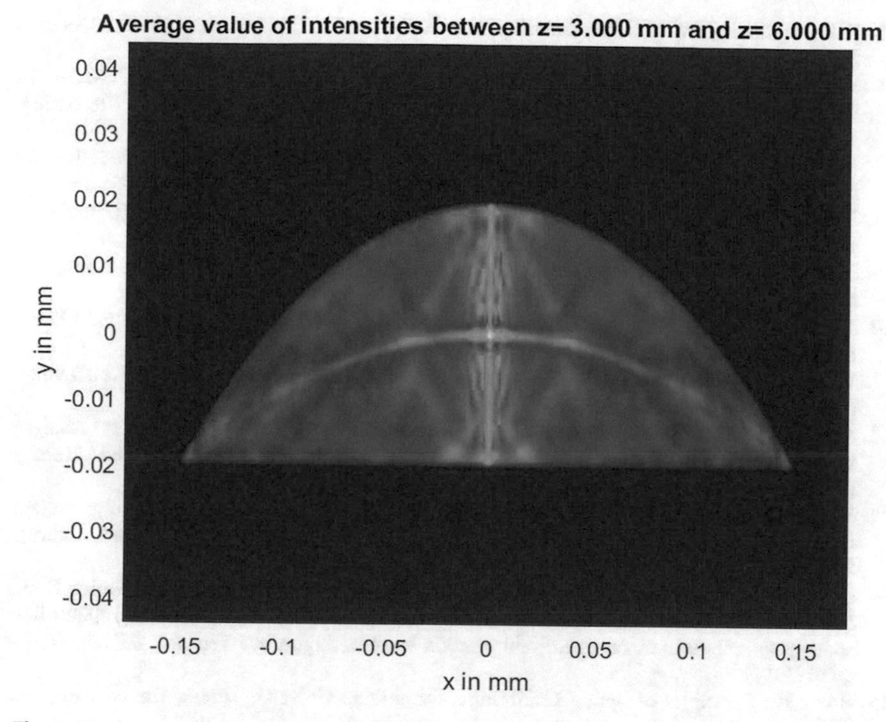

Fig. 3.45 Averaged intensity in the range z = 3 mm to z = 6 mm

propagated along the z-axis of a straight and defect-free waveguide like it is also the case for the ray tracing result. So, the results for ray tracing and for the WPM agree quite well if the intensity distribution obtained by ray tracing is compared with the axially averaged intensity distribution obtained by the WPM.

Acknowledgements We sincerely thank the Deutsche Forschungsgemeinschaft for funding the research group OPTAVER FOR 1660.

References

1. Bhagat, A., Clauson, L.: Zemax, LLC: design for manufacturability. SPIE Exhibition Product Demonstrations: International Society for Optics and Photonics (OP20EX), p. OP20EX12, (2020)
2. Jennato, S.: Synopsys, Inc.: Synopsys optical solutions. SPIE Exhibition Product Demonstrations: International Society for Optics and Photonics (OP20EX), p. OP20EX14 (2020)
3. Zhang, S. VirtualLab fusion: a physical optics simulation platform. A physical optics simulation platform. In SPIE Exhibition Product Demonstrations: International Society for Optics and Photonics (OP20EX), p. OP20EX (2020)
4. Jacobsen, D.A.: Lambda Research Corp: Start to finish optical design using OSLO, RayViz, and TracePro. SPIE Exhibition Product Demonstrations: International Society for Optics and Photonics (OP20EX). p. OP20EX0F (2020)

5. Kirk, D., Hwu, W.-M.: Programming massively parallel processors, 3rd edn. Morgan Kaufmann, Cambridge, MA (2017)
6. Shih, M., Chiu, Y.-F., Chen, Y.-C., Chang, C.-F. Real-time ray tracing with CUDA. In ICA3PP; International Conference on Algorithms and Architectures for Parallel Processing. Taipei, Taiwan (2009)
7. Moreland, K., Angel, E.: The FFT on a GPU. In Proceedings of the ACM SIGGRAPH/EUROGRAPHICS conference on graphics hardware, San Diego, California, (2003)
8. Brenner, K. H., Singer, W.: Light propagation through microlenses: a new simulation method. Appl. Opt., AO (Applied Optics) **32**(26), 4984–4988 (1993)
9. Purcell, T.J., Buck, I., Mark, W.R., Hanrahan, P.: Ray tracing on programmable graphics hardware. ACM Trans. Graph. **21**(3), 703–712 (2002)
10. Kawano, K., Kitoh, T.: Introduction to optical waveguide analysis: solving Maxwell's equation and the Schrödinger equation, 1. ed. Wiley (2001)
11. Yee, K.: Numerical solution of initial boundary value problems involving Maxwell's equations in isotropic media. IEEE Trans. Antennas Propag. **14**(3), 302–307 (1966)
12. Stallein, M.: Einkopplung in multimodale Lichtwellenleiter – Wellentheoretische Analyse und ein Vergleich zur strahlenoptischen Modellierung. Dissertation, Paderborn University (2010)
13. Schmidt, S., Tiess, T., Schröter, S., Hambach, R., Jäger, M., Bartelt, H., Tünnermann, A., Gross, H.: Wave-optical modeling beyond the thin-element-approximation. Opt. Express **24**(26), 30188–30200 (2016)
14. Lorenz, L., Nieweglowski, K., Al-Husseini, Z., Neumann, N., Plettemeier, D., Wolter, K.-J., Reitberger, T., Franke, J., Bock, K.: Asymmetric optical bus coupler for interruption-free short-range connections on board and module level. J. Lightwave Technol. **35**(18), 4033–4039 (2017)
15. Born, M.: Principles of optics, Cambridge. Cambridge University Press, United Kingdom (2020)
16. Richards, B., Wolf, E.: Electromagnetic diffraction in optical systems, II. Structure of the image field in an aplanatic system. Proc. R. Soc. Lond. A (Proceedings of the Royal Society of London. Series A. Mathematical and Physical Sciences) **253**(1274), 358–379 (1959)
17. Spencer, G.H., Murty, M.V.R.K.: General ray-tracing procedure. J. Opt. Soc. Am. **52**(6), 672–678 (1962)
18. Kuper, T. G.: Optical design with nonsequential ray tracing. In Optical Design Methods, Applications and Large Optics (1989)
19. „Photonics Buyers Guide - Optical design," SPIE Europe Ltd, 7 September 2020. [Online]. https://optics.org/buyers/category/67. Zugegriffen: 7 Sept. 2020
20. Lindlein, N.: Simulation of micro-optical systems including microlens arrays. J. Opt. A: Pure Appl. Opt. **4**, S1–S9 (2002)
21. Lindlein, N.: Simulation of micro-optical array systems with RAYTRACE. Opt. Eng (Optical Engineering) **37**(6), 1809–1816 (1998)
22. Lindlein, N.: Analyse der Aberrationen von holographischen Elementen mittels Raytracing. Erlangen (1991)
23. Lindlein, N.: Optik-Design unter Einschluss von diffraktiver Optik und Mikrooptik. Erlangen (2002)
24. Backhaus, C., Dötzer, F., Hoffmann, G.-A., Lorenz, L., Overmeyer, L., Bock, K., Lindlein, N.: New concept of a polymer optical ray splitter simulated by Raytracing with a new Bisection-Algorithm. In Frontiers in Optics, Washington DC (2019)
25. Backhaus, C., Hoffmann, G.A., Reitberger, T., Eiche, Y., Overmeyer, L., Franke, J., Lindlein, N.: Analysis of additive manufactured polymer optical waveguides: measurement and simulation of their waviness. In Photonics West, San Francisco (2020)
26. Karioja, P., Howe, D.: Diode-laser-to-waveguide butt coupling. Appl. Opt., AO (Applied Optics) **35**(3), 404–416 (1996)

27. Burden, R.L., Faires, J.D.: 2.1 The bisection algorithm. In Numerical Analysis, 3 ed., PWS Publishers (1985)
28. Backhaus, C., Lindlein, N., Zeitler, J., Franke, J.: Beeinflussung der optischen Eigenschaften von Polymer Optischen Wellenleitern durch das Druckpfad-Design. In DGaO-Proceedings 2019, Darmstadt (2019)
29. Backhaus, C., Vögl, C., Zeitler, J., Reithberger, T., Lindlein, N., Franke, J.: Optical simulations of printed Polymer Optical Waveguides (POWs): search for their optical limitations caused by fabrication and application geometry. In 2019 CLEO Europe and EQEC, München (2019)
30. Förner, J.: Simulation der Einflüsse von Produktionsfehlern auf das Dämpfungsverhalten in Polymer Optischen Wellenleitern. FAU Erlangen-Nürnberg, Erlangen (2020)
31. Loosen, F.: Design und Simulation unkonventioneller optischer Systeme basierend auf strahlen- und wellenoptischen Methoden. FAU Erlangen-Nürnberg, Erlangen (2019)
32. Loosen, F., Backhaus, C., Zeitler, J., Hoffmann, G.-A., Reitberger, T., Lorenz, L., Lindlein, N., Franke, J., Overmeyer, L., Suttmann, O., Wolter, K.-J., Bock, K.: Approach for the production chain of printed polymer optical waveguides-an overview. Appl. Opt. **56**(31), 8607–8617 (2017)
33. Halbe, K., Griese, E.: A modal approach to model integrated optical waveguides with rough core-cladding-interfaces. 2006 IEEE Workship on Signal 2006, pp. 133–136 (2006)
34. Bierhoff, T. Griese, E., Mrozynski, G.: An efficient monte Carlo based ray tracing technique for the characterization of highly multimode dielectric waveguides with rough surfaces. In Microwave Conference (2000)
35. Bierhoff, T., Wallrabenstein, A., Himmler, A., Griese, E., Mrozynski, G.: An approach to model wave propagation in highlymultimodal optical waveguides with rough surfaces. Proceeding X. International Symposium on Theoretical Electrical Engineering (ISTET' 99), p. 515–520 (1999)
36. Backhaus, C.: Monte-Carlo-Raytracing zur Simulation der Streuprozesse in gedruckten Wellenleitern auf Polymerbasis. Erlangen (2016)
37. Cooley, J.W., Tukey, J.W.: An algorithm for the machine calculation of complex fourier series. Math. Comput. **19**(90), 297–301 (1965)
38. Brenner, K.: A high-speed version of the wave propagation method applied to micro-optics. *2017 16th Workshop on Information Optics (WIO)*, pp. 1–3 (2017)

Conditioning of Flexible Substrates for the Application of Optical Waveguides

4

Gerd-Albert Hoffmann, Alexander Wienke, Stefan Kaierle and Ludger Overmeyer

4.1 Wetting Control by Conditioning

The wetting of solid surfaces in the technical field is a topic that receives such enormous inspiration from nature as is the case with only a few others. One of the best-known examples of this is the lotus plant. The hierarchical micro- and nanostructures on the surface of the lotus plant's leaves exhibit superhydrophobic behavior and thus cause water to be absorbed. The dewetting of water drops causes a self-reliant cleaning of the leaf and prevents the spreading of fungi or other organisms on its surface. Imitating this surface structure to influence wetting behavior has implications in many areas of science and engineering today. Historically, Thomas Young was the first to describe the peculiarities of wetting and contact angles in 1805. It was not until 1936 and 1949, respectively, that Wenzel explained the influence of roughness on wetting and thus made a technical implementation conceivable. At the beginning of the 2000s, a large number of surfaces with wetting properties were produced and their possible application in areas such as self-cleaning, oxidation protection or anti-fogging was investigated. In the

G.-A. Hoffmann (✉) · L. Overmeyer
Institut für Transport- und Automatisierungstechnik, Leibniz Universität Hannover, Garbsen, Germany
e-mail: g.hoffmann@lzh.de

L. Overmeyer
e-mail: ludger.overmeyer@ita.uni-hannover.de

A. Wienke
Production and Systems Department, Laser Zentrum Hannover e. V., Hannover, Germany
e-mail: alexander.wienke@lzh.de

S. Kaierle
Laser Zentrum Hannover e. V., Hannover, Germany
e-mail: s.kaierle@lzh.de

J. Franke et al. (eds.), *Optical Polymer Waveguides*,
https://doi.org/10.1007/978-3-030-92854-4_4

following section, the status in the field of wetting and capillarity is presented and the technical implementation of geometric and chemical wetting control by means of locally resolved conditioning will be elaborated.

4.1.1 Wetting and Capillarity

To characterize the static wetting property of a surface, the contact angle is measured. This contact angle arises from interfacial tensions between the solid and gaseous (σ_{SG}), liquid and solid (σ_{LS}) and liquid and gaseous (σ_{LG}) phases and is established in the three-phase contact point (TCP) on the surface of a solid (Fig. 4.1).

This mathematical relationship is described in Young's equation:

$$cos\theta = \frac{\sigma_{SG} - \sigma_{LS}}{\sigma_{LG}} \tag{4.1}$$

Depending on the surface energy, this contact angle increases or decreases as a result of the otherwise prevailing force imbalance. In general, a surface with a contact angle to water of > 90° is considered to have poor wettability, while a contact angle < 90° indicates a surface with good wettability. The relationship shown in 4–1 assumes ideal initial conditions with a smooth uniform surface. In order to describe the relationships on rough surfaces, Wenzel has modified Young's equation by introducing the factor r. This factor forms the ratio between the actual and the geometrically projected surface and changes the contact angle to:

$$cos\theta_w^* = r \cdot cos\theta \tag{4.2}$$

with r > 1. Thus, it becomes clear that a rough structure enhances the wettability of a well-wettable surface even more, while a poorly wettable surface forms an even larger contact angle due to a rough structure. For other liquids, however, these relationships apply only to a limited extent and must therefore be considered separately. In addition to the static contact angle, consideration of the dynamic contact angle, as it occurs on moving surfaces, e.g., in manufacturing processes, is relevant. To characterize the dynamic contact angle, the volume of the droplet on

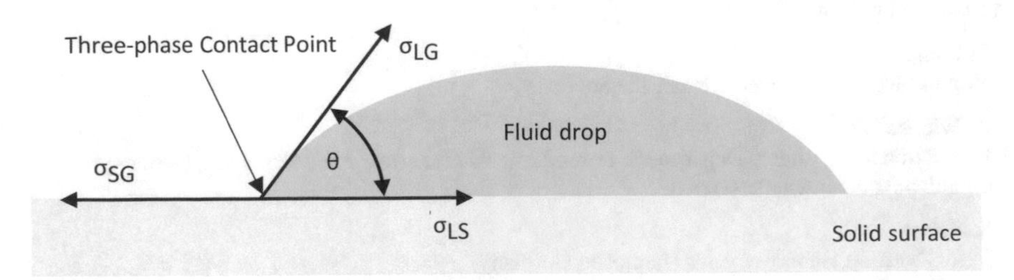

Fig. 4.1 Representation of the contact angle at the three-phase contact point of a liquid droplet

the surface is first slowly increased with additional liquid or reduced by suction. The resulting maximum contact angle (θ_{Adv}) as well as the minimum contact angle (θ_{Rec}) without displacement of the TCP is measured. The resulting span is referred to as hysteresis. A further distinction is made between forced and free wetting. In coating processes, the dynamic contact angle plays an important role. The higher the velocity, the greater this hysteresis or the difference to the static contact angle [1]. This difference is thus the result of forced wetting. Both the static and the dynamic contact angle are thus relevant for wetting control by locally resolved conditioning.

Another force that is important in wetting processes is capillarity. This force results from the difference between the pressure in the fluid and the surrounding atmosphere. Among other things, it describes the behavior of fluid in contact with capillaries (narrow gaps, cavities) in solids. Depending on the capillary geometry and the fluid properties, it outweighs the forces due to gravity. In this work it is relevant for balancing processes of the fluid structures in the fabrication of the conditioning and the later application of the optical waveguides. An important quantity is the so-called capillary length (κ^{-1}), which puts the surface tension σ in a relation to the gravitational force g (ρ: density of the considered liquid).

$$\kappa^{-1} = \sqrt{\frac{\sigma}{\rho g}} \tag{4.3}$$

For liquids that have a radius in the TCP below the capillary length ($r < \kappa^{-1}$), surface tension is the dominant variable and the effects of gravity can be neglected. For radii above the capillary length ($r > \kappa^{-1}$), liquid droplets are flattened and no longer exhibit a circular segment cross section (Fig. 4.2 left). Similarly, a wavy surface of a liquid film is smoothed.

If the wavelength of curvature is $\lambda < \kappa^{-1}$, then the surface tension is the dominant force. Depending on the direction of curvature of the surface, regions of higher or lower pressures are generated, inducing material flow (Fig. 4.2 right) [1]. If the ripple has a wavelength $\lambda > \kappa^{-1}$, gravity and hence the higher hydrostatic pressure in the elevated regions is the driving force of the equalization process [2]. In dynamic processes, the viscosity η of the fluid must be further considered. It is referred to as the property of a fluid to absorb a stress during deformation. The

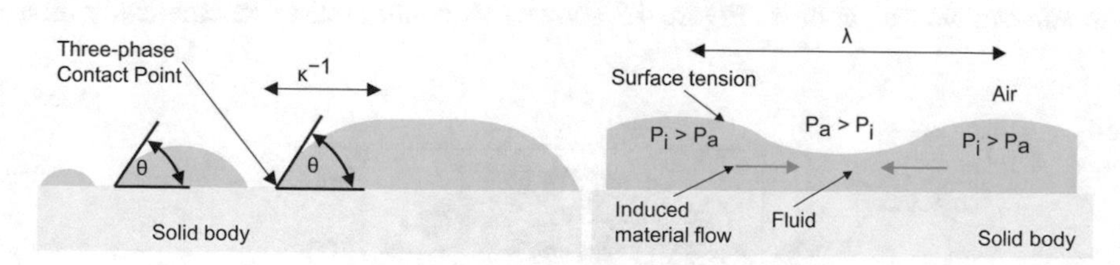

Fig. 4.2 left: Flattening of a liquid droplet as a result of the exceeded radius in TCP; right: Compensation processes at curvatures in liquid layers due to surface tension

ratio of this quantity and the characteristic process velocity V, to the surface tension σ, is called capillary number (Ca) and gives information about the influence on dynamic wetting processes.

$$Ca = \frac{\eta V}{\sigma} \tag{4.4}$$

For small capillary numbers ($Ca \ll 1$), the viscosity of the fluid is negligible. The behavior can consequently be compared with that of the static case, since here the surface tension always ensures the minimization of the surface area. In contrast, for large capillary numbers ($Ca > 1$), viscosity is the dominant force. The fluid cannot follow the deformation caused by the dynamic process, and contact angles and complex pressure distributions in the fluid layers can result that differ greatly from the static case [1].

Taking these findings into account, there is essentially two ways in which surfaces with locally adapted wetting properties can be produced. On the one hand, surface modification by geometric structures leads to a modified surface energy. The extent of this can be influenced by the shape and size of these structures. Secondly, wetting can be controlled by changing the chemical properties by means of a coating material if this has a surface energy that differs from that of the actual substrate. The latter mechanism can be implemented by means of printing technology and is therefore described in more detail below.

4.1.2 Wetting Control by Changing the Chemical Properties

When wetting is controlled by changing chemical properties through coating, two effects come into play. First, the application of a local coating creates an interface between the actual substrate and the material used for coating. This interface represents a defect site on the surface, which leads to the attachment of the contact line of a liquid droplet [3]. On the other hand, areas on the base substrate are spatially contained due to a lower surface energy of the coating material. Liquids subsequently adopt the higher contact angle of the coating in these areas on the base substrate [4].

By selectively conditioning substrate surfaces with fluids that have a lower surface energy in the cured state than the substrate itself, both effects can be exploited to achieve wetting control. Figure 4.3 shows this configuration schematically on a

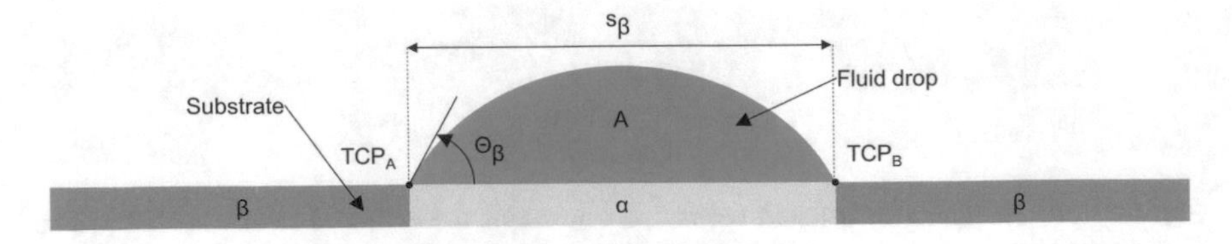

Fig. 4.3 Wetting on areas confined by zones with lower surface energy

substrate surface. The different regions have a specific static contact angle Θ_α and Θ_β. Here, the α region represents the base substrate and the β region represents the conditioned surface. A change of the contact line toward the center leads to an increase of the energy, provided that the contact angle Θ_c that is established lies between the specific contact angles Θ_α and Θ_β. Each contact angle in this range consequently forms its contact line at the same point (TCP_A or TCP_B).

Both the contact angle Θ and the distance s between the conditioned regions are shown as a function of the cross-sectional area A of the liquid droplet in Fig. 4.4. The cross-sectional area A results with the radius of the circular segment r to:

$$A = \frac{r^2}{2} \cdot \left(\frac{\Theta}{2} - \sin\left(\frac{\Theta}{2}\right)\right) \tag{4.5}$$

As the cross-sectional area A increases due to the wetting liquid droplet, the distance between the contact lines s (Fig. 4.4, Case 1) increases to the distance s_β until the conditioned regions are reached. Due to the lower surface energy of these areas, as the cross-sectional area A continues to increase, only the contact angle increases from Θ_α to Θ_β, while the distance between the two TCPs remains unchanged at s_β (Fig. 4.4, Case 2). Only by further increasing the cross-sectional area A, the conditioned area is also wetted with the liquid and the distance s between the contact lines increases, while the contact angle Θ_β remains the same (Fig. 4.4, case 3). This consideration results in the printing implementation of

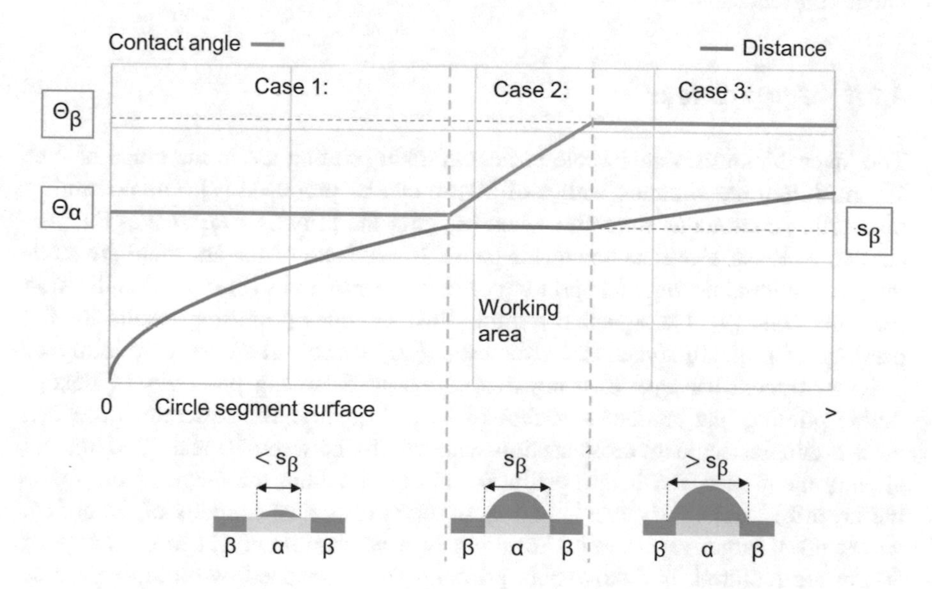

Fig. 4.4 Progression of the contact angle and the distance between the conditioned regions over the cross-sectional area of a liquid droplet

wetting control. The wetting behavior can be influenced by printed conditioning lines in such a way that after the application process between the print-conditioned areas, the waveguide core material undergoes self-alignment and assumes the shape shown in Fig. 4.3 [5].

4.2 State of the art for Functional Flexographic Printing

Printing is classified in DIN 8580 [6] of the manufacturing processes, in the 5th main group “coating” in the subgroup “coating from the liquid state.” Since the printing of books in the fifteenth century, printing has primarily been used as a manufacturing process for creating visual appearances on surfaces. In addition to the printing of documents, these days it is mainly used for printing packaging and design elements on almost any type of product.

If, in addition to the visual appearance, another benefit is created, this can be described as functional printing. From this benefit, technical requirements for the printing production can be defined. The most important requirement in visual printing is the appearance that the printed product leaves in the eye of the beholder. Humans are able to recognize structures that have a minimum distance of 0.15 mm to each other at a normal viewing distance of approx. 25 cm without the increase of aids. In addition to resolution, other properties can play an important role in functional printing, since they influence measurable physical variables. The weighting of the manipulated variables differs depending on the application under consideration.

4.2.1 Applications

The layer thicknesses achievable in flexographic printing are in the range of 3 to 10 μm. Minimum structure widths of 10 μm can be produced [7]. The possibility of creating continuous structures when imaging the printing form makes the process suitable for electrical conductor paths. In the state of the art, there are studies on functional flexographic printing of antenna structures [8] or of metallization for solar cells [9]. Flexographic printing also provides promising results for the printing of optically functional structures. [10] demonstrates direct printing of polymer optical waveguides using fluid-dynamic balancing processes of flexographic printing that produce a surface of optical quality. The structures produced have a circular segment cross section with widths between 10 and 1000 μm. As already mentioned, the height of the structures is within the range of the feasible layer thickness. For better handling, in terms of optical coupling of the optical waveguides, larger aspect ratios resulting in a waveguide height in the range of 50 μm are required. In flexographic printing, this is realized by multiple process cycles. However, layering impairs the productivity of manufacturing such systems. Furthermore, interfaces form between the individual printed layers. These can be caused by contamination with dirt particles or by stresses introduced during polymerization and as a result have a negative impact on the optical performance of the optical waveguides. The approach in this work is conditioning the substrate by flexographic printing in combination with subsequent application of the optical waveguide. In this way, the advantages of flexographic printing in terms of achievable resolution and the production of structures of optical quality can be combined with a high aspect ratio due to the locally modified wetting property in a downstream process step.

4.2.2 Flexographic Printing Process

The process flow in flexographic printing is shown schematically in Fig. 4.5 (left). The essential components of a flexographic printing unit are as follows:

The anilox roller in the flexographic printing unit is responsible for the constant and homogeneous supply of printing fluid. It consists of a cylindrical base body on whose surface a chromium oxide ceramic layer is applied in most cases according to the current state of the art. In this surface, subtractive processed depressions (the so-called cells) contain the printing varnish for the next printing cycle. These cells can differ in shape and size and thus influence the transfer volume and also the printing result. In addition to the geometry of the individual cells, their density on the anilox roller surface is also relevant. The number of cells is specified to a reference distance of 1 cm and is called line count. Common line counts range from 55 L/cm to 200 L/cm.

The doctor chamber contains the pressure fluid with which the anilox roller is refilled after each revolution. The doctor blade is used to fill the cells evenly across the width of the anilox roller and to wipe off the excess material.

The printing form (Fig. 4.5 right) is mounted on the printing cylinder. This can be slid on as an endless sleeve or attached as a printing plate. Due to the imaging by raised structures on the printing form surface, flexographic printing is defined as a direct letterpress printing process among the rotary printing processes. The imaging of the printing form is currently carried out in most cases by lithographic or laser structuring processes [11]. Depending on the imaging process,

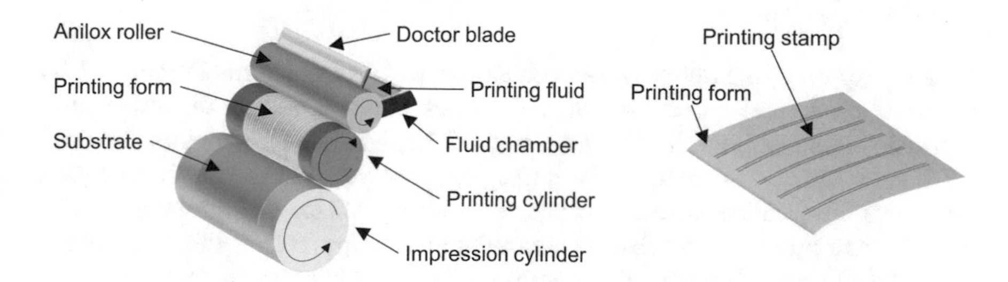

Fig. 4.5 left: Schematic representation of the process flow and components in flexographic printing; right: Schematic representation of the printing form with raised printing stamp structures

photopolymers or rubber of different hardness are used as the base material. The type of imaging also limits the minimum structure widths that can be produced on the printing stamps how the raised structures are also called. The stamp faces, directly involved in the printing process, are wetted with printing varnish by continuous rolling on the anilox roller. In the further process sequence, ink splitting then takes place as the printing stamps roll on the substrate. Due to the process and depending on the selected line count and the variable printing parameters, squeezing edges may appear in the printed image on the substrate [12]. The substrate is guided on the impression cylinder during ink splitting and then passed on for infrared or UV curing. The substrate material used in flexographic printing can be paper or polymer materials which, depending on the printing machine used, can be in sheet form or continuous.

In the case of printing form and anilox roller parameters, only the process parameters are variable in the printing process and thus influence the printing result. Essentially, these are the distance (feed) between anilox roller and printing form and between printing form and substrate, and the printing speed, which is set by the rotation speed of the cylinders.

4.3 Selection and Design of the Printing Form

Specific to flexographic printing plates is the nature of the elastic top layer, which contains an imaging and meets a rigid impression cylinder in the printing process. Depending on the intended use and the material, the imaging of the printing plates is produced using different technologies, which also differ in terms of the resolution that can be achieved. Digital imaging of printing plates, also known as "computer to plate," has been available in the flexographic printing market segment since the end of the 1990s. The layout of the image to be printed is available in digital form on a computer and is transferred to the corresponding substrate in the imaging step using various technologies:

- Laser exposure
- Direct imaging (by means of UV light)
- Laser engraving.

In laser exposure, a laser-sensitive coating on the printing form is removed by means of low-intensity laser radiation. The areas of the printing form surface thus exposed are then polymerized by means of UV light. A photopolymer is also used for direct imaging. In contrast to laser imaging; however, no masking is necessary, since here the printing areas are exposed locally by means of high-intensity UV light. In both processes, a subsequent development step removes the non-printing areas of the printing form and thus completes its manufacturing process.

Laser engraving uses a rubber as the printing form material. Imaging and development take place in a single process step by ablating and extracting the non-printing areas using high-intensity laser radiation.

The materials used differ in combination with the surface properties in terms of their surface energy (see Sect. 4.1). Furthermore, the hardness of the printing form and thus the stress occurring in the printing process is influenced by the material. The printing plates considered in this work are shown in Table 4.1 with the corresponding hardness, the achievable resolution of the imaging process and the surface energy.

The resolution of the respective printing forms refers to the manufacturer's specifications and relates to the size of the pixels that can be transferred in the imaging process. Freestanding structures must be many times this size to enable mechanical stability in the printing process. [10] describes the process window for flexographic printing and thus the minimum printable structure size. This is about 10 µm [7] (40 µm [13]), and thus, many times higher than the possible resolution of the production processes. Nevertheless, this property is decisive for the printing of conditioning lines, since it limits not only the minimum achievable distance between the printing form stamps but also the size for substructures on the stamp surface. In order to select a suitable printing form, evaluations were carried out by means of confocal microscopic images with regard to the surface properties and the quality of the edge waviness.

Figure 4.6 shows the surfaces of the printing plates used. The Kodak printing plate produced by laser exposure has a significantly lower surfacc roughness $S_a = 434$ nm ± 232 nm compared to the Continental printing plate produced by laser engraving ($S_a = 1930$ nm ± 245 nm). The intermediate area of the double stamps is also shown. Here, the Kodak printing form has significantly lower edge waviness at the stamp end faces too. In addition to the material, the surface energy of the printing forms is essentially defined by the surface roughness (as described in Sect. 4.1). As listed in Table 4.1, these are 12.26 ± 0.65 mN/m (Continental) and 19.59 ± 0.58 mN/m (Kodak). Lower surface energy also results in lower fluid transfer to the printing substrate for the printing form as a first approximation and thus in a changed line geometry.

The design of the printing form layouts plays an important role in the printing of the conditioning lines. Vector graphics of the print image are created on the computer using graphics software. The printing stamps and support structures at the edge of the printing form are colored as black areas (CMYK: 0/0/0). The widths of the printing form stamps and their spacing are dimensioned absolutely for this purpose. When using a full tone image, which does not allow any

Table 4.1 Overview of the considered printing forms and their properties

Printing form	Manufacturing technology	Minimum structure size in µm	Hardness (Shore A)	Surface energy in mN/m
Continental Laserline CSC	Laser ablation	10	68.77 ± 0.69	12.26 ± 0.65
Kodak Flexcel NX	Direct exposure	2.6	76.56 ± 0.51	19.59 ± 0.58
Flint Nyloflex Gold A	Laser exposure	100	80.01 ± 0.17	29.32 ± 0.73

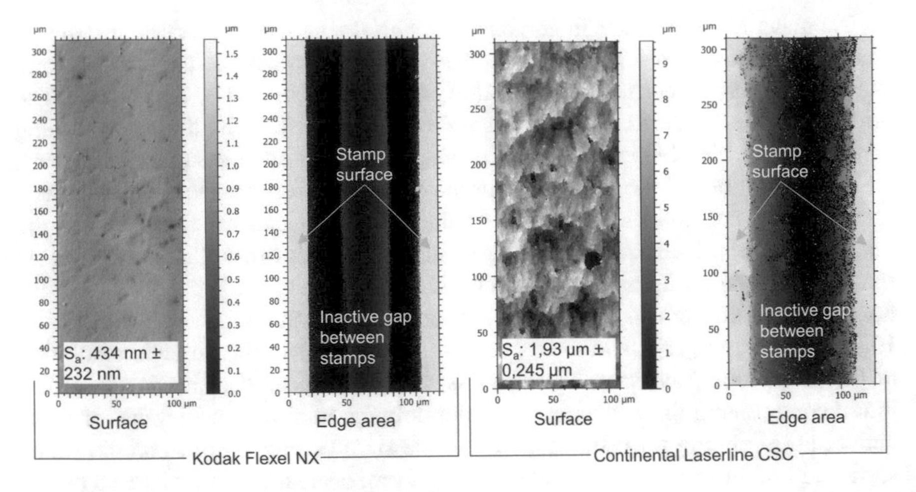

Fig. 4.6 Confocal microscope image of the surface as well as the border area of the stamp end faces

gradation between the stamp face and the printing form base, standard values are assumed for the angles of the printing stamps. These are usually 54°. By means of a gradation of the inked areas in the form of a grayscale image, this angle is another adjustable parameter that influences the stamp deformation in the printing process (Sect. 4.6). Figure 4.7 shows a layout of one of the printing plates used. The entire printing form has dimensions of 524 × 426 mm. To evaluate several printing parameters on one substrate, an anilox roller with four sections is used for this layout. The available line counts on this anilox roller are 55 L/cm, 100 L/cm, 160 L/cm and 200 L/cm and influence the ink transfer to the printing form. The higher the line count, the less printing fluid is absorbed by the printing stamps. In order to be able to guarantee uniform rotation over the entire print image, support structures are provided in the layout. These ensure that there is contact between the printing unit cylinders at all times, irrespective of the print image. A print mark containing different line thicknesses and spacing combinations is used to check the correctly set infeeds in the printing unit. The double stamps vary by widths between 100 µm and 1000 µm and have spacings between 50 µm and 500 µm. The labeling of the selected stamp parameters is included in the print layout and thus simplifies assignment during subsequent characterization.

4.4 Characterization and Selection of the Materials

The selection of materials for optical waveguides is limited by various prerequisites. On the one hand, these are optical parameters, such as the refractive index, the transmission behavior and the surface properties, and on the other hand,

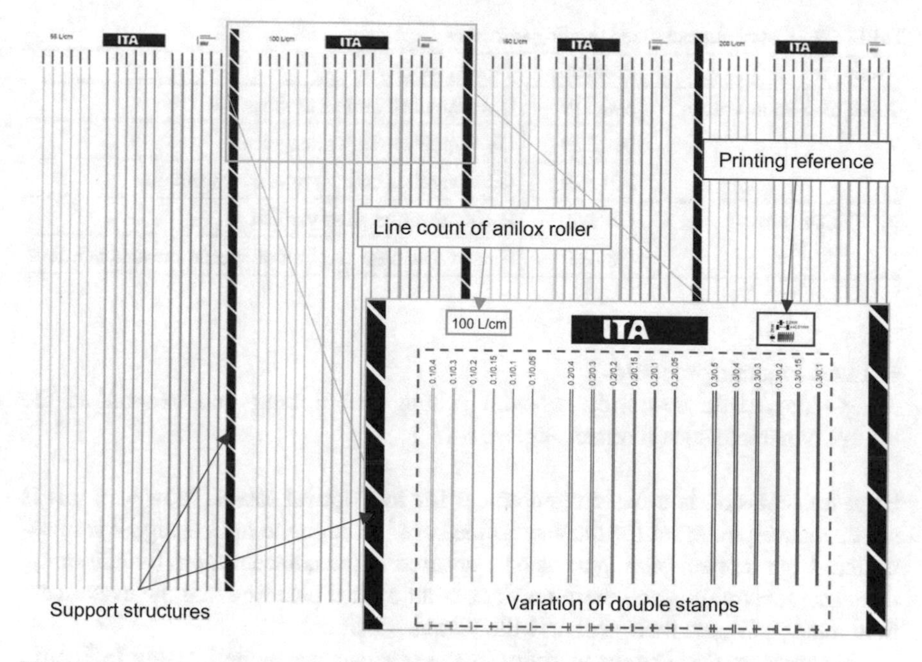

Fig. 4.7 Printing form layout for printing conditioning lines with banded anilox roller

these are parameters that have an influence on the manufacturing process and must therefore be evaluated. This may concern fluid-dynamic parameters such as viscosity and surface tension and energy, or mechanical properties in terms of strength and ductility. Subsequently, the printing varnish and the substrate are examined separately for this suitability and with regard to interactions.

4.4.1 Selection of the Printing Varnish

The selection of the printing varnishes for conditioning the film substrates and applying the optical waveguide is based on a characterization of commercially available UV-curing polymers. To be suitable for flexographic printing, the materials must be in a viscosity range of 50–500 mPas. Based on this property, the following fluids are further characterized:

The properties relevant to the printing varnishes are classified as follows for further characterization (Table 4.2):

- Optical properties
  - High transmissivity (> 90%) in the wavelength range used (850 nm)
  - Combination of refractive indices for waveguide core and conditioning line material (approx. n < 1.51 @850 nm)

Table 4.2 List of characterized printing varnishes

Producer	Product	Type / property
Jänecke & Schneemann	390,119	UV Supraflex varnish embossable
	390,120	UV Supraflex High gloss varnish
	391,629	Gloss varnish, embossable, overprintible
ACTEGA Terra	G8/372	Highly reactive gloss varnish
	G8/611	Gloss varnish for creation of special movement effect

- Fluid dynamic properties
  - Characteristic dewetting behavior of the conditioning line material in the polymerized state (contact angle $\Theta > 45°$)

High transmission is a basic requirement for low optical losses. However, this is not exclusively relevant for the waveguide core in which the optical signal is transmitted. High transmission must also be ensured in the adjacent areas (conditioning lines and substrate), since these may act as an optical cladding and the evanescent field must propagate there with as little loss as possible.

To characterize the transmissivity in the relevant wavelength range (850 nm), samples (5 mm thickness) are prepared from the respective UV-curing polymers and then measured using a spectrophotometer (LAMBDA 1050+UV/Vis/NIR). The wavelength range selected for the measurement is between 200 and 900 nm (Fig. 4.8). All printing varnishes exhibit complete absorption below 380 nm due to

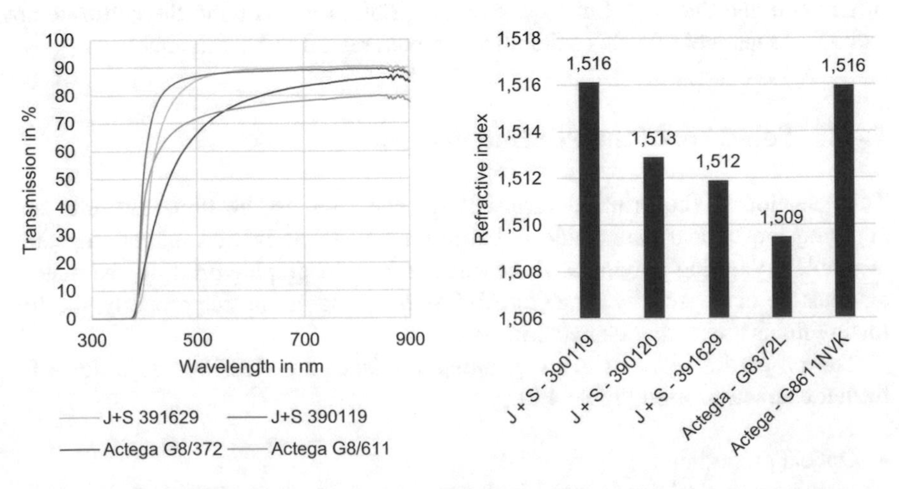

Fig. 4.8 Left: Transmission measurement on the selected print varnishes; right: Refractive index measurements on the selected print varnishes

the photoinitiators they contain, which are responsible for polymerization. Above this wavelength, which is visible to the human eye, light is transmitted. Compared to the other materials, the G8/611 printing varnish from Actega Terra exhibits a significantly increased absorption of more than 20% over the entire range and also at the relevant wavelengths and is therefore not further characterized. The manufacturer's other material (Actega Terra—G8/372) also exhibits slightly increased absorption. The print varnishes from Jänecke & Schneemann consistently exhibit high transmissivity and are therefore particularly suitable for use as waveguide core materials.

Depending on the application process used, the conditioning material can form an interface with the waveguide core material. Since these subsequently act as optical cladding material, a reduced refractive index compared to the waveguide core material is also necessary. Due to the interdependence, a characterization of the print varnishes to this effect is necessary. The refractive indices are measured using a refractive index profilometer (rinck-electronik profilometer) on previously prepared samples. The evaluation shows that all materials have a refractive index above 1.5. The Actega Terra G8/372 printing varnish has the lowest value of 1.509 and is therefore suitable for use as a coating material for any other printing varnish.

In order to achieve the defined objective, a characteristically dewetting behavior between the waveguide core material and the conditioning line material is an important property and will be investigated in the following. As described in Sect. 4.1, the wetting property of substrates by liquids can be obtained from the measurement of contact angles. The measurement is performed on a contact angle measuring station (Surftens, OEG) using the pendant drop method. First, a homogeneous layer of the material is applied to a film substrate with a film applicator (Erichsen Coatmaster 510) and subsequently polymerized. A contact angle measurement from the remaining combinations of the coating material with the print varnish is used to investigate the interaction in terms of wettability. The results are shown in Fig. 4.9.

It is noticeable that the print varnishes Jänecke & Schneemann 390119 and 391629 show a clearly increased contact angle in the range between 58 and 62°on the cured layers consisting of Actega Terra G8/372 and Jänecke & Schneemann 390120. This can be explained by the low surface energy of the cured layers and is thus a very good prerequisite for their use as conditioning line material.

Conclusion

Due to its sufficient transmissivity, its good suitability as an optical coating and its very good dewetting behavior, the Actega Terra G8/372 printing varnish is used to investigate the printing of conditioning lines. For the selection of the waveguide core material, the two print varnishes Jänecke & Schneemann 390,119 and 391,629 continue to be considered. A final evaluation is made after characterization of the interaction with the substrate materials.

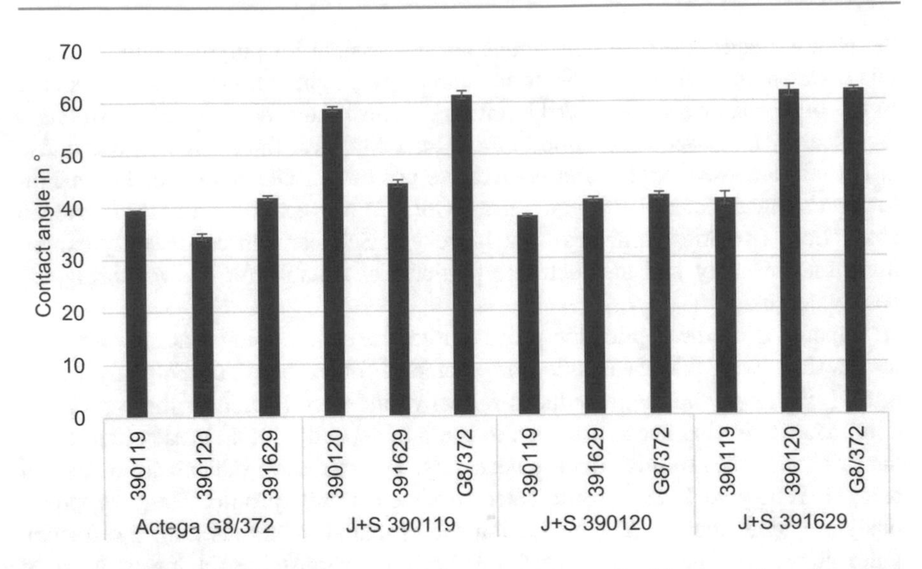

Fig. 4.9 Wetting properties of the material pairings, contact angle of the respective print varnish on a coating of the cured material. (Source for measurement data: Tim Wolfer, ITA Hannover)

4.4.2 Selection of the Substrate Material

The substrate material has to be usable in the system technology used (Sect. 4.5.1). Although the availability of the material in the required size and thickness is not a prerequisite for basic experimental investigations, it should be taken into account at the beginning of the selection process. Furthermore, bending flexibility is required for processing, since the substrate is guided around the impression cylinder during the printing process and is elastically deformed in the printing direction.

Based on these properties, the following substrate materials are further characterized (Table 4.3):

To select the substrate material, they are analyzed considering the following criteria:

- Optical properties
  - High transmissivity (> 90%) in the wavelength range used (850 nm)
  - Refractive index below the used waveguide core materials (approx. $n < 1.50$ @850 nm)
  - Surface roughness of optical grade ($S_a < \lambda / 10$, $S_a < 85$ nm)
- Fluid dynamic properties
  - Sufficiently high surface energy (contact angle $\Theta < 90°$).

To evaluate the optical properties of the different substrate materials, the transmission behavior is investigated first. This plays an important role, since the evanescent

Table 4.3 Characterized substrate materials for flexographic printing

Substrate material	Producer / product	Substrate thickness
PMMA (Polymethylmethacrylat)	Plexiglas XT 99,524, ThyssenKrupp Plastics	175 µm
PVC (Polyvinylchlorid)	Pentaprint PR M180/23, Klöckner Pentaplast	150 µm
PET (Polyethylenterephthalat)	Mylar A, Coloprint tech-films	130 µm
PET (Polyethylenterephthalat)	PET-GAG, Daams Kunststoffe	200 µm
PI (Polyimid)	Kapton HN, DuPont	130 µm

field transmitted in the optical cladding must be able to propagate as loss-free as possible. For this purpose, samples of the different materials are prepared by cutting them to the required size and measured in a spectrophotometer (LAMBDA 1050+UV/Vis/NIR) (Fig. 4.10).

The measurement range is set to a wavelength between 200 and 900 nm and the proportion of absorption is evaluated. The materials PET Mylar and PI Kapton exhibit an absorption of approx. 20% and 30%, respectively, in the relevant wavelength range and can be excluded from further consideration as substrate materials. The remaining substrate materials PVC and PMMA have an absorption of less than 10% and are therefore suitable for use with minimal optical losses.

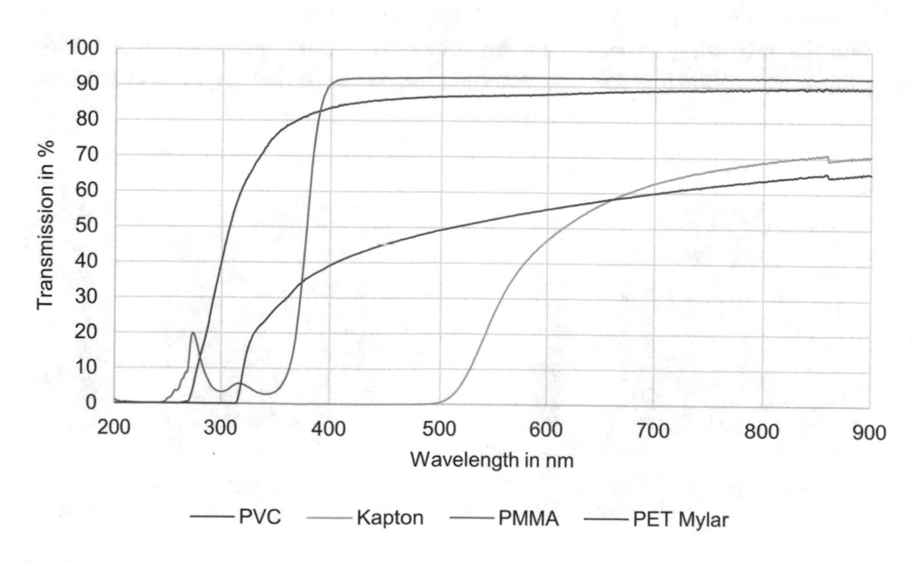

Fig. 4.10 Transmission measurement on the selected substrate materials

Each interface in the system of the planar optical waveguide contributes with its surface roughness to the optical losses and the roughness must therefore be minimized. According to the state of the art, a reference value of about 10% of the transmitted wavelength is given [14], below which the surface roughness does not generate any optical losses. The quality of the interface to the waveguide core is significantly influenced by the substrate surface. Confocal microscopic analysis was performed to measure the surface roughness of the two remaining substrates (Fig. 4.11). Here, it can be seen that both substrate materials investigated, with Sa = 65.42 ± 4.2 nm and 31.8 ± 0.1 nm, respectively, have a sufficiently low surface roughness to satisfy as a surface of optical quality.

In order to be suitable for the flexographic printing process, a low contact angle when the substrate is wetted by the waveguide material is advantageous, in contrast to the conditioning lines. A low contact angle indicates good printability and is therefore a prerequisite for its use (Fig. 4.12).

For this purpose, the two remaining materials that are suitable for use as waveguide core materials are measured with regard to the contact angle. The evaluation shows that the print varnish Jänecke & Schneemann 390,119 has a lower contact angle ($\theta < 45°$) and thus better wetting on the film surfaces.

The next property compared is the refractive index of the substrate materials. These values are taken from the literature or determined using a refractive index profilometer (rinck-electronik profilometer). To allow light to be guided in the waveguide core, the substrate acts as a lower optical cladding and must therefore have a lower refractive index. Considering that all characterized waveguide core materials have a refractive index $n > 1.500$, only the PMMA substrate with a refractive index of $n = 1.49$ @635 nm can be used without additional coating. The PVC film with the refractive index of 1.531 [15] is clearly above the print varnishes and can therefore not be used as substrate material.

For the following experimental investigations of the pressure of conditioning lines, the following material combination is defined in summary: substrate

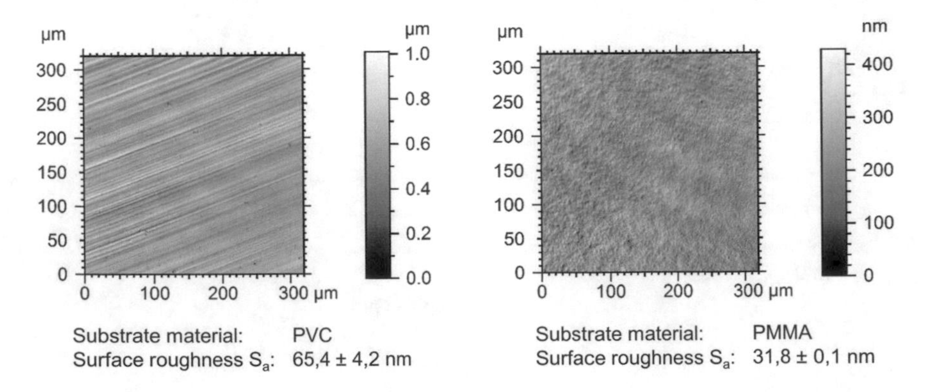

Fig. 4.11 Confocal microscopy measurement of surface roughness of substrate materials, left: PVC; right: PMMA

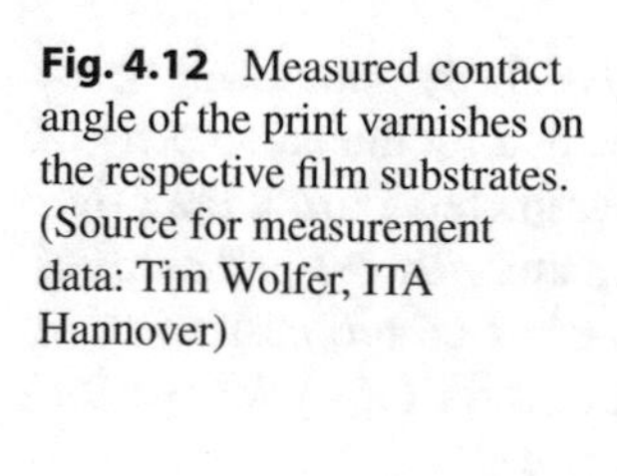

Fig. 4.12 Measured contact angle of the print varnishes on the respective film substrates. (Source for measurement data: Tim Wolfer, ITA Hannover)

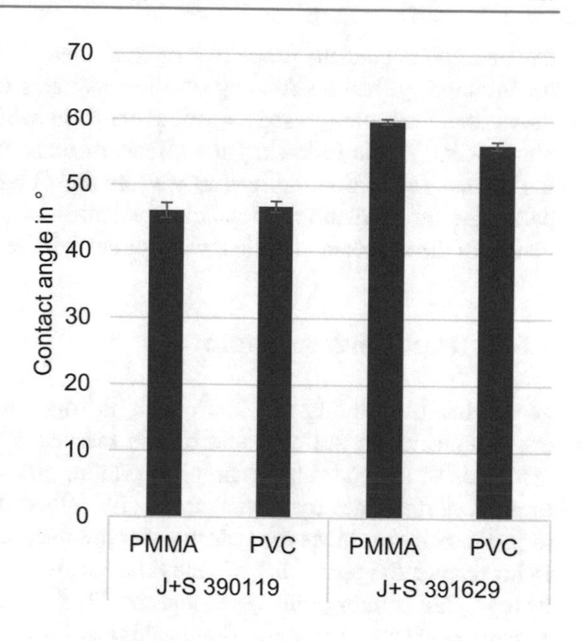

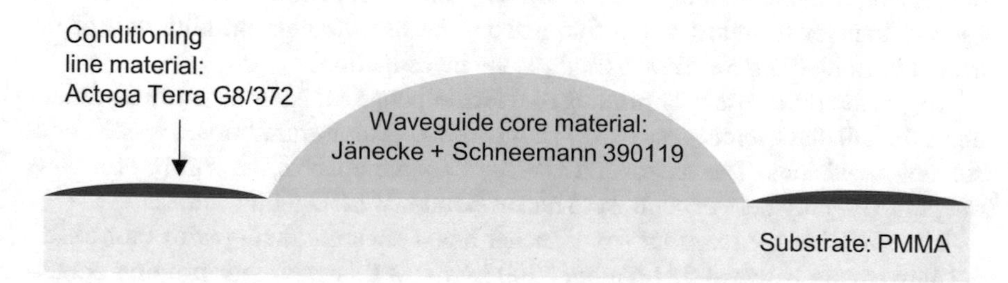

Fig. 4.13 Selected material configuration for experimental studies

(PMMA), conditioning line material (Actega Terra G8/372) and waveguide core material (Jänecke+Schneemann 390119). This configuration is shown in Fig. 4.13.

4.5 Experimental Studies of the Conditioning Line Printing Process

The geometry of printed optical structures, unlike that of optical fibers, is defined by the application on a substrate. This results in an asymmetric cross-sectional area depending on the process parameters and the materials used. From the state of the art, it is expected that the cross section of the structures can be approximated to a segment of a circle, varying in width as well as height. In addition to the material-dependent properties as described in Sect. 4.4, the parameter space for influencing the conditioning of film substrates is defined by the parameters of the system technology used as well as the adjustable parameters during the printing process. In the following, the effects of these parameters on the geometry of the forming conditioning lines as well as their characterization are described. In particular, the formation of topographical attributes relevant for the optical performance of the subsequent optical waveguides will be discussed.

4.5.1 Experimental Setup

A number of printing machines with flexographic printing units from different manufacturers are available on the market. They range from proof printing machines equipped with the simplest printing units for testing inks and coatings, to printing machines for laboratory use that allow the greatest possible variability, to production machines for industry that are used in 24-h shift operation. In order to investigate the possibility of industrial use, we use a sheetfed offset press with a flexographic printing unit (Speedmaster 52) from Heidelberger Druckmaschinen. A schematic cross section of the machine is shown in Fig. 4.14. The advantage of developing a manufacturing process directly on production machines is that it is easy to transfer to industry. For this reason, the flexographic printing unit of the available Heidelberg SM52 was used for the investigations.

To dry the fluids used for printing, the Heidelberg SM52 has an infrared emitter and a broadband mercury vapor UV light source. The infrared dryer is used with solvent-based inks. The broadband UV light source enables the curing of a wide range of UV polymers in contrast to narrowband UV LED light sources.

The available anilox roller from Zecher has a bank engraving with four different line counts between 55 L/cm and 200 L/cm. In this way, four printing parameters can be compared with each other on one substrate in a print test. Figure 4.15 shows the configuration of the anilox roller used, the geometry and the transfer volume.

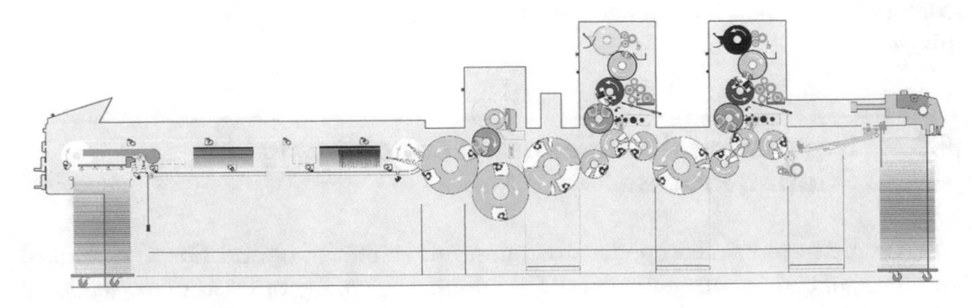

Fig. 4.14 Schematic illustration of Heidelberger SM52 in cross section

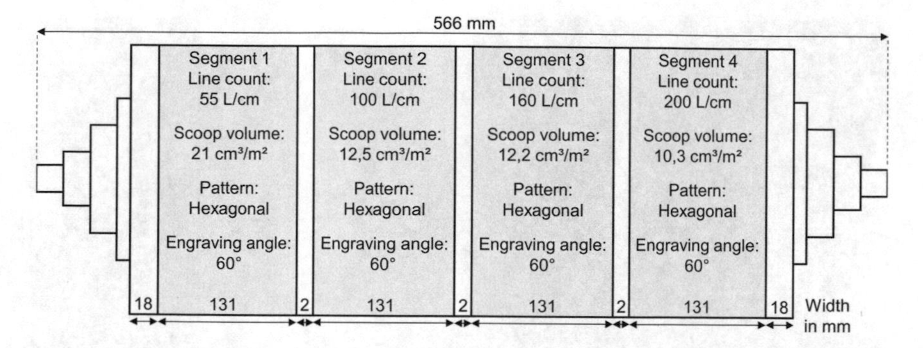

Fig. 4.15 Engraving parameters as well as different areas of anilox roller

The substrate is fed to the machine in stacks via a feeder. A gripper device is used to position the substrate in the plane. The substrate size is 350 mm in printing direction and 500 mm in width. Possible substrate thicknesses are between 50 and 500 μm.

4.5.2 Automated Evaluation of Geometric Quality

To characterize the properties of the conditioning lines described below, they are measured using a confocal microscope (μsurf custom, Nano-Focus). Subsequently, the three-dimensional measurement data are analyzed and evaluated within a MATLAB script. For this purpose, the actual line edges are automatically detected by means of image processing and a boundary line is generated (Fig. 4.16).

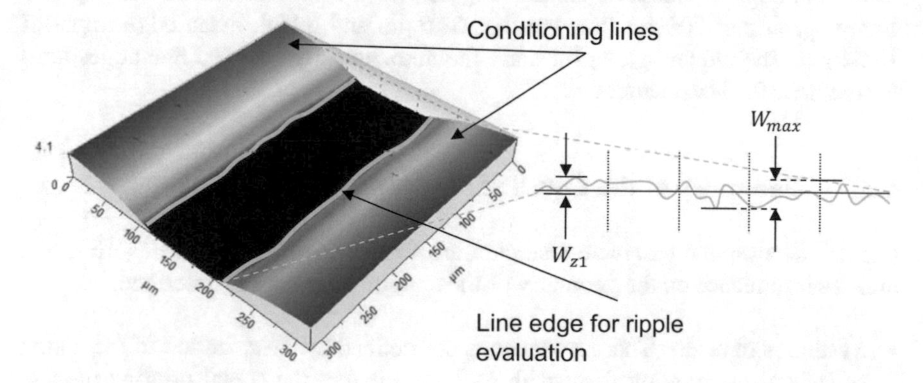

Fig. 4.16 Automated detection of conditioning line edges for calculation of shape tolerances (waviness WA) [DIN 4760] using MATLAB script

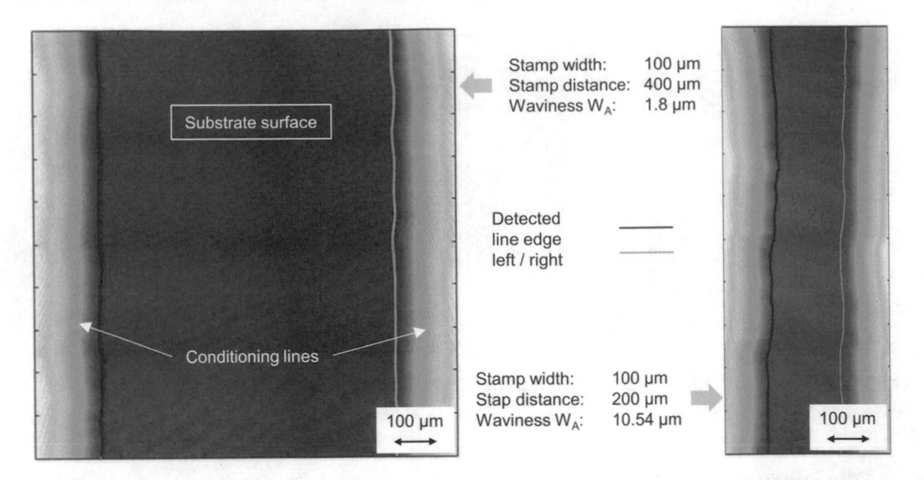

Fig. 4.17 Qualitative comparison of two interstices of conditioning lines with automatically detected line edge; left: large stamp spacing with low edge ripple; right: small stamp spacing with large edge ripple

The Douglas–Peuker algorithm provides the best results for this and is therefore used for the detection of the line edge [16]. For the conditioning lines, shape tolerances are selected according to known calculation bases (DIN 4760) for the evaluation of waviness widths W_A. Furthermore, the distance including the statistical error between the conditioning lines is determined by the script.

The following figure shows an exemplary qualitative comparison of two conditioning line configurations. For both conditioning lines, a printing stamp with a width of 100 μm and a line spacing of 100 L/cm is used. Figure 4.17 on the left shows the gap at a stamp spacing of 400 μm (mean line spacing: 275 μm) and a low mean edge waviness of 1.8 μm. Figure 4.17 on the right shows the gap at a stamp spacing of 200 μm (line spacing 68.5 μm and a high mean edge ripple of 10.54 μm. The red and green lines are the automatically detected line edges used for the described evaluations.

4.5.3 Geometry of the Conditioning Lines

For this section, the parameters studied are first divided into two categories, and then their influence on the geometry of the conditioning lines is described.

- System parameters: These parameters are defined by the selection of the anilox roller and the printing form with its layout before the actual printing process and therefore cannot be changed in the process.

- Process parameters: These parameters are set during the printing process independently of the system parameters. The process parameters identified as relevant from the state of the art are the infeed between anilox roller and printing form and the infeed between printing form and substrate.

The following Fig. 4.18 shows the characteristic geometry of the conditioning lines under investigation. From the preliminary work of [1] it is known that the cross section of printed lines after completed fluid dynamic balancing processes corresponds in the ideal case to a segment of a circle. By the exaggerated representation of the z-axis the predominant characteristics of the conditioning line cross sections are recognizable. These are, on the one hand, symmetrical lines with constant height and width (Fig. 4.18 left), and asymmetrical lines of variable height with squeezing edges (Fig. 4.18 right). For the subsequent application of the optical waveguide core, however, the width and the distance between the conditioning lines are likely to be the decisive factors and will therefore be further investigated.

Influence of the System Parameters

The line count of an anilox roller affects the ink transfer behavior to the printing form. The cells engraved in the surface of the anilox roller absorb the printing varnish from the ink reservoir. Excess material is wiped off by the doctor blade. With a constant cell type (hexagonal pattern, engraving angle 60°, Sect. 4.5.1), the

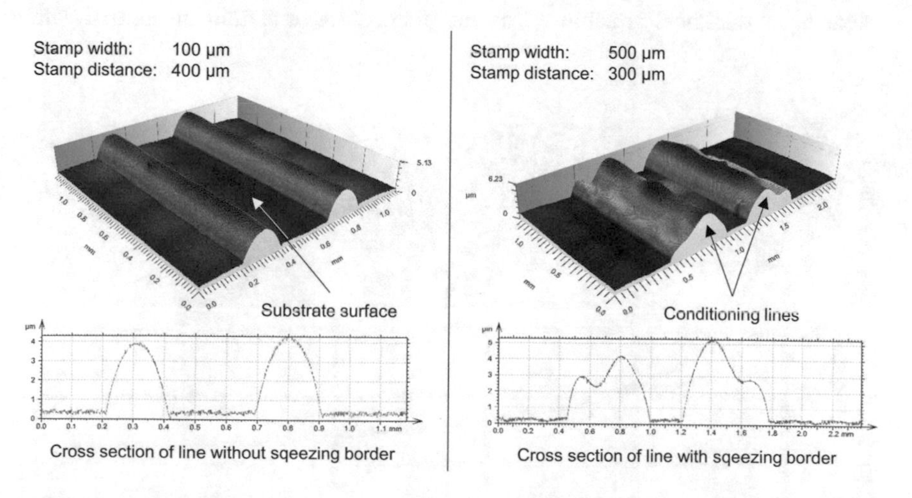

Fig. 4.18 Qualitative geometry of conditioning lines with the profile cross sections, left: Profile cross section without squeeze edges as well as the respective profile sections, right: Profile cross section with squeeze edges as well as the respective profile sections

depth and width are the decisive factors for the amount of varnish transferred. This geometrical property must therefore result in a larger volume of coating transferred to the printing form. For this purpose, printing tests were carried out with different anilox roller line counts and the resulting conditioning line geometry was measured. The cross-sectional area of the polymerized line structure allows to make conclusions about the layer thickness of the liquid coating film on the printing form before ink transfer. Since the printing varnish undergoes material shrinkage during polymerization, this difference was measured by weighing defined volumes of material before and after polymerization and determined to be 11.8%. Furthermore, it is known from the prior art that the ink splitting ratio approaches 50% with increasing printing speed [17]. The varnish still liquid on the printing form surface at the time of wetting therefore has twice the volume of the liquid varnish on the film substrate. Normalized to the stamp width of the printing form, the coating thickness on the printing form can be given as a function of the ratio of cell diameter to stamp width. Figure 4.19 shows this relationship.

This ratio of stamp width to cell diameter is shown on the horizontal coordinate axis, and the layer thickness on the stamp face during the printing process is shown on the vertical coordinate axis. The use of small stamp width / cell diameter ratios results in a significantly increased ink transfer, which, however, has a negative influence on the resulting line geometry in terms of waviness on the surface and necking on the side areas [1]. This can be explained by the fact that the stamp of the printing form is not only wetted with varnish on the front surface but also on the stamp edges by dipping into the cell of the anilox roller, thus leaving a defective image on the printing substrate. In the investigations, it can be seen that a consistent layer thickness is achieved on the punch face at a ratio of approximately

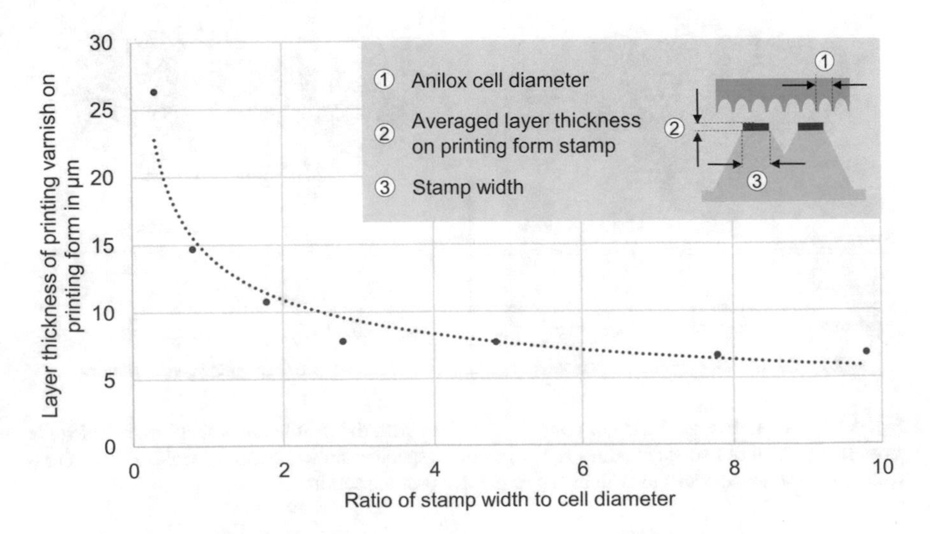

Fig. 4.19 Influence of the anilox roller's rulings on the coating thickness on the stamp surface

three and above. As the ratio increases, a constant layer thickness in the range of 6–7 μm is achieved. This result leads to the conclusion that this ratio should not be undercut even for printing functional coatings that require optimum surface quality.

In order to investigate the influence of the selected line count on the spacing of the conditioning lines, this is measured across all selected stamp widths (100, 200, 300, 500 μm) using confocal microscope images. In the diagram in Fig. 4.20, the mean value of this parameter is shown for different stamp spacing, including the statistical error. As expected, increasing the stamp space results in a linear increase in the conditioning line spacing. Furthermore, it can be seen that a finer line count also results in an increasing line spacing due to the decreasing transferred volume, since the decreased volume is distributed over the line cross section. Figure 4.20 (right) shows the linear regression degrees of the determined measured values for qualitative differentiation. Here, a constant increase of the line spacing between the line counts 100, 160 and 200 L/cm can be seen. Only the 55 L/cm line shows a reduced increase, which also indicates defective wetting of the stamp edges (compare Fig. 4.20).

In the following investigation, the conditioning line spacing of the respective stamp width is considered separately. Figure 4.21 shows the conditioning linc spacing including the statistical error again as a function of the stamp spacing but for the different stamp widths. The anilox line spacing used for this study is 100 L/cm. Here, the linear relationship between the two values can be seen too. The large error bars at the largest stamp width of 750 μm are noticeable. Also in comparison with the previous work of [1], this large variation can be explained

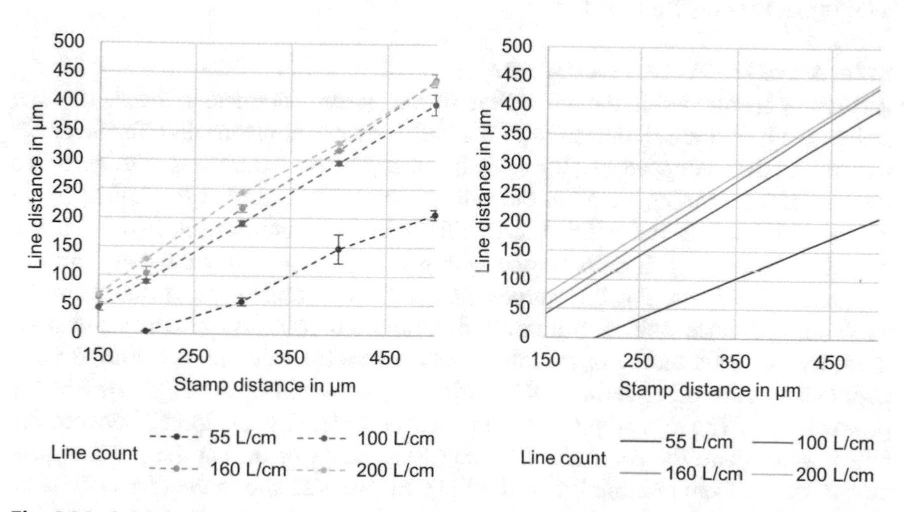

Fig. 4.20 left: Conditioning line spacing as a function of punch spacing incl. error bars for the different lineatures of the anilox roll; right: regression lines for the resulting conditioning line spacing as a function of stamp spacing for the different lineatures of the anilox roll, $R^2=0.989$ (55 L/cm)

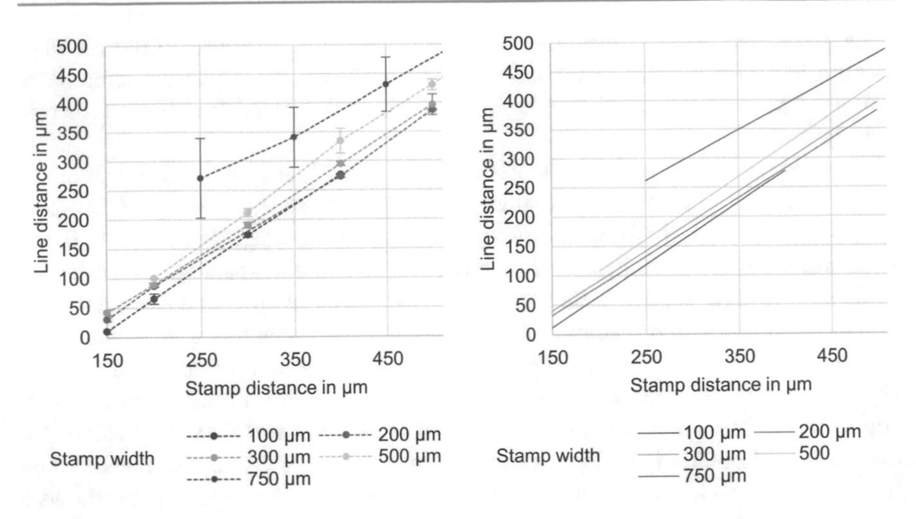

Fig. 4.21 Left: Conditioning line distance as a function of punch spacing incl. error bars for the different punch widths; right: regression lines for the resulting conditioning line distance as a function of stamp spacing for the different stamp widths, $R^2 = 0.9998$ *(300 µm)*

by viscous finger formation. When plotting the regression line in Fig. 4.21 (right), the distinction between the other stamp widths (100, 200, 300 and 500 µm) can be seen better. This analysis is also in line with the state of the art and shows that the widening of the printed lines results in a deviation from the printed image and that a reduction of the stamp spacing below 150 µm (stamp width 100 µm) results in a merging of the conditioning lines.

Influence of the Process Parameters

As already mentioned in the description of the system parameters, the ratio of the stamp width on the printing form to the cell diameter is responsible for incorrect wetting of the stamp edges. By selecting the process parameters, the defective wetting of the printing stamp in the cells of the anilox roller is intensified by an increased infeed. In the printing tests, this infeed is therefore selected in such a way that the printing form is wetted with printing varnish over the entire area at the stamp end faces. For this purpose, the infeed is gradually reduced from the maximum distance until a uniform and continuous print image can be seen on the substrate. This setting of the infeed between anilox roller and printing form is kept constant for all experimental tests in order to avoid incorrect wetting of the printing form. The second process parameter considered is the infeed between the printing form and the substrate. The dewetting behavior during the printing process is known from preliminary work of [1]. Figure 4.22 shows the relevant phases of ink splitting. The first phase shows the contact of the printing varnish on the stamp face with the substrate. From the second phase onward, the printing varnish is squeezed outward from the gap between the printing stamp and the substrate, causing maximum line widening. The infeed changes the minimum distance

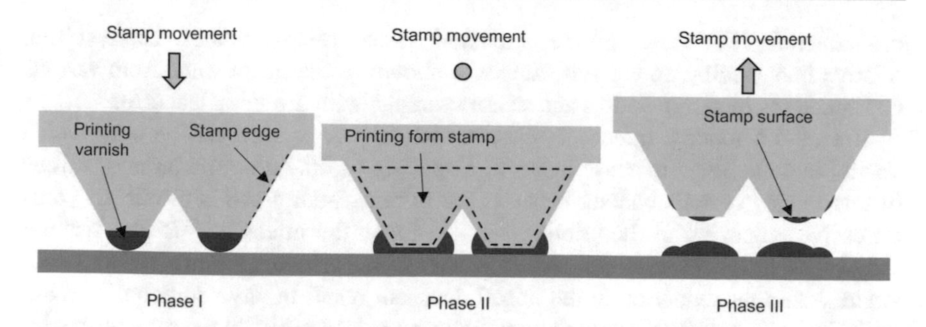

Fig. 4.22 Schematic representation of the three phases of the flexographic printing process, Phase I: Contact of the printing varnish with the substrate surface; Phase II: Resulting line widening due to displacement of the printing varnish; Phase III: Defective wetting of the stamp flanks after completion of the printing process

between the printing stamp and substrate in this phase. This also influences the widening of the printed structures and consequently the resulting distance between the conditioning lines after the print stamp is removed from the substrate. In phase 3, the ink splitting and equalization processes are completed, as described in [1]. Both the stamp face and the stamp edge are wetted with residual printing varnish after completion of the printing process, which influences the printed image during subsequent process cycles.

Figure 4.23 shows the previously described influence of the minimum distance between the printing stamp and the substrate from phase 3 on the resulting line spacing. Since the infeed is influenced by the thickness of the printing form and

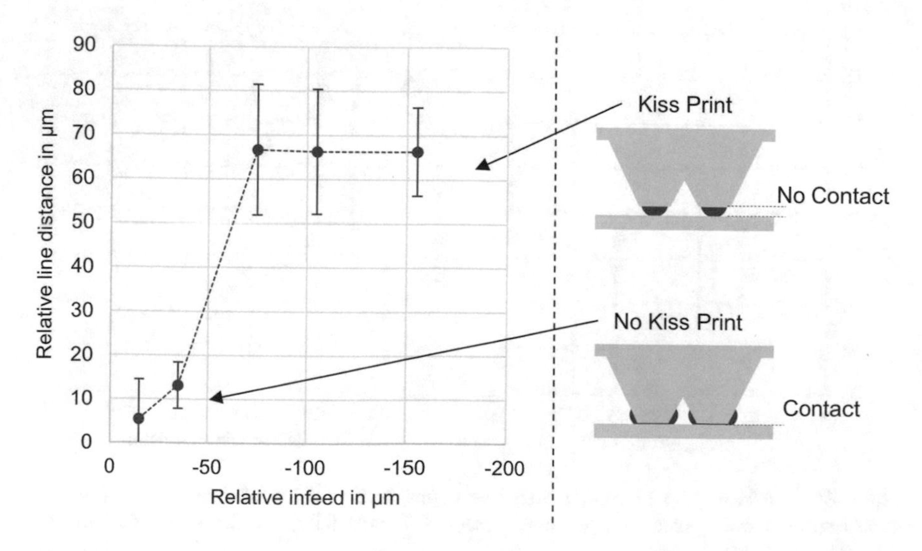

Fig. 4.23 Change of line spacing depending on the infeed (printing form/substrate)

the substrate, it is shown on the horizontal axis in relative values. The resulting relative line spacing on the vertical axis is shown as the mean value from various combinations of stamp widths and stamp spacings with the statistical error.

Between a relative infeed of—35 µm and–75 µm, a jump can be seen which results in an average increase in the line spacing of 66.7 µm. The reduced infeed in this range prevents contact between the printing stamp and substrate and thus excessive squeezing of the printing varnish. From the infeed of—75 µm, wetting only takes place through the contact between the printing varnish and the substrate. A further reduction in the infeed does not result in any additional increase in the line spacing. This configuration, known as "kiss print" in the state of the art, results in lower squeeze edges and is therefore identified as the target value for the experimental investigations.

4.5.4 Formation of Edge Ripples and Topography

The target value for the resulting line gap and the resulting line width is defined on the one hand by the selected application process and on the other hand by the application. In order to define a process window, the edge ripple of the resulting gap between the conditioning lines is investigated. In the following figure, the influence of the printing parameters on the edge waviness (W_A) is shown as a mean value with statistical error. Figure 4.24 (left) shows the resulting edge waviness for different stamp widths averaged over all stamp spacings. Due to the large error bars, however, no dependence can be seen here. The stamp width of 500 µm exhibits the lowest statistical errors and is therefore defined as the target value in

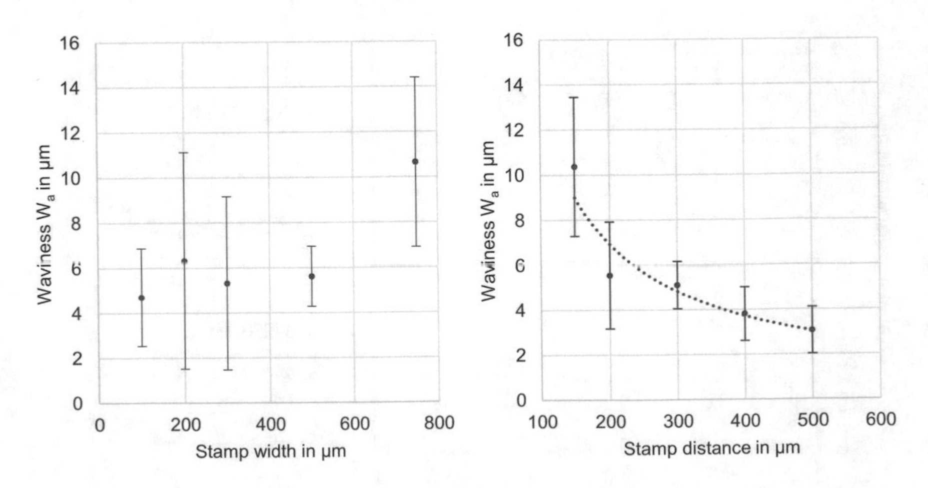

Fig. 4.24 *Waviness (W_a) of conditioning line edges as a function of stamp width (left) and* stamp spacing incl. potential regression "f(x)=816.14*x^{-0.9}, coefficient of determination R^2 = *0.9133 (right)*

the following. Figure 4.24 (right) shows the ripple of the different stamp distances averaged over all stamp widths.

It can be seen that with increasing spacing, a lower ripple at the edge occurs with likewise low statistical deviations. The value approaches asymptotically toward a constant value at large punch distances, since here there is no longer any mutual fluid-dynamic influence of the conditioning lines and these are to be regarded as individual lines.

From the relationships between the system parameters and the resulting line spacing described in Sect. 4.5.3, the relative widening of the line spacing can be represented as a function of the stamp width. This relationship has been determined for a line count of the anilox roll at 100 L/cm. With smaller stamp widths, the proportion of squeeze margins is clearly disproportionately included in the resulting line spacing. This is shown by the automatic logarithmic trend line

$$g(x) = 25.963 \cdot \ln(x) - 114.4 \tag{4.6}$$

with a coefficient of determination of $R^2 = 0.8886$ (Fig. 4.25 left). From this function, an estimation of the resulting line spacing can be made from the combination of stamp width and stamp spacing (Fig. 4.25 right). This prediction can additionally be supplemented by the model of the resulting edge ripples (Fig. 4.25 right) as a function of the stamp spacing. This waviness is shown by coloring the corresponding line spacing, where red equals a high waviness and green equals a low waviness.

This consideration results in a process window where, however, the influence of the edge ripple on the optical performance of the optical waveguide has to be reviewed afterward.

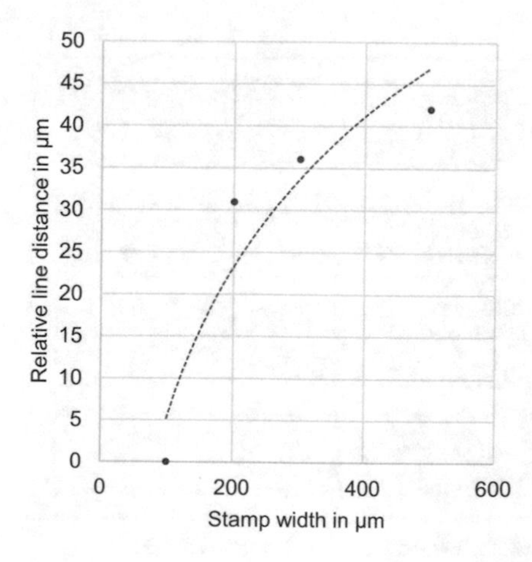

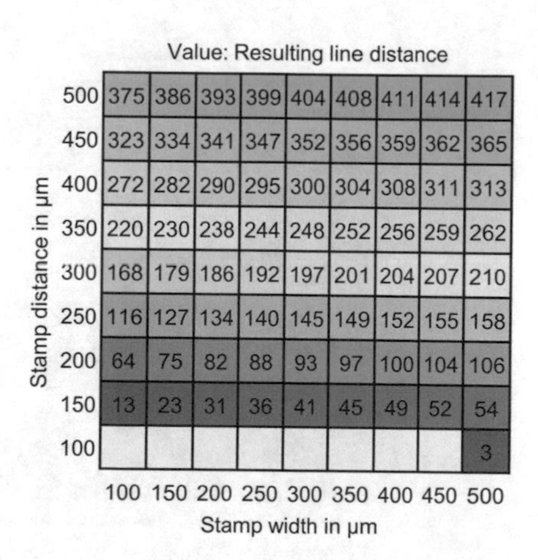

Stamp distance in µm	100	150	200	250	300	350	400	450	500
500	375	386	393	399	404	408	411	414	417
450	323	334	341	347	352	356	359	362	365
400	272	282	290	295	300	304	308	311	313
350	220	230	238	244	248	252	256	259	262
300	168	179	186	192	197	201	204	207	210
250	116	127	134	140	145	149	152	155	158
200	64	75	82	88	93	97	100	104	106
150	13	23	31	36	41	45	49	52	54
100									3

Fig. 4.25 Left: Relative line spacing for different stamp widths incl. logarithmic regression g(x); right: Process window for printing conditioning lines with estimation of the line spacing to be set, green = low edge ripple, red = high edge ripple

4.6 Modeling of Stamp Deformation in the Printing Process

In addition to the analytical investigation, a qualitative investigation of the conditioning line geometry is also carried out. Figure 4.26 shows the profile cross section of the resulting conditioning lines with different combinations of stamp widths and stamp spacings. It is noticeable that large stamp widths can lead to the formation of squeeze edges (300 μm and 500 μm) and that the symmetry of the individual conditioning lines increases with increasing stamp spacing (profile cross sections marked in red). This characteristic suggests that a non-uniform loading case of the printing stamps prevails during the printing process. Since the stamp spacing as described in 4.5.4 also has an influence on the ripple of the interspace and thus affects the quality of the optical waveguides, this source of error is investigated further. For this purpose, the deformation characteristics of the printing stamps in the printing process are modeled and analyzed using the finite element method.

For the geometric modeling of the printing stamp, an idealized design is used, which is created on the basis of the confocal microscopic measurement data (Fig. 4.27). In the simulation of the deformation behavior, parameters are varied. These are, with respect to the printing stamp geometry, the width, the spacing, and the edge angle. Furthermore, different materials are modeled, which differ from manufacturer to manufacturer. The properties of the stamp deformation investigated here are the widening of the stamp face and the stress distribution in the printing stamp.

Simulation using the finite element method is performed by describing the physical behavior of individual geometric elements of the overall model. To evaluate the system behavior, the behavior of each individual element is calculated for

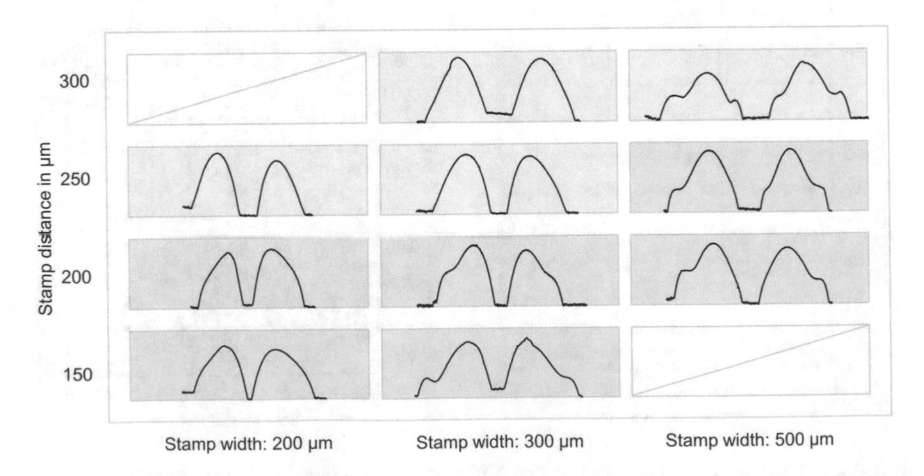

Fig. 4.26 Qualitative influence of punch width and punch spacing on the geometry of the conditioning line cross section, green: symmetrical profile cross sections, red: asymmetrical profile cross sections

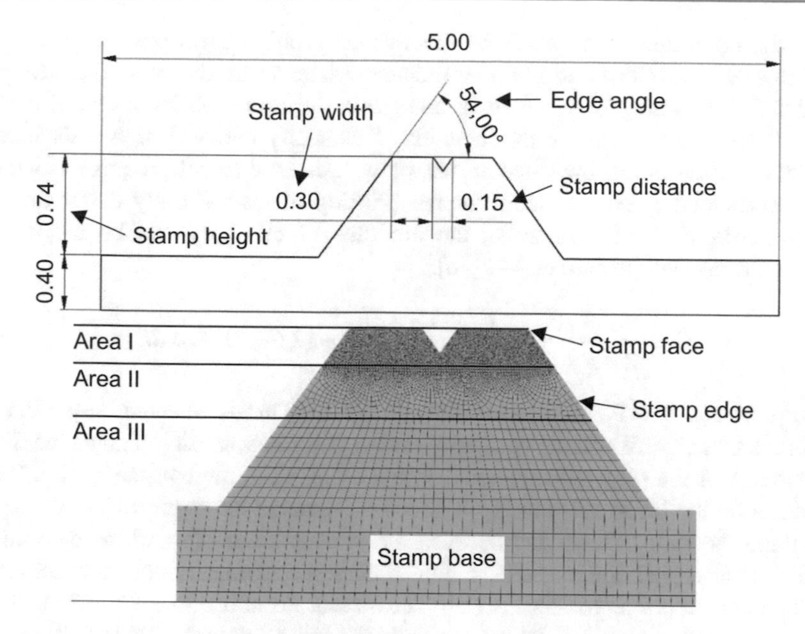

Fig. 4.27 Above: idealized geometry of the modeled printing stamp in mm; below: Division of the printing stamp into different meshing zones

the action of forces and loads as well as existing boundary conditions and their interactions with each other. The partitioning into these elements is referred to as "meshing." The geometric shape of the individual elements is selected in such a way that the overall model can be represented as ideally as possible by stringing them together. For the printing stamp shown in Fig. 4.27, square basic shapes are therefore used. The number of individual elements must be sufficiently large to enable the most realistic modeling possible and thus to obtain meaningful simulation results. However, this number also increases the required computing power, which necessitates a division of the overall model into areas to be prioritized. These areas are shown in Fig. 4.27 (bottom). Area I is defined as the area below the punch face as well as the upper area of the stamp edges, where direct contact with the printing varnish is expected and which, due to its small cross section, has the proportionally lowest stiffness during stamp deformation. Here, the node spacing defining the width of the individual elements is selected to be 5 μm. Area III represents the base area as well as the stamp base. This area only requires a coarse mesh, since no critical forces and deformations are expected there. A node spacing of 50 μm is selected here. Area II forms the transition between the fine mesh of individual elements in the upper Area I and the lower Area III with larger elements. This is defined at a position from 50 μm below the lowest point in the intermediate area of the individual stamps. In this area, a node spacing of 30 μm is selected in order to represent deformations and stresses still sufficiently accurately and to satisfy the low deformation to be expected.

In the definition of materials by the various printing form manufacturers, the modulus of elasticity is an important characteristic value that describes the proportional relationship between stress and strain during the deformation of a solid body. Thus, a high value of the modulus of elasticity means a high resistance to elastic deformation. A low value means a low resistance to deformation. However, this information is not available for the printing plates used. By converting the Shore hardness A of a material, the modulus of elasticity can be determined approximately with the aid of 4–7 [18].

$$E = \frac{1-\mu^2}{2 * R * C_3} * \frac{C_1 + C_2 * Sh_A}{100 - Sh_A} * (2{,}6 - 0{,}02 * Sh_A) \quad (4.7)$$

Here μ denotes the Poisson's ratio defined above. This is selected as $\mu = 0.5$ for materials that exhibit volume constancy during deformation. Sh_A denotes the Shore hardness A of the respective material. In the calculation, the constants C1, C2 and C3 describe the linear relationship between the indentation depth of the test specimen and the spring force, and between the indentation depth and the determined Shore hardness [18]. R denotes the radius of the test specimen defined for measuring the hardness according to Shore A [19]. The measurement results are shown together with the respective result of the conversion in the following table (Table 4.4).

Due to the rotary printing process, the load profile of the printing stamp is not linear. The movement profile of the printing process depends on the system technology used. The diameters of the individual cylinders in the printing machine, printing form cylinder (diameter 180 mm), impression cylinder (diameter 360 mm), together with the selected printing speed result in the kinematic relationship shown in Fig. 4.28.

The motion profile in Fig. 4.28 (right) is shown for an infeed of 150 μm. By selecting a lower infeed, the time to the reversal point of the deformation is shortened. The press is operated with a throughput W = 3000 sheets/h for all tests. With a substrate length of 500 mm, the angular velocities ω_1 and ω_2 from Fig. 4.28 (left) result as follows:

$$\omega_1 = \frac{1}{2}\omega_2 = 2 \cdot \pi \cdot W \quad (4.8)$$

The vertical distance between points 1 and 2, S_y, is given by Fig. 4.28 (right):

$$S_y = r_1 \cdot (1 - \cos(\alpha)) + r_2 \cdot (1 - \cos(\beta)) \quad (4.9)$$

Table 4.4 Results of the conversion from Shore hardness to modulus of elasticity of the different compression molding materials

Producer	Shore Hardness A in $\mathbf{Sh_A}$	Young's modulus E in N/mm^2
Conti Laserline	68.77 ± 0.69	8.82
Kodak Flexcel NX	76.56 ± 0.51	10.69
Flint Gold A	80.01 ± 0.17	12.45

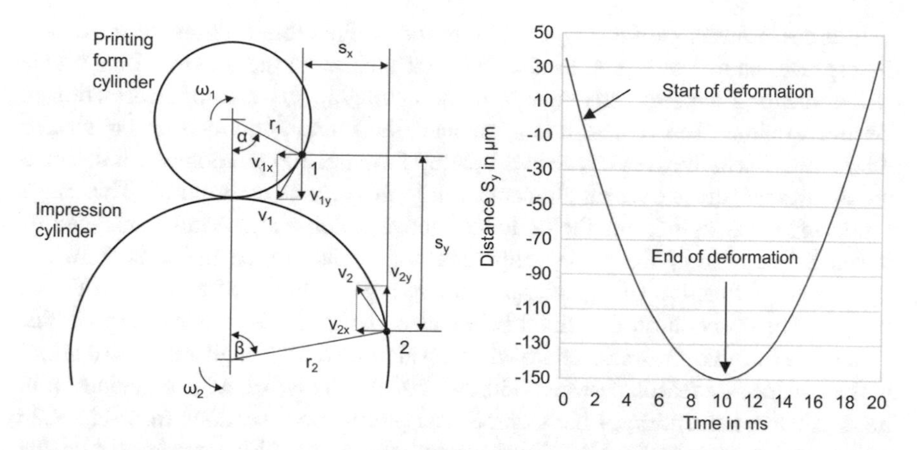

Fig. 4.28 left: Cross section of printing unit; right: Kinematics in printing unit, view rotated about Z-axis [1]

This consideration enables the calculation of the motion profile and is used for the investigations in the FEM simulation.

For a first qualitative consideration of the stamp deformation in Fig. 4.29, the geometry from Figure Fig. 4.27 (top) is used. The infeed selected here is 50 µm. The outer and inner stamp edges are curved outward by the compression. The resulting von Mises stresses in the printing stamp exhibit local maxima at the respective edges of the stamp face. In addition, they are unevenly distributed over the width of the stamp. In each case, the region of high stress is more pronounced at the stamp exterior and is a first indication of the asymmetric profile cross section of the conditioning lines (Fig. 4.26). For the further simulations, only one half of the stamp is considered, since the behavior can be transferred to the other half due to the geometric symmetry.

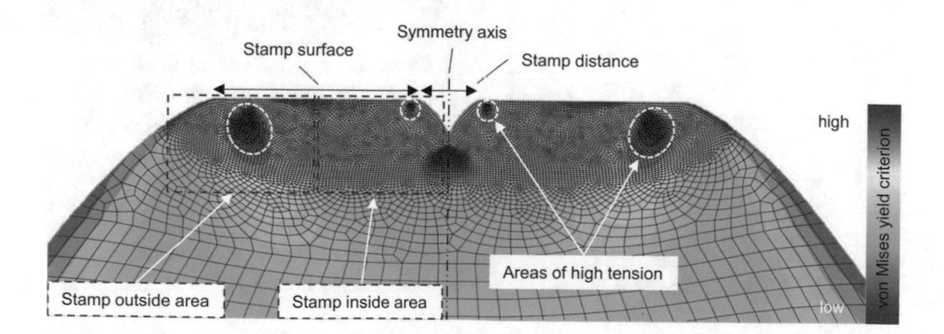

Fig. 4.29 Qualitative representation of stamp deformation in the printing process with characteristic curvature of the stamp flanks and uneven distribution of stress between the outer and inner sides of the stamp; stamp width: 300 µm, stamp spacing: 150 µm, infeed: 50 µm

In order to investigate the asymmetry of the profile cross sections of the conditioning lines shown in Fig. 4.26, the stress ratio between the outer and inner side of the stamp is subsequently analyzed by simulating the stamp deformation at varying spacings. To evaluate this imbalance, the stamp is divided into outer and inner sides. Then, the area between the stamp face and 200 μm below it is selected and an average stress over all elements within this area is determined. This range is selected because it forms the boundary above which a stress imbalance can be detected. The ratio of the mean value from the outside to the inside is shown in Fig. 4.30 as a function of the stamp spacing, representative of a stamp width of 300 μm. The observation described below also applies to the other stamp widths. It can be seen that the ratio approaches a value around 1 (uniform distribution) as the distance increases between 350 and 500 μm. This behavior coincides with the qualitative perception of the symmetrical profile cross sections from Fig. 4.26 with increasing stamp spacing. Furthermore, this result also correlates with the edge ripple of the conditioning lines described in Sect. 4.5.4, which decreases with increasing distance and approaches a limit value.

Another parameter that has an impact on the printing result in the printing process is the stamp widening due to the compression caused by the applied load. Unlike process monitoring, this deformation behavior can be evaluated in the simulation environment at any time during the process. Figure 4.31 shows the percentage stamp widening at the infeeds 10 μm, 50 μm and 150 μm. At low infeeds of 10 μm and 50 μm, the punch widening remains constant in a range of 0.6% and 5% with increasing punch width (Fig. 4.31 left). Only at a significantly increased infeed of 150 μm, the percentage stamp widening is significantly greater at smaller stamp widths (38.06% at 200 μm stamp width). This suggests that the relative line

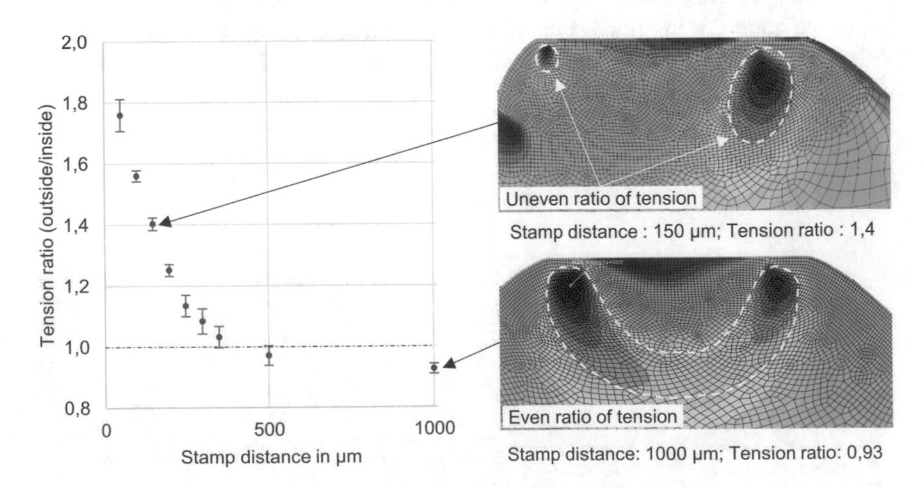

Fig. 4.30 left: Stress ratio of the punch deformation during the printing process as a function of the punch spacing incl. error bars for the different infeeds (50 μm, 100 μm, 150 μm) at a punch width of 300 μm; right: representation of the stress distribution at 150 μm and 1000 μm punch spacing, respectively

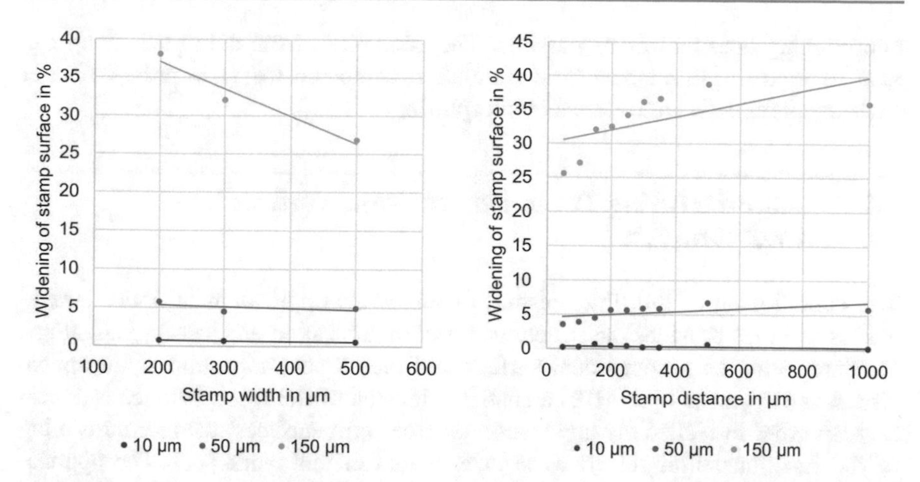

Fig. 4.31 left: Percentage widening of the punch face as a function of punch width including linear regression for different infeeds: 10 µm, 50 µm, 100 µm; right: Percentage widening of the punch face as a function of punch spacing including linear regression for different infeeds: 10 µm, 50 µm, 100 µm;

spacing resulting from a changed infeed will be subject to an even larger variation at smaller stamp widths and therefore the state of the "kiss print" has an even greater relevance for specifically predicting line spacing. Figure 4.31 on the right shows the percentage of stamp widening for the same infeeds as a function of the stamp spacing. Here, it is noticeable that at low infeeds (10 µm and 50 µm) the stamp spacing does not have a major influence and the stamp widening is almost constant at 0.6% and 5%, respectively. The result for an infeed of 150 µm is consistent with the findings from the dependence on the stamp width. Here, an increase in the percentage widening is seen at a stamp spacing of up to 500 µm, which can be explained by a lower stiffness of the single structure compared to the double stamp. A further increase in the stamp spacing does not result in an increase in the percentage stamp widening, since there is no longer any mutual influence between the individual stamp structures.

There is no dependence on the choice of the printing form manufacturer and the material-related hardness for both the stamp widening and the stress ratio. The influence of these materials with regard to their fluid dynamic properties can nevertheless have a relevant influence.

Conclusion

The simulative modeling of the stamp deformation provides an indication of the resulting asymmetry of the conditioning lines at the different combinations of stamp width and stamp spacing. Also, the uneven stress ratio between the inside and outside of the printing stamp at low stamp spacings correlates with the high waviness of the conditioning lines. FEM simulation thus offers a way to make predictions for a stamp geometry in order to keep the resulting waviness in conditioning lines as low as possible. The geometry of the deformed stamp can also be used for an adapted fluid simulation compared to [1] in order to better understand fluid dynamic effects in ink splitting.

4.7 Manufacturing Three-Dimensional Optical Interconnects

The conditioning of the film substrates enables the application of optical waveguides and thus provides the structures for communication and sensor technology. For integration on a component surface and thus functionalization as an optical circuit carrier or 3D-opto-MID, a spatial extension of the foil substrates is necessary. In order to realize the application on free-form surfaces, the thermoforming of the functional film substrates is investigated in this work [20]. The application and material-specific findings obtained in the process are explained in the following.

4.7.1 Studies on the Thermoforming of Conditioned Film Substrates

The 686PT thermoforming machine from Formech is used for forming the film substrates. This is a manual floor-standing machine in which forming takes place by means of a vacuum. Infrared quartz heaters with six adjustable heating zones are used to heat the film substrate to the forming temperature. In order to qualitatively assess the heat distribution during the heating and forming process, the film substrate is measured using a thermal imaging camera (Flir T360). Figure 4.32 (left) shows the experimental setup used for this purpose. The measurement image of the thermal imaging camera with the temperature scale and the current measured value in the marked measurement range can be seen in Fig. 4.32 (right). The lower limit of the measuring range is defined as the glass transition temperature of the PMMA film (TG = 105 °C), which is required for forming in the thermoforming process.

In addition to the measurement parameters, which are defined by material properties and environmental conditions, the active measurement range with the contour of the forming tool can be seen in the measurement image. As soon as contact with the tool surface occurs during the forming process, the film substrate cools to a value in the lower measuring range. As soon as the glass transition temperature falls locally below this value, the forming process is completed at this position. For spatial extension, the forming behavior of the conditioned film substrates is investigated. In this way, a process window is to be determined without causing damage to the functionalized film substrates. The damage results from the material selection made for the manufacturing process according to Sect. 4.4. For example, the print varnish used for the conditioning lines consists of acrylates which are cured by means of crosslinking radical UV polymerization. Thermal forming

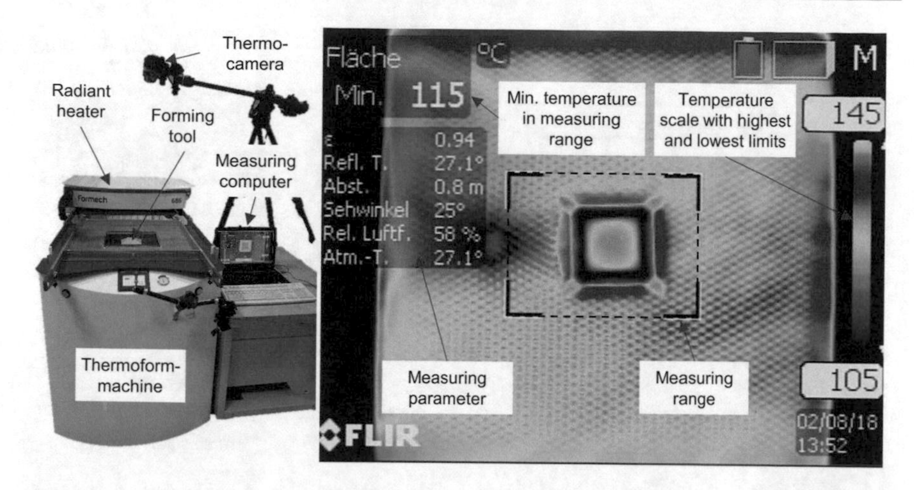

Fig. 4.32 Left: Experimental setup for measuring the temperature distribution of the film substrates by means of a thermal imaging camera; right: Measurement image of the thermal imaging camera with temperature scale (lower limit glass transition temperature PMMA: TG = 105 °C) immediately after the forming process

of the conditioned substrates results in elongation of the thermoplastic substrate. The imprinted structures are subject to the same elongation. Due to the thermoset material, however, the permissible elongation is limited. If the limit is exceeded, cracks appear in the conditioning lines and waveguides as a result of this elongation, which negatively affects the subsequent application process or the functionality of the optical waveguides themselves. A measure with which the locally occurring strain after the forming process can be determined is the wall thickness of the substrate [21]. A truncated pyramid is selected as the base body for the analytical investigation. Figure 4.33 (left) shows the tool geometry with the variable parameters (tool height and base angle). The radius of curvature r was kept constant at 10 mm.

Furthermore, the measuring points at which the wall thickness is determined by micrometer screw are shown. Figure 4.33 (right) shows the resulting wall thickness at the respective measuring point along the mold surface as a function of the selected mold height (10, 20, 30 mm).

Here, it can be seen that an uneven wall thickness distribution occurs. The minimum wall thickness is determined for each tool height at the measuring point in the lower radius of curvature and increases in each case to the measuring points on the upper surface and at the base of the truncated pyramid back to the range of the original wall thickness (175 µm). At a tool height of 10 mm, there is even an increase in wall thickness on the upper plane at the middle measuring points. Since cracking in the printed structures due to forming always occurs at the point of lowest wall thickness, only this measuring position in the lower radius of curvature is considered in the following. The resulting wall thickness is considered in Fig. 4.34 by combining the respective tool height with a variable base angle. Here,

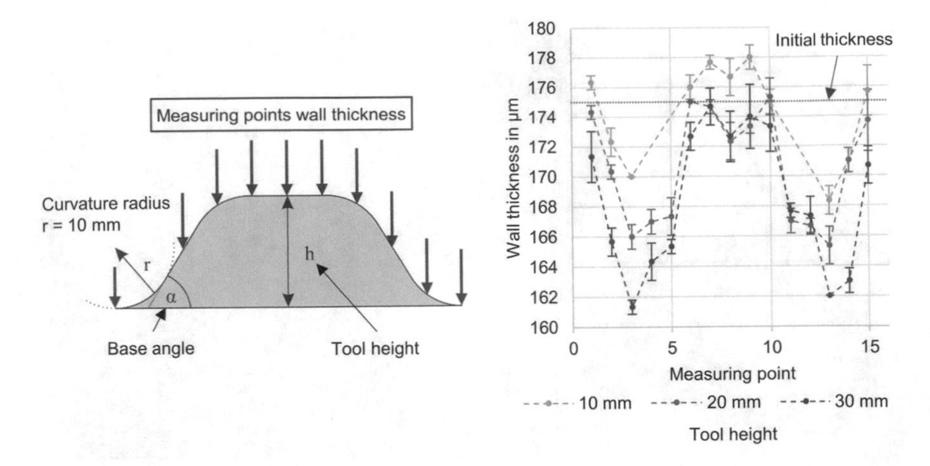

Fig. 4.33 Left: Basic geometry of the forming tools with the variable parameters base angle, radius of curvature and tool height as well as positions along the tool surface for measuring the wall thickness; right: Resulting wall thickness at the respective measuring points along the tool surface incl. error bars depending on the tool height (10, 20, 30 mm)

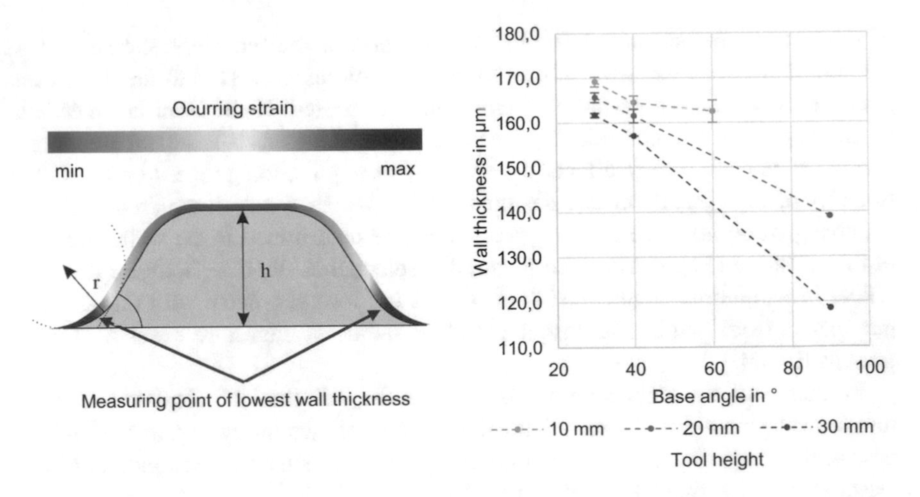

Fig. 4.34 Left: Occurring strain at the tool surface incl. the measuring points at which the smallest wall thickness is to be expected; right: Resulting wall thickness at the measuring point of the largest strain (lower radius of curvature) incl. error bars as a function of the base angle and the tool height

it can be seen that the wall thickness also decreases with increasing base angle and drops to up to 70% of the original substrate thickness (175 µm) at a tool height of $h = 30$ mm and a base angle of $\alpha = 90°$.

In order to assess the effect of the resulting wall thickness on the functional structures, the damage caused by the forming process is then characterized. This characterization is carried out by categorizing the type of crack and plotting the number of cracks as a function of the conditioning line geometry.

Figure 4.35 (left) shows an example of a thermoformed conditioned film substrate. The film substrate is illuminated from the side to identify the position of the cracks. The cracks represent scattering centers and can subsequently be clearly identified. For categorization, the cracks in the area of the lower radius of curvature are examined using microscope images Fig. 4.35 (right). The evaluation identifies two types of damage, cracks on the surface of the conditioning lines (crack type I) and cracks that result in local delamination of the conditioning line (crack type II). The number of cracks can subsequently be evaluated using microscope images (Fig. 4.36 top) and characterized as a function of the mold geometry and the geometry of the conditioning lines. Figure 4.36 (left) shows the number of cracks as a function of the conditioning line cross section and the selected line count of the anilox roll. As described in Sect. 4.5.3, the cross-sectional area of the printed structures increases with increasing cup volume on the anilox roll.

This behavior can also be seen in this analysis. Furthermore, an increasing number of cracks in the conditioning lines correlates with increasing line cross section. This value rises to over 250 cracks for the largest line cross section in the 3000 μm^2 range. The proportion of type II cracks is shown in Fig. 4.36 (right). Here, the correlation of the number of cracks with increasing cross-sectional area of the conditioning lines can be seen too. With approx. 30 cracks at the largest line cross section of approx. 3000 μm^2, this is at a significantly lower level, but is subject to large percentage fluctuations. This shows that conditioning lines with a smaller cross-sectional area are better suited to prevent the formation of cracks.

In order to correlate the crack formation in the conditioning lines, as a result of the occurring strain, with the resulting wall thickness and thus with the geometry of the forming tool, a further investigation takes place. For this purpose, the number of cracks is again documented, which occurs with constant conditioning line

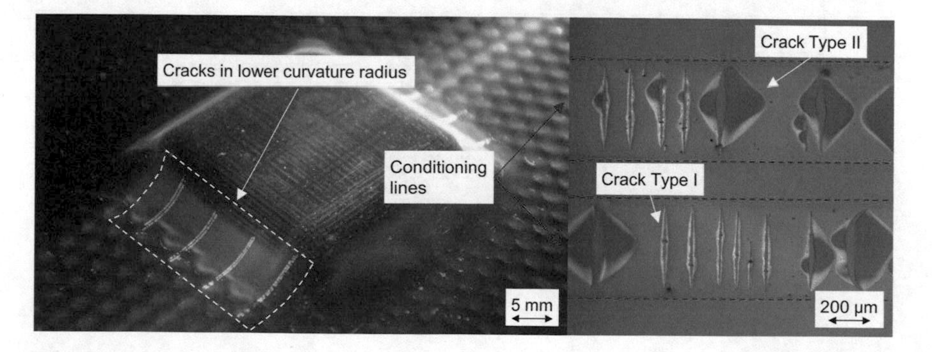

Fig. 4.35 Left: Identification of the area of crack initiation by the scattering of laterally coupled light; right: Categorization of crack types by microscopic examination into type I and type II cracks

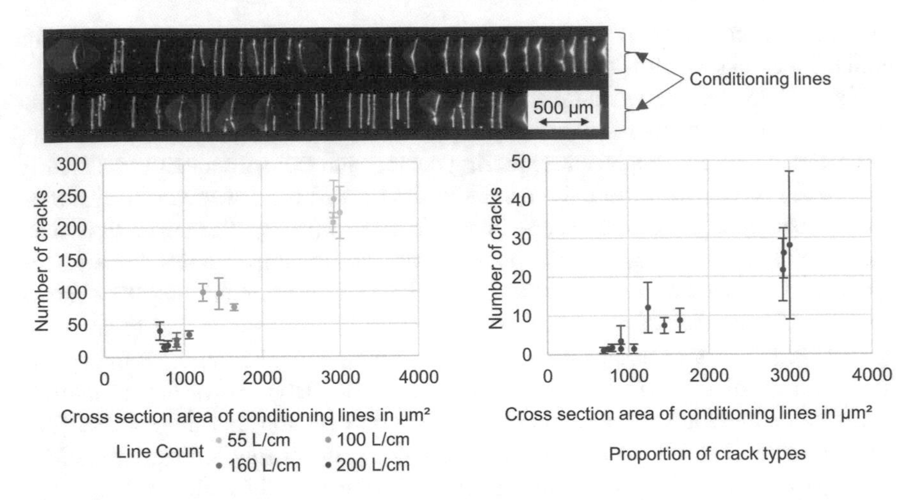

Fig. 4.36 Top: Evaluation of the number of cracks in the conditioning lines based on microscopic images; left: Number of cracks in the conditioning lines incl. defect bars as a function of their cross-sectional area and the anilox roller ruling used; right: Proportion of type 2 cracks in the conditioning lines incl. defect bars as a function of their cross-sectional area

geometry but with varying tool geometry (Fig. 4.37). With increasing mold height, the number of cracks in the conditioning lines also increases. Type I cracks occur with statistical significance (mean number of cracks = 14.6) only from a mold height of 20 mm.

Type II cracks occur at this height with an average frequency of 0.4. Only at a mold height of 30 mm, a statistically significant average number of cracks (13)

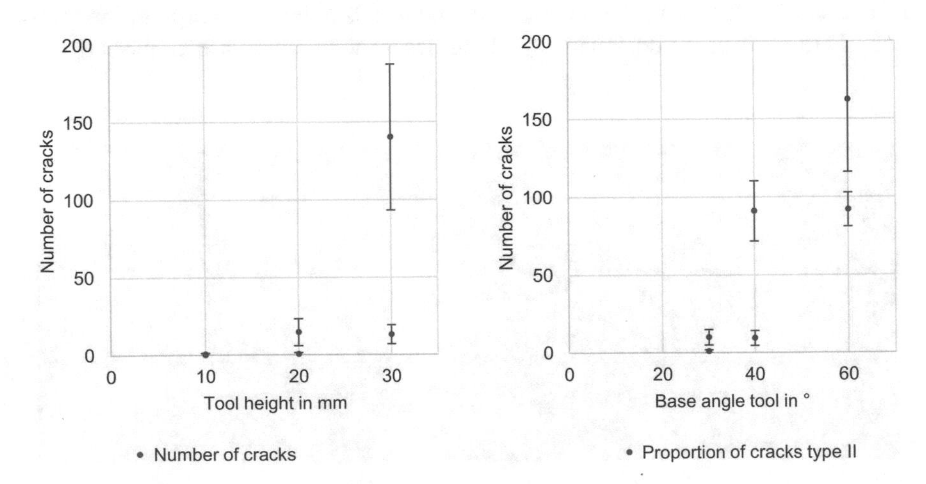

Fig. 4.37 Left: Number of cracks in the conditioning lines incl. error bars as a function of the mold height; right: Number of cracks in the conditioning lines incl. error bars as a function of the mold base angle

is recorded. A similar behavior is also seen with increasing base angle (Fig. 4.37 right). Here, cracks occur from 30° (mean number of cracks type I = 10) and from 40° (mean number of cracks type II = 9).

Conclusion
In both crack types, the edges of the conditioning lines are not interrupted and can thus potentially continue to locally control the subsequent wetting. The characteristics of the application processes for the waveguide cores are the occurring overspray (aerosol jet) and the flooding of the conditioning lines (dispensing). The influence of these process characteristics depending on the crack type as well as their frequency will be investigated with regard to the quality of the resulting optical waveguides in the following.

4.8 Surface Functionalization for Wetting Behavior Adjustment of Flexographic Printing Forms

The term functionalization is generally not clearly defined, so the first section will describe the meaning in more detail in the present context.

In flexographic printing, the printing form has the task of transferring the material to the substrate (see Sect. 4.2). Based on a conventional printing form, it can be modified through further process steps in such a way that certain properties can be achieved or improved in the printed result. The changes in the print form are intended to fulfill a specific function.

Following examples illustrate this:

1. Fig. 4.38 shows schematically printed conditioning lines (red) of different quality, between which there is a waveguide inserted (yellow). One of the most important criteria of the conditioning lines is their waviness, because the shape of the conditioning lines directly determines the shape of the waveguides. As explained in 3.2.4, the damping of the waveguides correlates significantly with their outer geometry. Therefore, the printing form should be

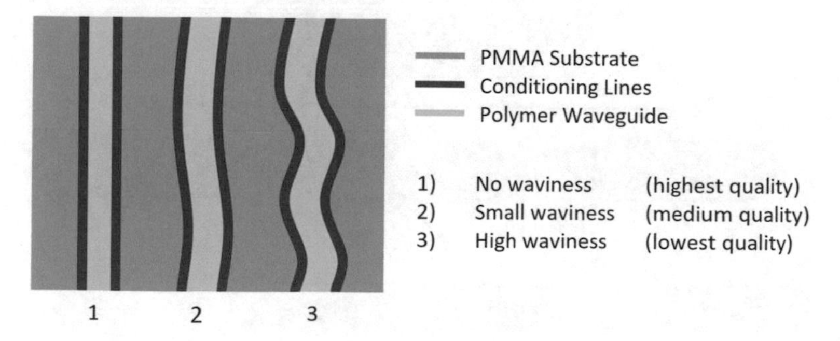

Fig. 4.38 Schematic representation of substrate, conditioning lines and waveguide

functionalized / processed in such a way that defects, waviness, etc. of the conditioning lines decrease, so that a waveguide with lower attenuation can be manufactured.

2. To create three-dimensional optical waveguides with the OPTAVER process, the PMMA substrate is thermoformed after the conditioning lines are applied. Since the aerosol process is 3D capable, thermoforming either can happen before or after the waveguide is applied. If the waveguide has not yet been applied, it is mandatory that the conditioning lines remain intact even at high forming rates. Higher formability can be achieved by various changes in the print image. For this purpose, the printing form is in turn adapted / functionalized.

In general, however, it can be said that the functionalization of flexographic printing forms as well as functional printing itself has not yet been sufficiently investigated. Although there are initial approaches, such as the Kodak Digicap [22]—uniform groove on a printing form, which are intended to ensure optimum ink flow and solid ink coverage for specific ink types and anilox roller volumes—however, the introduction of the microstructure is limited to just one application. Furthermore, the comprehensive degrees of freedom using laser structuring is not remotely being utilized.

There are various ways to functionalize a conventional printing form. An overview of these, which will be discussed below, can be seen in 9. In summary, three main methods will be considered in more detail:

1. Functionalization through the insertion of microstructures
2. Functionalization by application of a coating
3. Functionalization through a chemical change of the surface

Each main method has specific advantages and disadvantages. Combinations of the three main methods are also possible, e.g., when areas are first removed and then refilled with another material (Fig. 4.39).

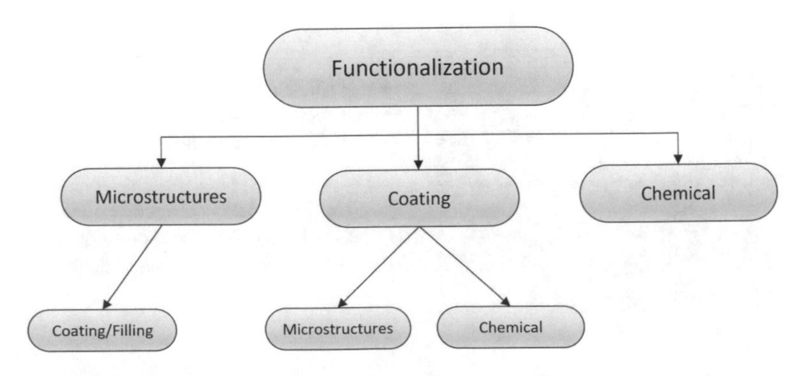

Fig. 4.39 Overview of the functionalization methods for flexographic printing forms

4.8.1 Functionalization by Laser-Induced Microstructures

The micro- and macrostructures can be inserted into the printing form with different laser systems. In order to identify a suitable laser system, first various material properties must be investigated. One of the most important parameters is the absorption property of the printing form material, which is usually a wavelength-dependent variable. During absorption, the incident photons interact with electrons in the work piece [23, 24]. Accordingly, the absorption properties are also material-dependent. If a photon hits an electron, the electron is transferred to an energetically higher state, if the photon possesses the right amount of energy. The energy of a photon is described by $E = h * f$, where f and h are the frequency and the Planck constant. If the absorbed energy is greater than the ionization threshold, ionization occurs, i.e., a bound electron is removed from an atomic shell and a positively charged atom or molecule remains (photoionization). The detached electron is thus present as a free electron and can absorb further photons. It is further accelerated until it either recombines with an ion and / or interacts with phonons or hits another electron. If the latter is the case, electron-induced ionization / impact ionization can occur. In this case, the kinetic energy of the free electron exceeds the ionization potential / binding energy of the impacting particle [25, 26].

Linear Absorption

For the interaction between laser radiation and matter, the fluence (F)—the energy received per unit area—is a characteristic quantity in linear absorption, provided that the pulse duration is constant and long enough so that nonlinear effects such as two-photon absorption are negligible. Since this quantity cannot be measured directly, it is determined from the energy, the spatial beam profile and the temporal intensity profile [27]. The assumption underlying linear absorption is described by Lambert's law:

$$\frac{dF}{dz} = -\alpha F \tag{4.10}$$

where F is the fluence, z the propagation direction and α the wavelength-dependent absorption coefficient. The solution of this differential equation yields the sought fluence profile.

$$F(z) = F * e^{-\alpha z} \tag{4.11}$$

As already mentioned, at least two conditions must be fulfilled for an atom/molecule to be excited. On the one hand a photon must hit an electron and on the other hand the energy of the photon must be sufficient to lift the electron to a higher energetic level. Since the energy of a photon is determined by $E = h * f$ respectively $E = h * \frac{c}{\lambda}$, it can be seen that the higher the frequency of a photon, the higher its energy, or that the lower its wavelength, the higher its energy. Therefore, linear absorption is more likely to occur in or below the UV wavelength range for several materials. Since the printing form consists of a dielectric (polymer), the energy required between two energy levels is relatively large.

Simplified, the absorption can be estimated by means of transmission measurement. Assuming that the total intensity is composed of the individual intensities of reflection, absorption, scattering and transmission (cf. Eq. (4.12)), an approximate value for the absorption is obtained as a complementary component to the transmission—since the values of scattering and reflection are expected to be very low in the application discussed here, so that they can be neglected.

$$I_{Total} = I_{Reflection} + I_{Absorption} + I_{Scattering} + I_{Transmission} \quad (4.12)$$

Figure 4.40 shows the transmission data of the three introduced printing form materials including the typical wavelengths of a Yb:YAG / Nd:YAG laser—including their second and third harmonic frequency, respectively—which are commonly used for pulsed laser systems in micromachining processing. The materials exhibit distinctly different absorption characteristics. The Conti Laserline material (green) absorbs almost 100% of wavelengths between 200 and 2000 nm at the surface. The Flint printing form material (yellow) shows a similar transmission curve, but this does not have much significance, due to its multi-material structure. The substrate is an aluminum layer onto which a photopolymer is cured [28]. Most of the radiation is now reflected or absorbed by the aluminum substrate, while the photopolymer is expected to show rather a transmission behavior more like the Kodak material (blue). This has a polymer as a substrate, which transmission properties are similar to the photopolymer. The absorption of the Kodak material drops sharply at approx. 355 nm as transmission increases. At the second harmonic frequency (515/532 nm), the absorption is only about 30% while the transmission increases to around 70%. At the fundamental wavelength (1030/1064 nm), the absorption is down to just over 10%.

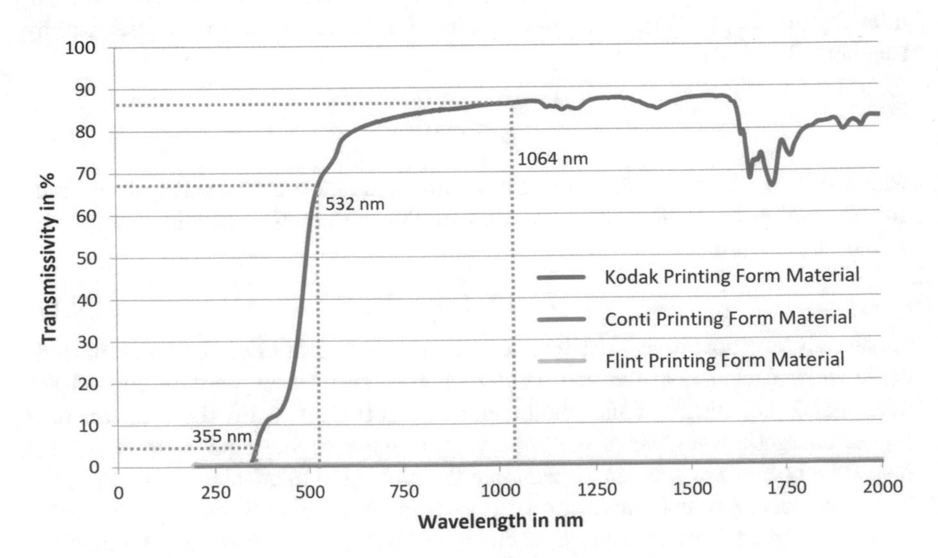

Fig. 4.40 Transmission measurement of the printing form materials

The absorption properties of the Kodak material can be well explained by the effects of linear absorption mentioned above. The good absorption of the Conti Laserline material at most optical wavelengths (200–2000 nm), despite its polymeric structure, is related to its complex composition. Similar to a car tire, soot particles, which are also responsible for the black color, and other additives provide sufficient free electrons, so that linear absorption occurs.

Nevertheless, it may also be useful to work with wavelengths (λ>450 nm) at which low absorption is to be expected from the transmission measurement. This has several reasons, e.g., that one can work with the fundamental wavelength. As a result, frequency conversion is no longer needed, which reduces the complexity and the expenses for the system. Additionally, there are no conversion losses when working with the fundamental wavelength, which results in a better overall efficiency. Therefore, one does not necessarily have to rely on the short wavelengths and linear absorption, because at high intensities ($>10^{11}\frac{W}{cm^2}$), more absorption effects occur between laser beam and material. The best-known and most used theories for nonlinear laser-material interactions are based on the "strong field approximation" (SFA) also known as "KFR approach," named after the developers Keldysh, Faisal and Reiss [29–31]. However, also in this case, the absorption remains material-dependent, so that there is a specific ionization energy for each atom or molecule. The basis of the theory is the Keldysh parameter γ (cf. equation (4.13)), which describes a generally valid case. By means of a case distinction (cf. equation (4.14)), various nonlinear absorption phenomena such as multiphoton ionization, tunnel or field ionization can occur.

$$\gamma = \sqrt{\frac{I_p}{(2U_p)}} = \frac{\omega\left(2m_e I_p\right)^{1/2}}{E_0 e} \tag{4.13}$$

where I_p represents the ionization potential of an atom, U_p represents the ponderomotive potential of the laser, ω represents the angular frequency, E_0 represents the amplitude of the electric field of the laser, e represents the elementary charge and m_e represents the mass of an electron.

$$\gamma \begin{cases} \ll 1 \rightarrow \textit{Tunnel or field ionization} \\ \gg 1 \rightarrow \textit{Multiphoton absorption} \end{cases} \tag{4.14}$$

Tunnel or Field Ionization

In a first approximation, the magnetic field of the laser beam can be neglected in the case $\gamma \ll 1$. Thus, this can be described as a time-varying electric field. When this becomes very strong, it suppresses the Coulomb field that binds the electrons to the positively charged atomic nucleus. If the electric field of the laser is strong enough, the Coulomb field can be suppressed to such an extent that the probability increases that a valence electron tunnels through a potential barrier and is thus present as a free electron. This effect dominates at strong electric fields, which occur at high laser intensities (approx.$>10^{13}\frac{W}{cm^2}$), depending on the material to be ionized. Since these regimes are not achieved in the here presented functionalization process of flexographic printing forms, it is assumed that there is only a very low probability of tunneling or field ionization, and therefore, this will not be discussed further. For additional information, please refer to the current technical literature [32–34].

Multiphoton Absorption

At high laser intensities ($>10^{11}\frac{W}{cm^2}$), as seen for example in ultrashort pulsed laser systems, simultaneous absorption of several photons can occur. The probability that multiphoton absorption occurs depends not only on the material to be processed, but likewise on the laser parameters used. Unlike tunnel ionization, multiphoton absorption is more likely to occur at higher laser frequencies, which results in shorter wavelengths and therefore in higher photon energy. The energy of a single photon would not be sufficient to ionize an atom or a molecule. By absorbing multiple photons simultaneously, the combined energy of the absorbed photons can exceed the ionization potential of the material being processed and release electrons from their bonds. Mathematically, the formula (4.11) now changes to:

$$\frac{dF}{dz} = -\sum_k \alpha_k * F^k \tag{4.15}$$

k now represents the order of multiphoton absorption. This differential equation can only be solved analytically for special cases, such as for k = 1, because then the formula would again be identical to the formula (4.11) [35, 36]. Otherwise, numerical models can only approximate the solution of this equation.

In summary, either lasers with short wavelengths (UV or shorter) or lasers with ultra-short pulse durations and high intensities are suitable for processing printing forms.

Of the three printing form materials Conti Laserline, Kodak Flexel NX and Flint nylofl ex Gold A, only the first two will be investigated further. This is because the Flint Group's printing form consists, as already mentioned above, of a multi-material structure. As a result, it is likely that only the substrate will be laser processed instead of the actual printing form material. In addition, the Kodak printing form material, which also consists of a photopolymer, is examined more closely. The results are therefore to a certain extent transferable to the Flint printing form material. Based on the above presented theory, a UV nanosecond and an IR femtosecond laser system are used for the following studies.

Laser System 1: Coherent Avia 355–7000.
Laser System 2: Coherent Monaco 1035–60.

Ablation trials [37]

Based on the above theoretical principles, removal tests were carried out. In the following, the ablation tests of the Conti printing form will be presented as an example. The tests were carried out with an ultrashort pulse laser from Coherent Inc. (Monaco 1035–60). The nominal wavelength of the laser is 1035 nm. The maximum average optical power is 60 W. The M2 factor is 1.16. The repetition rate ranges from single pulse to 50 MHz. The pulse duration can be set from 252 fs to 10 ps but was set to the shortest pulse duration for all experiments. A Gentec Maestro power meter and a UP19-W detector were used to measure the average power. To ensure a flat surface, the compression mold material was positioned on a vacuum table. A cross-jet and suction system were used to minimize interaction with the ablated particles. After beam expansion, the raw beam had a diameter of 10 mm, which resulted in a focal diameter of 16.5 µm @ e^{-2} after the telecentric F-theta objective with a focal length of 59.7 mm.

Ablation was studied at 100 kHz and 250 kHz, while peak beam fluence was varied from 0.5 up to 50 J/cm². Percussion drilling with 10, 25 and 50 pulses was used for ablation. After measuring the depth with a laser scanning microscope (model: VX-1000 from Keyence), the ablation depth per pulse is calculated by dividing the measured depths d_m by the number of pulses n_p.

$$d_p = \frac{d_m}{n_p} \tag{4.16}$$

The fluence (F) for a classical Gaussian beam can be calculated according to (2) with E_p as pulse energy and r as beam radius (@ e–2).

$$F = \frac{2 * E_p}{\pi * r^2} \tag{4.17}$$

The results of the ablation experiments at 100 kHz and 250 kHz are shown in Fig. 4.41 with a logarithmic x-axis. The results from both repetition rates look qualitatively similar, because in both cases, the ablation depth increases with increasing fluence and the ablation depth per pulse is higher when fewer pulses arc applied. This indicates that the removal efficiency is higher during the first pulses.

In contrast to metals, there is no clear boundary between sublimation and thermal ablation in the case of pressure forming material (EDP). The ablation depth increases significantly with increasing fluence. The fit in the logarithmic plot is

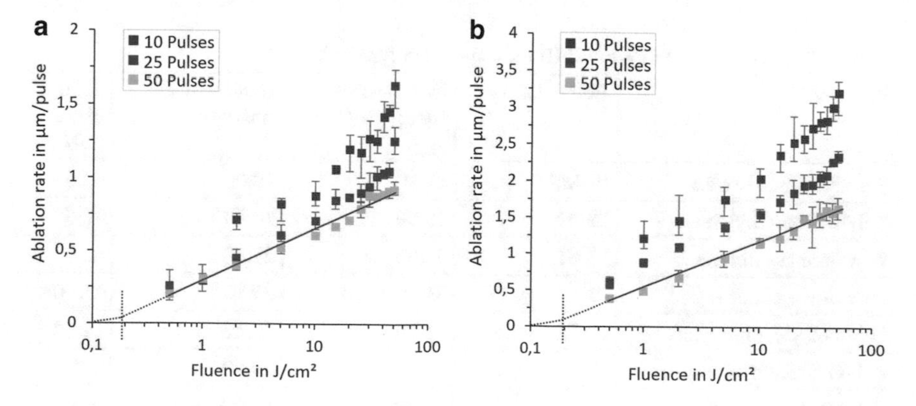

Fig. 4.41 Ablation rate (µm/pulse) as a function of fluence (J/cm²) a) at 100 kHz and b) at 250 kHz [37]

nearly linear for the 25 pulse and 50 pulse series of measurements, while the fit for 10 pulses shows slightly exponential trends. For the 100 kHz tests, the ablation rate in the 50 pulse measurement series appears to level off at 0.9 μm / pulse. The highest measured ablation depth per pulse is 1.7 μm and occurs at the maximum fluence in the measurement series with the lowest number of pulses.

In the test series at 250 kHz, the highest measured ablation depth per pulse is 3.4 μm and occurs at the maximum fluence in the measurement series with the lowest number of pulses. The higher ablation depth per pulse at a higher repetition rate can be well explained by the fact that more heat is introduced into the process, which increases the absorption and thus the ablation rate. The ablation threshold detectable here is about 0.5 J/cm^2. Below this fluence, no significant ablation could be detected. Nevertheless, as with common polymers, the ablation threshold is probably well below the 0.5 J/cm^2 limit. If a fit is made through the measured values and a further flattening in the lower range is taken into account, an ablation threshold around 0.1 J/cm^2 is expected. This could be verified by further investigations, in which the number of pulses was significantly increased. Depending on the laser repletion rate, ablation could be detected even below 0.1 J/cm^2.

4.8.2 Functionalization by Coating Mechanisms

The second method to be considered for the functionalization of printing forms is coating. The basic idea of this method is that a coating medium is applied to the printing form and cured. Due to the fact that the coating has a different chemical composition than the printing form, it also differs in its surface energy. As previously described, this is a critical factor in material transfer in flexographic printing. By selecting a suitable coating medium, the material transfer can thus be specifically manipulated. Table 4.5 shows various coating media with their free surface energies:

In order to achieve the greatest possible change in surface free energy and its effect on transfer behavior, polydimethylsiloxane is examined in more detail below.

Table 4.5 Surface free energies (SFE) for common polymers [38]

Polymer	Abbreviation	Free surface energy [mJ/m^2]	Dispersive part [mN/m]	Polar part [mN/m]
Polydimethylsiloxane	PDMS	19.80	19.00	0.80
Polytrifluoroethylene	P3FEt	23.90	18.80	4.10
Polytetrahydrofurane	PTHF	31.90	27.40	4.50
Polyvinylacetate	PVA	36.50	25.10	11.40
Polymethylmethacrylate	PMMA	41.10	29.60	11.50
Polyvinylchloride	PVC	41.50	39.50	2.00
Polyethyleneterephthalate	PET	44.60	35.60	9.00

Polydimethylsiloxane (PDMS)

Polydimethylsiloxane is a polymer of the silicone class and is a widely used organic material in industry and research. It consists of a repeating siloxane unit (Si–O) and methyl groups (CH_3 or H_3C) attached to it. Between the end groups, the C_2H_6OSi-unit can repeat any number of times. The structural formula can be seen in Fig. 4.42.

Polydimethylsiloxane is characterized by both its attractive physical and chemical properties. Thus, on the one hand, it exhibits high elasticity and optical transparency. On the other hand, it is chemically inert, non-toxic and thus biocompatible. Therefore, it is used in numerous biomedical [39, 40], optical [41, 42], lithography [43, 44] and microfluidic applications [45, 46]. A UV-polymerizing version has been developed and is available from the manufacturer Shin-Etsu Chemical for a few years [47]. This two-component material is characterized by having photo initiators that trigger the exothermic curing reaction upon UV exposure. Furthermore, the physical properties can be influenced by the addition of additives. For example, the PDMS can be diluted with silicone oil so that it has a lower viscosity, which is a decisive factor for the coating process.

Experimental setup:

In order to achieve the goal of generating the thinnest (~1 μm) and most homogeneous PDMS layer on a printing form, several aspects had to be taken into account. On the one hand, the printing form is much larger than the coating thickness, which resulted in a challenging manageability. On the other hand, the surface of the raised stamp structures had to be coated, so that common coating methods such as spin coating did not need to be considered. It was decided to use a wide slot nozzle system for coating, as shown schematically in Fig. 4.43.

First, the pressure vessel is filled with the coating fluid (PDMS). Afterward, it is then pressurized (max. 1 bar). This presses the fluid through the system so that the system is vented and the medium pressure module / syringe module is filled. As soon as there is no more air in the system, valve 2 is closed. Now, the syringe module can be used to adjust the flow rate down to single-digit μl/min range. As soon as the coating has been applied, the axis system moves with the coated substrate under the galvometer scanner, which cures the coating using a UV laser.

In order to achieve sub 50 PDMS coating thicknesses, a wide slot nozzle was designed for the present coating case by the company FMP Technology GmbH. First, the coating process is simulated. From this, all relevant coating parameters like the slot width, coating speed, volume flow, distance nozzle—substrates, etc.

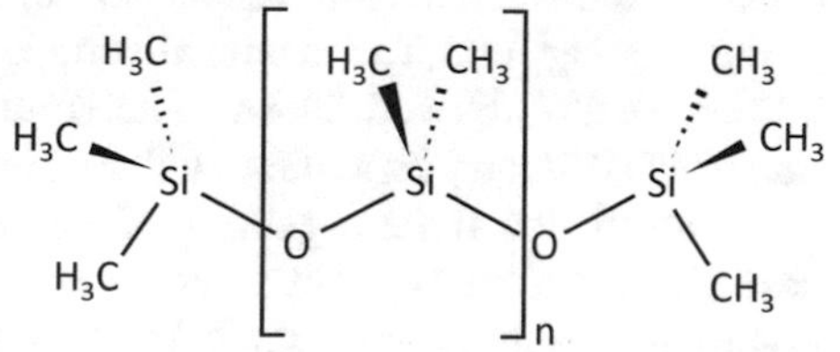

$$H_3C-Si(CH_3)_2-[O-Si(CH_3)_2]_n-O-Si(CH_3)_3$$

Fig. 4.42 Chemical structural formula of Polydimethylsiloxane (PDMS) [39]

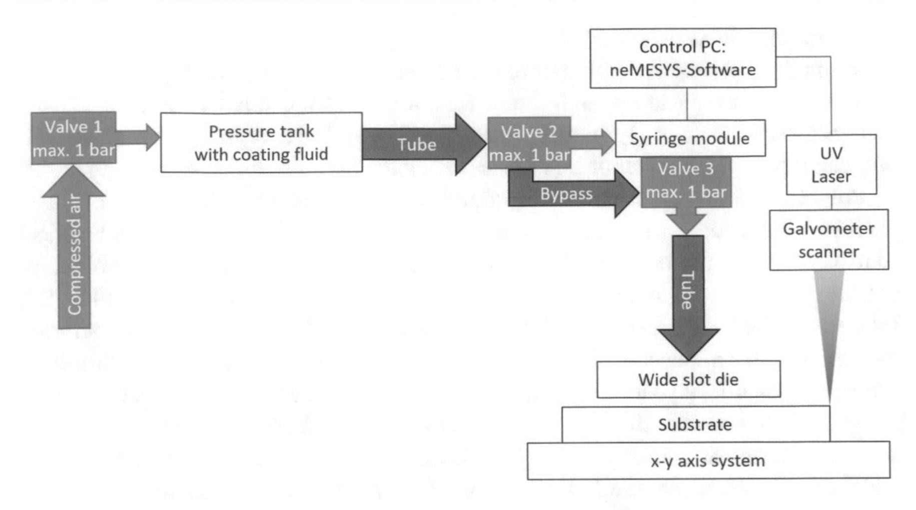

Fig. 4.43 Schematic layout of the wide slot nozzle coating system

are obtained as a function of the viscosity. The following table shows the coating parameters resulting from the simulation for an assumed coating length of 420 mm (Table 4.6).

The two-component material used here is from ShinEtsu (UV-PDMS KER-4690-A & KER-4690-B), which is mixed together before coating. The mass ratio is 50:50, and for the diluted version, 10 mass% silicone oil is added so that the mass ratio is 45:45:10. Figure 4.44 shows the coating process.

The PDMS coating resulted in no material being transferred at all. Although this is an interesting observation, it does not help in the present application—the production of conditioning lines. Therefore, this approach will not be pursued at the moment.

4.8.3 Functionalization by Chemical Modification

Even though the coating functionalization will not be pursued at the moment, further functionalization options will be briefly discussed here. In the present case of chemical printing form functionalization, the functionalization approach is based on the following idea: the composition and chemical structure of the printing form material is a well-kept secret of the printing form manufacturers. This makes chemical functionalization difficult to implement. If a thin PDMS layer is applied to the printing form, the chemical surface properties and functionalization possibilities are already well known. Therefore, a brief overview of the functionalization of PDMS/polymer surfaces will be given below.

As mentioned at the beginning of the chapter, there is no clear definition of the term functionalization. Therefore, one must always differentiate from application to application. For example, in order to apply a certain coating, it can also be

Table 4.6 Simulatively determined coating parameters for PDMS coating using a wide slot nozzle

No	Fluid description	Slot width slot nozzle [µm]	Coating width of the slot nozzle [mm]	Layer thickness wet [µm]	Coating speed [m/min]	Distance of the nozzle to the substrate [µm]	Volume flow [ml/min]	Coating time per unit [min]	Fluid volume per unit [ml]	Viscosity of the coating material [Pas]
1	PDMS (diluted)	100	290	1	0.0035	50	0.0010	120.0	0.12	2.1
2	PDMS (diluted)	100	290	2	0.00765	50	0.0044	54.9	0.24	2.1
3	PDMS	200	290	20	0.0306	100	0.1775	13.7	2.44	3.2
4	PDMS	200	290	30	0.04985	100	0.4337	8.4	3.65	3.2

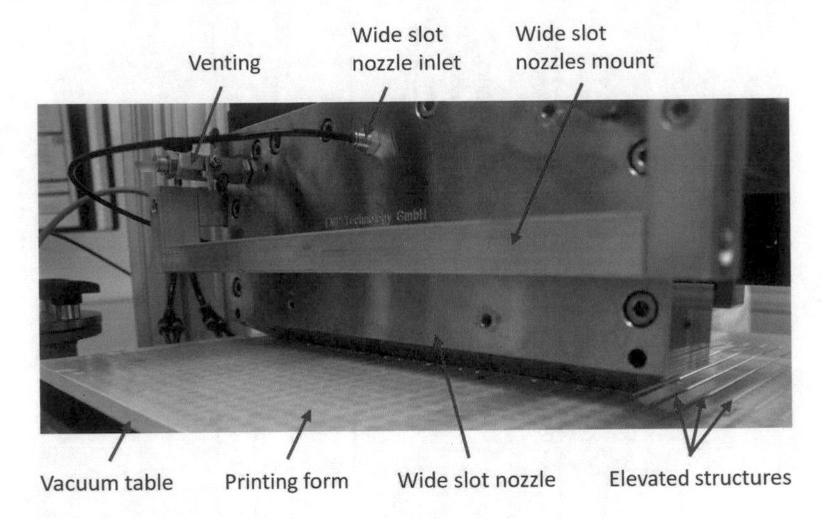

Fig. 4.44 Experimental setup for coating by using a slot die

useful to temporarily increase the surface energy, as for example on the printing of packaging material [48, 49]. Here, the surface energy was adapted by introducing polar reactive functional groups. Other use cases include creating superhydrophobic surfaces to provide a stain-resistant effect [50], or ensuring biocompatible surfaces. In order to carry out the modification of the PDMS/polymer surface, a wide variety of methods are available [51]:

- Low pressure plasma
- Corona discharge
- Flame treatment
- Radiation modification (UV, γ-, electron beams)
- Ozone
- Wet chemical processes (treatment with acids and bases).

Further, there is an approach to add additives to the polymer layer. In this way, for example, doped PDMS thin films can be created, which results in a further reduction of the surface energy. This has been shown for example in [52] using the example of a perfluoroether allylamide doping.

Another approach is to use the PDMS coating as a new substrate material. For example, a self-assembled monolayer (SAM) can be deposited on the PDMS. Here, organic monomers arrange themselves on a surface to form a self-assembled monolayer. The monomer has an anchor group with which it is coupled to the substrate and a head group which is responsible for the chemical surface properties. These self-assembled monolayers have the advantage that they are extremely thin (down to about 0.1 nm, depending on the molecule used), at the same time very stable, and the head group can be further functionalized.

4.8.4 Application of Functionalization Mechanisms on Flexographic Printing Forms and its Impact on the Printing Results

As already described above in the introduction to Sect. 4.8, the functionalization of printing forms can be realized by different approaches. This section will take a closer look at laser-based functionalization. First, the shape and structure of the printing form will be examined in more detail [53].

Printing form

Figure 4.45 shows an example schematic cross section of a flexographic printing form used here. This is divided into three types of areas, which are examined in terms of their functionality in the printing process:

A1: Raised area that transfers the material.
A2: Rising edge areas.
A3: The transition area between A1 and A2. This is 150 μm, of which 75 μm protrude into A1 and 75 μm into A2.

On the one hand, this functionalization can be achieved by a subtractive removal on the stamp surface, whereby a reduced area is available for wetting by the anilox roller (area control). On the other hand, surface modification, chemical surface modification or additive coating of the stamp surface can reduce the surface energy, for example, so that there is less wettability. As a result, the printing stamp picks up less material to be transferred and the volume of the transferred material is reduced (volume control). A strict separation of these two control variants is usually not possible, since, for example, laser-based material removal also results in at least a local chemical change of the material. Nevertheless, in both cases,

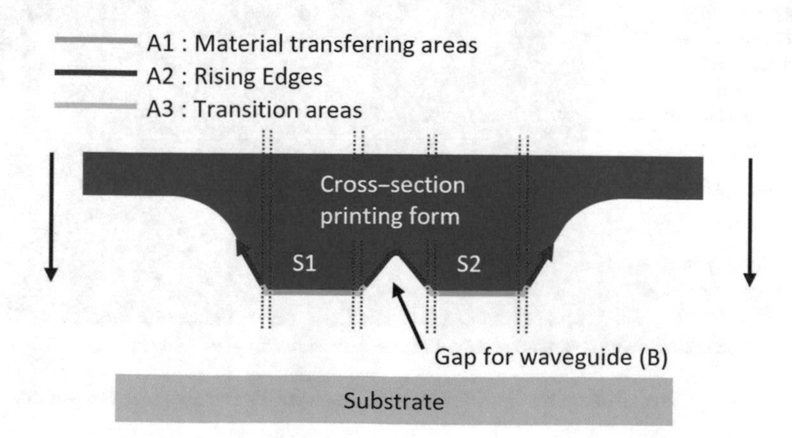

Fig. 4.45 Schematic cross section and its classifications of a flexographic printing form [53]

the print result is adjusted by a functionalized printing form without changing the printing process itself.

Investigation of the Influence of Areas A1–A3:

For the following investigations, as for the ablation tests presented above, the femtosecond laser system "Monaco 1035–60" from Coherent Inc. is used. The machining setup used in the following can be seen in Fig. 4.46.

In order to ensure a level working plane, the printing form is fixed on a vacuum table. This is located on an x–y axis system. This, together with a camera system (CCD camera, optical zoom, filter, deflection mirror, tube, magnifying lens), is used to align the printing form. The optical setup (beam expansion, telecentric F-Theta lens—focal length of 59.7 mm) results in a focus diameter of 16.5 µm at e^{-2}.

A pattern of microstructures was inserted in each area. The pattern consists of squares (75 µm x 75 µm). The ablation between the squares is 31 µm wide and 50 µm deep. Figure 4.47 shows a laser scanning image with the patterned area A1. Further, special attention was paid to area A3 (the transition between the material-transferring areas (A1) and the flanks (A2)). Since only two squares â 75 µm can be introduced into this area anyway, it is additionally changed geometrically. Figure 4.47 b) shows this change. Instead of an abrupt drop with an angle (organic line) of about 98°, there is an area rounded over 110 µm (black line). Contrary to what Fig. 4.47 a) implies, the examinations were performed independently, i.e., either the square structures were introduced or the edge was rounded. The marking in Fig. 4.47 a) shows only the position of the cross section, which, however, was rounded at a different position.

After the effects of the individual areas had been examined, further investigations were carried out. On the one hand, line ablations of different widths were introduced

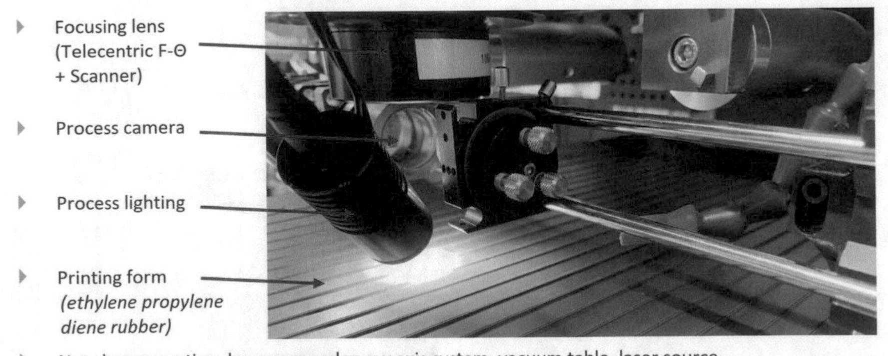

Fig. 4.46 Experimental setup for inserting laser-induced microstructures into the printing form [53]

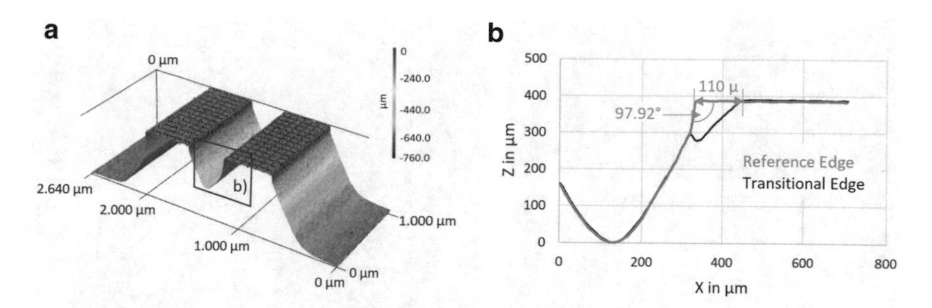

Fig. 4.47 **a** Laser scanning microscope image of the printing form with a structured A1 area **b** Cross section of the edge of an untreated printing stamp S1 and a transitional version [53]

into the material-transferring stamp area in order to investigate the effects on the print result more closely. The focus here was on the correlation between the line width removed and the resulting gap on the substrate. Furthermore, the square structures were varied in distance and depth and the effects on the print result were also analyzed.

After the defined ablations had been applied, printing tests were carried out with the modified printing forms on the Speedmaster 52 printing machine from Heidelberger Druckmaschinen AG presented in the previous section (Sect. 4.5.1). A laser scanning microscope (model: VX-1000 from Keyence) was mainly used for analysis.

4.8.5 Impact of the Functionalization on the Printing Results

Fig. 4.48 shows the print results for the investigations into the influence of the different areas. 4–48-0) shows a reference print without functionalization, while Fig. 4.48 1) to Fig. 4.48 3) each show a printed conditioning line with one structured area each. In Fig. 4.48, 1) area A1 was patterned, in Fig. 4.48, 2) area A2 was patterned, and in Fig. 4.48, 3) area A3 was patterned. Fig. 4.48 4) shows two printed conditioning lines between which a waveguide can subsequently be applied. While the left conditioning line was printed with an untreated edge, the right line was printed with a transitional edge as shown above in Fig. 4.47 b).

While a continuous conditioning line can be seen in Fig. 4.48 0), the print result of the conditioning line with the introduced structure on the material-transferring area (A1) differs significantly. Individual dots can be seen in the print image. Each micro-square of the printing form material leads to one transferred dot in the printing result. Apparently, the width of the ablation is chosen so that already some dots run into each other. In Fig. 4.48 2), where the transition area was structured with two rows, only the row that was introduced in the

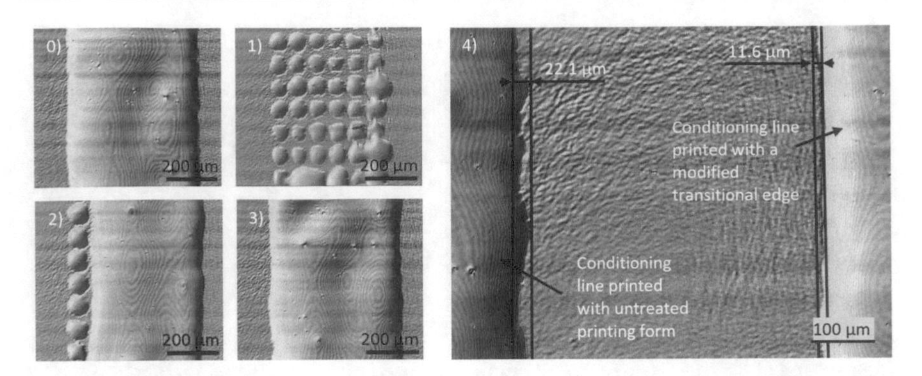

Fig. 4.48 DIC images of conditioning lines printed with an unprocessed printing form (0) and functionalized printing forms (1–3 each correspond to one previously defined areas A1-A3). 4) DIC image of two conditioning lines printed with an untreated and a modified printing form edge [53]

material-transferring area can be seen. The second row, which is located on the rising flank, does not seem to have any influence. The conditioning lines in Fig. 4.48 0) and Fig. 4.48 3) are very similar, which suggests that the structuring of the rising flanks has no or a negligible effect on the print result. In contrast, however, Fig. 4.48 4) shows that the shape of the transition from the material-transferring surfaces to the rising flanks has a large effect. Comparing the edge ripples of the two conditioning lines, it is clear that the one with the rounded edge has a much lower edge ripple. Comparing the peak-to-valley values with each other, the ripple values are lower by almost half for the rounded edge. Considering that the waveguide takes the edge shape of the conditioning lines, the edge ripple of the waveguides can thus also be greatly reduced.

In summary, it can be derived from this experiment that the structuring of the material-transferring areas have a strong influence on the transferring material (area control). Not only the geometric shape of the stamps, but in particular that of the transition area (A3) also has a decisive influence on the print image. In the present case, this makes it possible to reduce the edge waviness of the conditioning lines and thus also improve the quality of the waveguides to be applied. If microstructures are applied to surfaces which, in this case, are 100 µm below the material-transferring plane, they have no effect on the print image or the printing process.

The investigations into the microstructures on the material-transferring surfaces were further extended by varying both the side length of the square structures and the depth and width of the ablated structures. If we first look at a typical cross section of a conditioning line printed with a 500 µm wide stamp, as shown in Fig. 4.49, we first notice the material accumulations on the sides. These are so-called squeeze edges, as described in more detail in Sect. 4.5.3. These occur

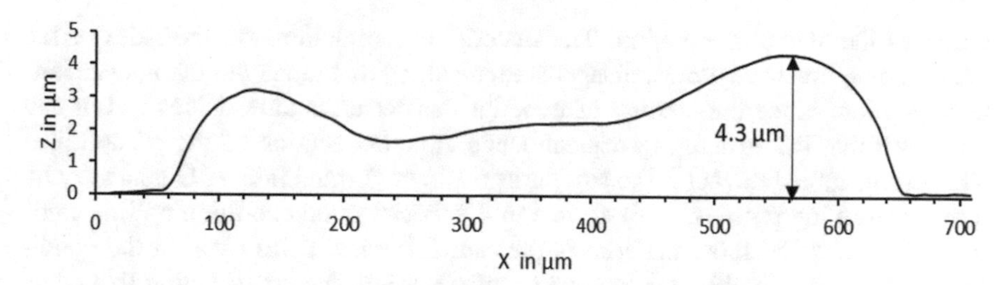

Fig. 4.49 Cross section of a printed reference conditioning line printed with an unprocessed printing form [54]

because the material is pressed out to the right and left of the stamp during the impact in the printing process.

If we now reduce the edge length of the square pattern to 50 µm, the depth and width of approx. 17 µm, we now give the material the opportunity to spread into the ablated lines during the impact instead of being pushed out to the right and to the left. Figure 4.50 shows this effect. Figure 4.50 a) shows a laser scanning image of the printing die, Fig. 4.50 b) the corresponding cross section, Fig. 4.50 c) the conditioning line printed with the die and Fig. 4.50 d) the corresponding cross-section of this line.

Comparing this cross section of the conditioning line (Fig. 4.50 d)) with a reference line (Fig. 4.49) printed with a printing form without microstructures

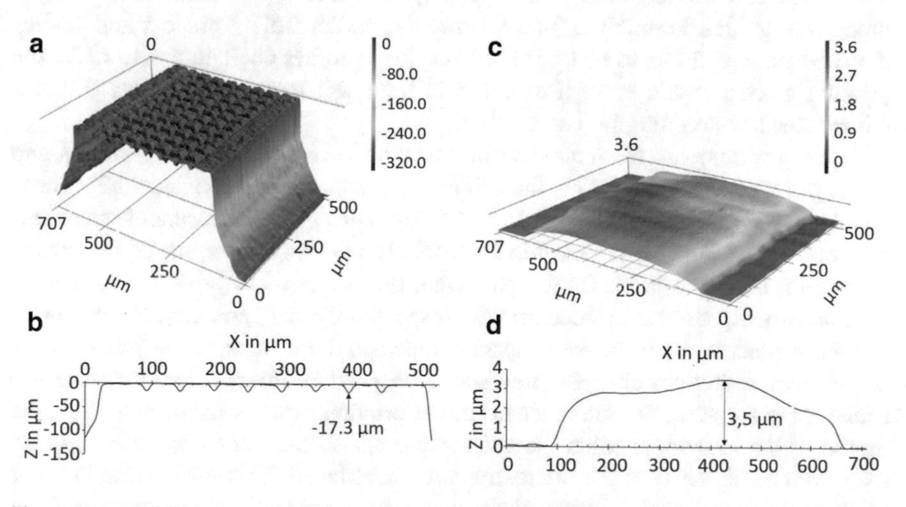

Fig. 4.50 a–b LSI image of a printing stamp with functionalized structures and its cross-sectional profile **c–d** LSI image of the corresponding printed conditioning line and its cross-sectional profile [54]

confirms the above assumption. The material accumulations on the sides of the conditioning line have almost disappeared, resulting in a much more homogeneous cross section. Since the amount of material transferred is almost identical, it can be shown that the structuring can influence the cross section of the printed line. This can be quite relevant if the structures are to be formed into a 3D geometry by a thermoforming process, as described in 4.7, because the conditioning line tends to crack where the most material is present. Likewise, cracks within the conditioning line are tolerable, while cracks that reach into the conditioning line edge, which is also the boundary of the waveguide, are very problematic. This is because as soon as the waveguide material pulls into the cracks material accumulation occurs and the entire waveguide becomes unusable.

Finally, investigations were carried out to study the effects of line removal at different widths on the print result [37]. For this purpose, different line widths were ablated centrally into the material-transferring areas. All in all, line removal of 31, 99, 185, 278 and 364 μm width was varied in five different trials. After print form modification and printing, the printed conditioning lines were again measured with a laser scanning microscope. The results can be seen in Fig. 4.51. While Fig. 4.51 0) shows the cross section of a reference line, Fig. 4.51 1) to Fig. 4.51 5) show the cross sections printed with the above-mentioned modified printing forms. It is noticeable that in the case of cross section Fig. 4.51 1), material accumulation has formed at the point where the 31 μm wide line ablation was located. In the following Fig. 4.51 1–5, a gap appears in the conditioning line at the position where the line ablation is located on the printing form. If the ablated line widths on the printing form is just under 100 μm, a gap of approx. 8 μm wide appears in the printed conditioning line. This is exactly the distance needed to insert single-mode waveguides there in the future. However, the aerosol jet process that applies the waveguide still has to be optimized for this. Another challenge is to make the printing process stable enough to maintain the small gap over a longer distance without the lines running into each other.

If we now compare the removed line width on the stamp with the resulting gap in the printed conditioning line, the following graph is obtained (Fig. 4.52). These results agree well with the results from 4.5.3, where similar experiments were carried out. As a result, it is possible to precisely adjust the spacing of the conditioning lines using a laser. On the one hand, this offers many advantages, such as significantly higher design freedom with respect to the die geometry. For example, in a laser process, both the exact spacing between the stamps, the edge geometry of the stamps, and possibly also microstructures can be inserted into the material-transferring surfaces. On the other hand, the printing process itself, with, e.g., the infeed of the individual rollers to each other, also has a very high effect, which may even be higher than the microstructures introduced. This means that in order to carry out successful printing, the printing form production, the printing form processing and the printing process itself must be coordinated with each other in a proper manner.

We sincerely thank the Deutsche Forschungsgemeinschaft for funding the research group OPTAVER FOR 1660.

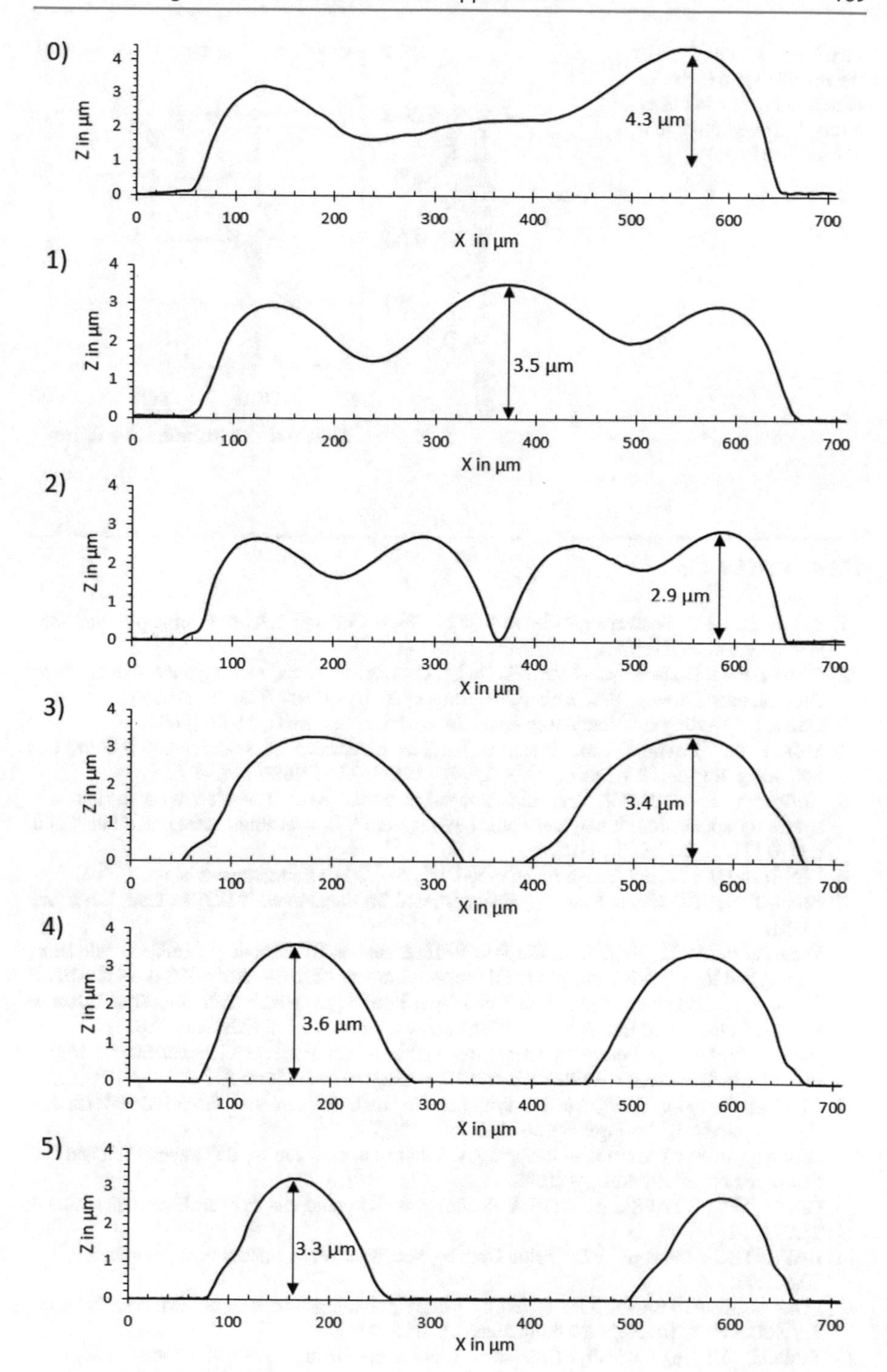

Fig. 4.51 Cross sections of printed conditioning lines depending on the ablated line width on the flexographic printing form. 0) is a reference print 5) is the print with the widest ablation

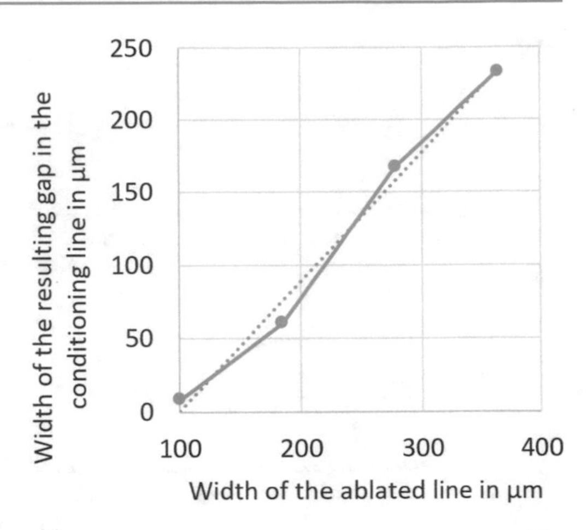

Fig. 4.52 Width of the resulting gap in the conditioning line in µm as a function of width of the ablated line in µm

References

1. S.F. Kistler S.F., Schweizer P.M. (ed.) Liquid Film Coating: Scientific principles and their technological implications, Dordrecht: Springer (1997).
2. Gennes, P.-G. D., Brochard-Wyart, F., Quéré, D.: Capillarity and wetting phenomena: Drops, bubbles, pearls, waves, New York, NY: Springer (2010)—ISBN: 9780387216560
3. Quéré, D.: Wetting and Roughness, Annu. Rev. Mater. Res. **38**(1), 71–99 (2008)
4. Yost, F. G., Hosking, F. M., Frear, D. R.: The Mechanics of Solder Alloy Wetting and Spreading. Boston, MA, Springer US (1994)—ISBN: 978-1-4684-1442-4
5. Hoffmann, G.-A., Wolfer, T. et. al.: „Improving partial wetting resolution on flexible substrates for application of polymer optical waveguides", Optical Engineering. dOI https://doi.org/10.1117/1.OE.56.10.103109
6. DIN Deutsches Institut für Normung e. V.-DIN 8580:2003: Fertigungsverfahren. (2003)
7. Klein, T. Dr.: HD Flexo: Neuer Qualitätsstandard im Flexodruck. VDD Seminar, Darmstadt (2010)
8. Vena, A., Perret, E., Tedjini, S.: Chipless RFID Based on RF Encoding Particle: Realization, Coding and Reading System, s.l.: ISTE Press—Else-vier (2016)—ISBN: 978-1-78548-107-9
9. Clement, F., Lorenz, A. et al.: High Throughput Printing for Highly Effi-cient Cost-Effective Si Solar Cells, e 33rd European PV Solar Energy Conference and Exhibition (2017)
10. Wolfer, T.: Additive Fertigung integrierter multimodaler Polymer-Lichtwellenleiter mittels Flexodruck. Garbsen, TEWISS Verlag (2020)—ISBN: 978-3-95900-377-3
11. Kipphan H. (ed.): Handbuch der Printmedien: Technologien und Produk-tionsverfahren. Berlin, Heidelberg, Springer Berlin Heidelberg (2000)
12. Podhajny, R. M.: The halos of flexography: What in heaven can we do? Paper, Film and Foil Converter, Nr. Bd. 72 Ausg. 8 (1998)
13. OE-A: WHITE PAPER (ed): OE-A Roadmap for Organic and Printed Electronics, Stand: 21.04.2021.
14. LAYERTEC GMBH (ed.): Precision Optics—Specification of surface form tolerance, Stand: 20.04.2021.
15. POLYMERDATABASE.COM (ed.): http://polymerdatabase.com/polymer%20physics/Ref%20Index%20Table2%20.html. Stand: 23.04.2021.
16. Saalfeld, A.: Topologically Consistent Line Simplification with the Doug-las-Peucker Algorithm, Cartography and Geographic Information Science **26**(1), 7–18 (1999)

17. VOß, C.: Analytische Modellierung, experimentelle Untersuchungen und dreidimensionale Gitter-Boltzmann Simulation der quasistatischen und in-stabilen Farbspaltung, Dissertation, Bergische Universität Gesamthoch-schule Wuppertal (2002)
18. KUNZ, J. UND STUDER, M.: Druck-Elastizitätsmodul über Shore-A-Härte er-mitteln, Kunststoffe, Carl Hanser Verlag, https://www.kunststoffe.de/a/article/article-258844 (2006)
19. American Society for Testing and Materials-ASTM D 2240:2015: Härteprüfung an Gummi (2015)
20. Hoffmann, G.-A., Wienke, A., et. al.: "Thermoforming of planar polymer optical waveguides for integrated optics in smart packaging materials", Journal of Materials Processing Tech. doi https://doi.org/10.1016/j.jmatprotec.2020.116763
21. Schwarzmann, P.: Thermoformen in der Praxis (2016)—ISBN: 978-3-446-44403-4
22. Weiss, D.A.: A minimum of ink enhances shelf appeal and improves productivity, Kodak (2017)
23. Seltzer, S.M.: Calculation of photon mass energy-transfer and mass energy-absorption coefficients. Radiat. Res. **136**(2), 147–170 (1993)
24. Takabe H.: The Physics of Laser Plasmas and Applications—Volume 1: Physics of Laser Matter Interaction, Springer (2020)
25. Becker, U., Shirley, D. A.: VUV and Soft X-Ray Photoionization—Theory Of Photoionization, Springer (1996)
26. Drake, G.: Springer Handbook of Atomic, Molecular, and Optical Physics, Springer (2006)
27. Finger, J.-T.: Puls-zu-Puls-Wechselwirkungen beim Ultrakurzpuls-Laserabtrag mit hohen Repetitionsraten. Dissertation, Apprimus Verlag, Aachen (2017)
28. Datasheet nyloflex® Gold A, [Online]. Available: https://www.flintgrp.com/media/644704/nyloflex_gold-a_us.pdf
29. Keldysh L. V.: Ionization in the field of a strong electromagnetic wave. Soviet Physics Jetp **20**(5) (1965)
30. Faisal, F.H.M.: Multiple absorption of laser photons by atoms. Journal of Physics B **6**, L89 (April 1973)
31. Reiss, H.R.: Effect of an intense electromagnetic field on a weak bound system. Physical Review A **22**(5) (1980)
32. Schaffer, C.B. : Andr´e Brodeur and Eric Mazur, "Laser-induced breakdown and damage in bulk transparent materials induced by tightly focused femtosecond laser pulses" Meas. Sci. Technol. **12**, 1784–1794 (2001)
33. Klaiber, M., Hatsagortsyan, K.Z., Keitel, C.H.: Tunneling Dynamics in Multiphoton Ionization and Attoclock Calibration. Physical Review Letters **114**, 083001 (2015)
34. Klaiber, M., Briggs, J.S.: "The cross-over from tunnelling to multiphoton ionization of atoms. Physical Rev A **94**, 053405 (2016)
35. Diels, J.C., Rudolph, W.: Ultrashort laser pulse phenomena. Academic Press, San Diego (1996)
36. Korte, F., Nolte, S., Chichkov, B.N., Bauer, T., Kamlage, G., Wagner, T., Fallnich, C., Welling, H.: Far-field and near-field material processing with femtosecond laser pulses. Appl. Phys. A **69**, 7–11 (1999)
37. Wienke, A., Hoffmann, G.-A., Koch, J., Jäschke, P., Overmeyer, L., Kaierle, S.: Ablation and functionalization of flexographic printing forms using femtosecond lasers for additively manufactured polymer-optical waveguides. Procedia CIRP **94**, 846–849 (2020)
38. Gooch, J.: Encyclopedic Dictionary of Polymers. Springer (2011)
39. Takai, M., Shirai, T., Ishihara, K.: Surface Functionarization of Polydimethylsiloxane by Photo-Induced Polymerization of 2-Methacryloyloxyethyl Phosphorylcholine for Biodevices. J. Photopolym. Sci. Technol. **24**, 597 (2011)
40. Ishikawa, N., Hanada, Y., Ishikawa, I., Sugioka, K., Midorikawa, K.: Femtosecond laser-fabricated biochip for studying symbiosis between Phormidium and seedling root. In: Appl. Phys. B **119**, 503 (2015)
41. Lu, D., Zhang, Y., Han, D., Wang, H., Xia, H., Chen, Q.: Solvent-tunable PDMS microlens fabricated by femtosecond laser direct writing. J. Mater. Chem. C **3**, 1751 (2015)

42. Ward, J., Yang, Y., Chormaic, S.: PDMS quasi-droplet microbubble resonator. In: Proc. SPIE 9343, 934314 (2015)
43. Lin, Y., Chou, J.: Fabricating translucent polydimethylsiloxane (PDMS) super-hydrophobic surface greenly by facile water-dissolved fillers. J. Adhes. Sci. Technol. **30**, 1310 (2016)
44. Nargang, T., Brockmann, L., Nikolov, P., Schild, D., Helmer, D., Keller, N.: Liquid polystyrene: a room-temperature photocurable soft lithography compatible pour-and-cure-type polystyrene. Lab Chip **14**, 2698 (2014)
45. Waheed, S., Cabot, J., Macdonald, N., Lewis, T.,Guijt, R., Paull, B., Breadmore, M.: 3D printed microfluidic devices: enablers and barriers. Lab Chip **16**, 1993 (2016)
46. Kant, M., Shinde, S., Bodas, D., Patil, K., Sathe, V., Adhi, K., Gosavi, S.: Surface studies on benzophenone doped PDMS microstructures fabricated using KrF excimer laser direct write lithography. Appl. Surf. Sci. **314**, 292 (2014)
47. Datasheet UV-curable liquid silicone rubber / UV-PDMS, [Online]. Available: https://www.microresist.de/produkt/uv-curable-liquid-silicone-rubber-uv-pdms/
48. Lahti, J., Savolainen, A., Räsänen, J.P., Suominen, T., Huhtinen, H.: The role of surface modification in digital printing on polymer-coated packaging boards. Polym. Eng. Sci. **44**, 2052–2060 (2004)
49. Mesic, B., Lestelius, M., Engström, G.: Influence of corona treatment decay on print quality in water-borne flexographic printing of low-density polyethylene-coated paperboard. Packag. Technol. Sci. **19**, 61–70 (2006)
50. Lee, H.J., Michielsen, S.: Preparation of a superhydrophobic rough surface. J. Polym. Sci., Part B: Polym. Phys. **45**, 253–261 (2007)
51. Roth, J.: Funktionalisierung von Silikonoberflächen. Dissertation, Technischen Universität Dresden (2009)
52. Vudayagiri, S., Junker, M.D., Skov, A.L.: Factors affecting the surface and release properties of thin polydimethylsiloxane films. Polymer Journal. **45**, 871–878 (2013)
53. Wienke A. et al.: Characterization and functionalization of flexographic printing forms for an additive manufacturing process of polymer optical waveguides. Journal of Laser Applications. **33**, 012017 (2021)
54. Wienke A. et al.: Surface functionalization of flexographic printing forms using a femtosecond laser for adjustable material transfer in MID production processes. In: SPIE Proceedings. 11267 (2020)

Aerosol Jet Printing of Polymer Optical Waveguides

5

Mohd-Khairulamzari Hamjah, Thomas Reitberger, Lukas Lorenz and Jörg Franke

5.1 State-of-the-Art Fabrication of the Polymer Optical Waveguide

Fabrication of polymer optical waveguides (POWs) has been routinely made by conventional selective removal of patterned materials processes such as etching, rib cladding, diffusion polymerization and photobleaching. These processes have shown that they can produce the desired quality of the POWs. However, this fabrication method cannot meet the current demand in POWs fabrication on desired substrates, particularly on 3D substrates. State of the art of POW fabrication methods have shown that additive manufacturing (AM) technologies such as inkjet printing (IJP), dispensing and aerosol jet printing (AJP) are promising approaches. AM technologies have shown a plethora of advantages and possibilities to encounter limitations in the conventional fabrication of POWs.

It is essential to address the research question on "How possible and reliable the POWs material can be additively fabricated on the spatial surfaces?". Hence, this work aims to study reliability of POW fabrication via aerosol jet printing onto desired geometrical surfaces. Figure 5.1 shows an overview of the POW

M.-K. Hamjah (✉) · T. Reitberger · J. Franke
Institute for Factory Automation and Production Systems,
Friedrich-Alexander-Universität Erlangen-Nürnberg, Erlangen, Germany
e-mail: mohd-khairulamzari.hamjah@faps.fau.de

J. Franke
e-mail: joerg.franke@faps.fau.de

L. Lorenz
Institut für Aufbau- und Verbindungstechnik der Elektronik, Technische Universität Dresden, Dresden, Germany
e-mail: lukas.lorenz@tu-dresden.de

J. Franke et al. (eds.), *Optical Polymer Waveguides*,
https://doi.org/10.1007/978-3-030-92854-4_5

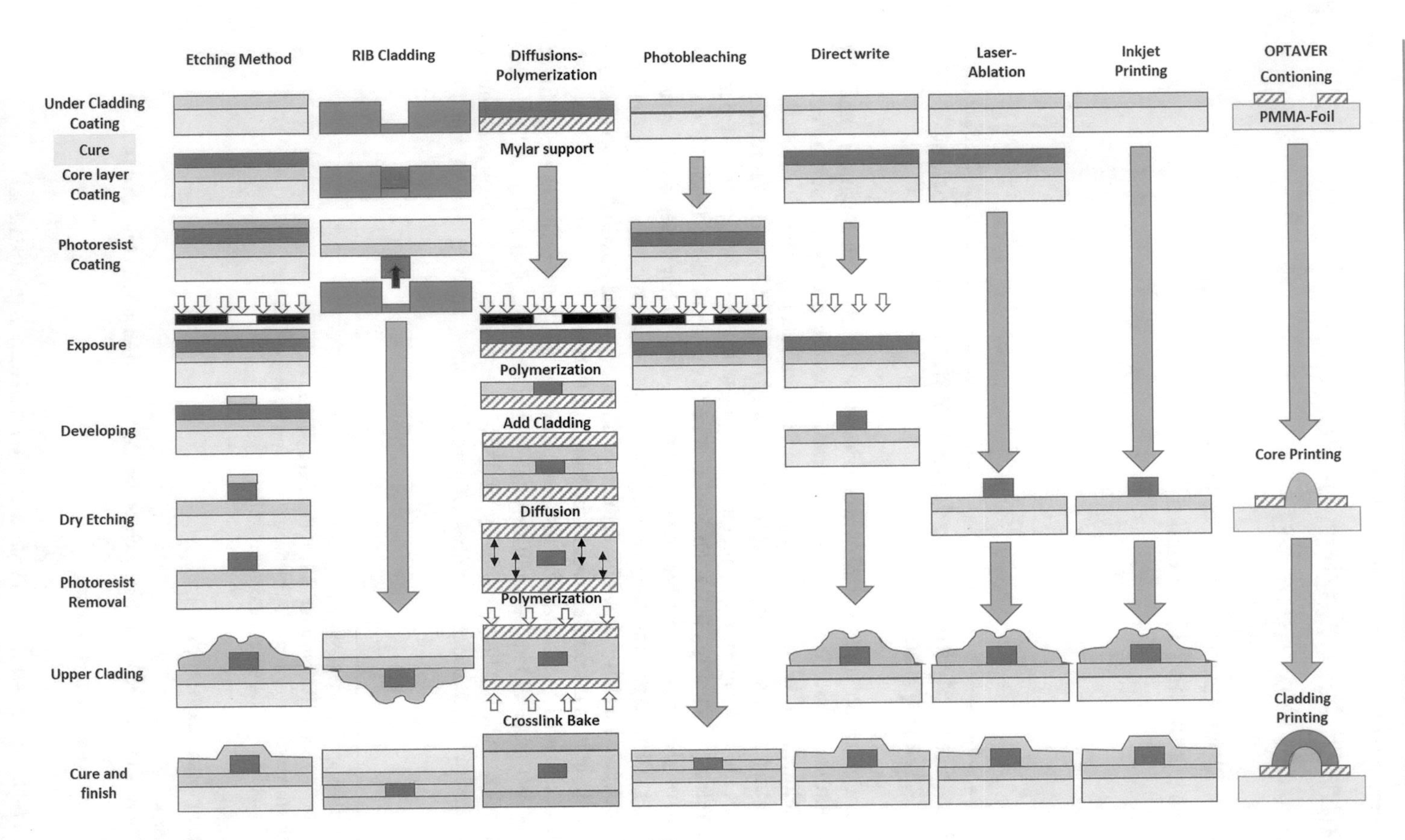

Fig. 5.1 State of the art of polymer optical waveguides process chains [1]

fabrication technologies. In the following section, existing processes and achievements in POW fabrications are explained in brief.

5.1.1 Photolithography

In this process, optical waveguides are produced layer by layer from the liquid phase. First, liquid cladding material is applied to a carrier substrate over a large area, for example, via spin coating process and then cured by UV exposure. In the next step, the subsequent core material is applied and selectively exposed using suitable masks. The remaining non-consolidated core material is chemically removed using suitable solutions. Finally, the large-area application of cladding material is again carried out, as well as its curing process. OrmoCore material has successfully been used in this method, and attenuation values as low as 0.04 dB/cm are achieved. Even though this process has shown promising results, direct fabrication on 3D substrates is not possible and the process also produces material waste during the etching process [2].

5.1.2 Photolysis

Photolysis is a process that is limited to the use of PMMA material. PMMA is applied to the substrate material and cured. Alternatively, PMMA can also be used directly as substrate material, which saves the first process step. By using ionizing radiation, the refractive index is increased locally, using a mask, thus enabling subsequent light conduction. Finally, the PMMA cladding material is applied and cured. This process can be performed for single-mode waveguides. However, this process requires masking steps and is currently restricted to PMMA material. Higher attenuation values of 0.9 dB/cm is reported [3].

5.1.3 Photolocking

First, the lower cladding material is applied and cured, or else a suitable substrate can be used directly as the cladding material. Subsequently, a polymer matrix with a high proportion of monomers and photoinitiators is applied and selectively exposed via a mask. The polymerization of the photoinitiators with the monomers occurs, resulting in local areas of increased refractive index, which represent the subsequent core structures. In contrast to other exposure processes, the unexposed areas of the core material are not removed. Finally, the cladding material is applied and cured. Attenuation values of 0.2 dB/cm $\pm$ 0.05 are archived in the process [4].

5.1.4 Photobleaching

This process is based on the use of photosensitive, dye-doped polymers. These are applied over a large area and selectively exposed via a mask. Unlike comparable processes, the refractive index is lowered in the exposed areas. These areas can subsequently be used as cladding layers. The waveguide itself is thus created indirectly. In the last step, the cladding layer is applied and cured. The attenuation values of 0.8 dB/mm is successfully achieved [5].

5.1.5 Reactive Ion Etching (RIE)

After the lower cladding material is applied and cured, a reactive polymer layer is applied, which serves as the subsequent core. Masked irradiation of the reactive polymer layer with ions occurs under a plasma atmosphere, which successively etches off the areas not protected by the mask. Finally, the application and curing of the upper cladding take place. This process can produce an excellent waveguide resolution of <100 nm. It is also possible to produce single-mode waveguides at extremely low attenuation values as low as 0.1 dB/cm [6]. Despite the significant output, it cannot be processed directly on 3D surfaces. Furthermore, the need for special masks and the costly plasma atmosphere plant technology are also disadvantages of the process [7, 8].

5.1.6 Laser Ablation

The lower cladding layer is deposited and cured, followed by applying the optical layers using spin coating. Larger boards require other techniques such as dip coating, roller coating or spray coating. The layer is exposed to UV light for the curing process. Then, the waveguide is formed using laser micro-structuring before deposition of the upper cladding material. An average propagation loss of 0.13 dB/cm at 850 nm wavelength has been reported [9].

5.1.7 Dispense Printing

Dispense printing is among the direct-writing process and has the capability of depositing various types of material. Typical materials considered in this process are pasty materials such as metals, semiconductors, polymers, ceramics and composites. During the dispensing process, the material is applied to the substrate through a mechanical air-pressure pump or extrusion mechanism. This process forces the continuous flow of the material exiting the syringe tip or needle via capillaries action. The size of the tip varies according to specified sizes and shapes. The syringe is attached to the maneuver system in x and y orientation to have pattern printing. After deposition, the material is let dried or sintered on demand. It is also possible to use different printing heads simultaneously. Thus, components with integrated functionalities made of different materials can be manufactured.

In the fabrication of POWs, polymer material is used. First, the bottom cladding material is prepared and properly cured. Next, the core material is then dispensed on top of the cladding material continuously. The gap between the nozzle tip and the substrate is usually half of the nozzle size. This gap is essential to ensure the deposited material is in contact between the nozzle and the substrate. Studies show that optical transparency and optical losses are achieved at 98% and below 0.22 dB/cm, respectively [10].

5.1.8 Mosquito Method

The mosquito method utilizing a micro-dispenser is a straightforward fabrication technique for polymer waveguides [11–16]. This method has the same working principle as the dispensing system, whereas the core monomer material is dispensed except on the cladding material. The cladding material is prepared in advance from the sol–gel material and placed in the mold.

Here, a viscous core monomer material is dispensed in the middle of the cladding material from a thin needle of a syringe connected to a dispenser. The needle tip remains inserted into the cladding monomer material while still in the sol–gel state when the core monomer is being dispensed. The needle is moved horizontally within the cladding monomer and forms a waveguide structure. Parallel arrays of circular cores are fabricated by repetitive parallel scans of a single needle. When the two monomers are miscible, after the core monomer is dispensed, the two monomers diffuse slightly into each other to form a concentrated distribution of the material. After dispensing multiple cores, both the cores and the cladding are exposed to UV-LED light for curing.

The mosquito method proves to be advantageous in maintaining the cross-sectional area of the waveguides. Study shows that the mosquito method is capable in the fabrication of graded-index (GI) circular core POW [12].

5.1.9 Inkjet Printing

The inkjet process jets a stream of material droplets onto the substrate to form the desired layout pattern. There are two types of inkjet printing, namely, drop-on-demand (DOD) and continuous mode (CIJ) [17]. Studies show that inkjet printing is used to deposit Truemode® core material and Truemode® cladding material on the functionalized glass substrates [18]. Klestova et al. [19] have inkjet printed multiple layers of optical waveguides for single-mode operation. In their studies, an attenuation loss of 3.52 dB/cm at 1.55 μm wavelength is achieved. Even though this process yields significant printed results, direct printing on the curvature surfaces is hard to be obtained. The inkjet printing requires a gap between the nozzle tip and the substrate surfaces in the range of 0.1 mm to 1 mm, making this printing method unsuitable for 3D surfaces.

5.1.10 Aerosol Jet Printing

Based on what had been briefly explained, POWs' current manufacturing processes are well suited to build 2D structures. In some processes, it is theoretically possible to manufacture 3D optical waveguides. However, associated costs and additional process steps (e.g., 3D masks, process control) become a limitation to further use in 3D applications.

AJP has shown its possibilities to functionalize 3D surfaces with light-conducting structures. Besides, the AJP process also excels in material diversity, the elimination of a mask and the unnecessary use of chemicals. In this chapter, a complete fabrication process of printed POWs via AJP technology is presented. The AJP respective process parameters and their effects on the quality of the light-guiding structures are also discussed. For POW fabrication via the AJP process, the selection of the suitable POW material and its compatibility is discussed. Next, the challenges in determining the best printing strategy toward reliable printed POW quality are explained. Besides, simulation studies of the velocity flow of the AJP-printhead are also briefly explained. An overview of the POW manufacturing technology in terms of its advancements and limitations is shown in Table 5.1.

5.2 Polymer Optical Waveguides Fabrication Through Aerosol Jet Printing

5.2.1 Principles of Aerosol Jet Printing

Aerosol jet printing technology consists of two types of systems, namely ultrasonic and pneumatic. The ultrasonic system refers to the use of ultrasonic waves for the atomization of the desired material. In contrast, the pneumatic system is the method, whereas the shear force effect is used to atomize the material. Both systems physically change the liquid to become atomize droplet size of approximately 1 to 5 μm. Figure 5.2 shows the schematic diagram of the ultrasonic and pneumatic systems of the AJP. Direct fabrication onto 3D surfaces is the key advantage of the AJP technology, which is an ideal solution for many potential application areas, such as 3D-integrated optical devices and 3D-Opto.MID.

In this research work, the pneumatic system is used as the ultrasonic system is not capable of atomizing the POW material with high viscosity. Hence, this section explains in detail the pneumatic system rather than the ultrasonic system. The AJP-pneumatic system consists of specific functional units, namely pneumatic, virtual impactor and deposition head. The operation principles started with high-velocity compressed carrier gases (nitrogen) passing through the atomizer jet and

Table 5.1 Comparison of the state of the art of the POW fabrication technology

Process technology	3D surface process ability	Attenuation	Single-mode	Masking process	Material diversity	Chemical process	Process time
Laser ablation	No	0.0035 dB/cm (840 nm) [9]	Yes	No	Very high	No	High
Reactive ion etching	No	0.01 dB/cm [6]	Yes	Yes	Low	No	Very high
Photobleaching	No	0.8 dB/mm [5]	No	Yes	Low	Yes	Very high
Photolocking	No	0.2 dB/cm [4]	Yes	Yes	Low	Yes	Very high
Photolysis	No	0.9 dB/cm [3]	Yes	Yes	PMMA	Yes	High
Photolithography	No	0.04 dB/cm [2]	No	Yes	Low	Yes	Very high
Mosquito method	No	0.033 dB/cm (850 nm) [16]	No	No	Low	No	Low
Dispensing	Possible	0.22 dB/cm [10]	No	No	Very high	No	Very low
IJP	Possible	3.52 dB/cm [19]	No	No	High	No	Very low
AJP	yes	0.22 dB/cm	No	No	High	No	Low

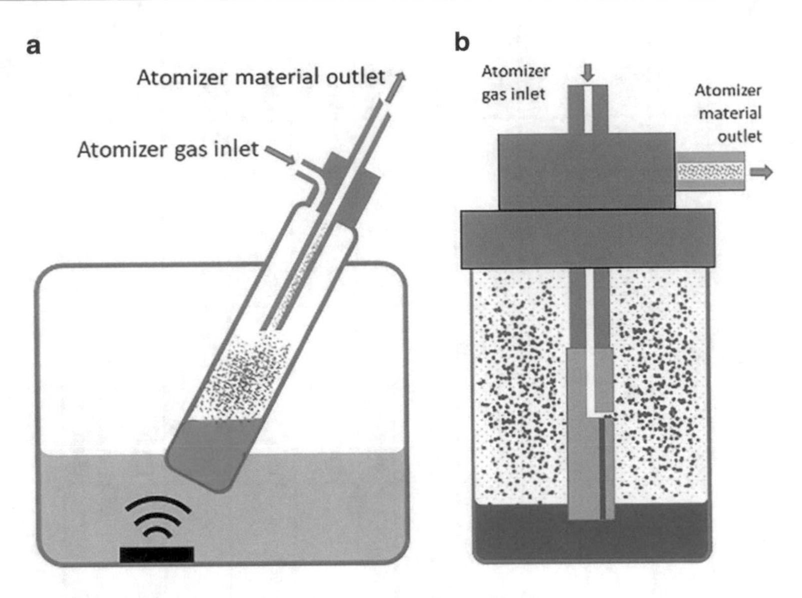

Fig. 5.2 AJP atomization systems **a** ultrasonic and **b** pneumatic

producing atomized droplets of the liquid material (optical polymer). Big and heavy droplets are collected back into the vial during this process, while small and light droplets stay in the gas form. Subsequently, the carrier gas channeled the droplets to the virtual impactor unit. This impactor unit functioned to further refine the droplets by removing the smaller and lighter droplets by the exhaust system, while the dense and high-inertia droplets remain at the center of the streaming gas flow. Next, the fine droplets are transported to the deposition head via a hose. Inside the deposition head, the droplets are further focused with the annular sheath gas to form a net output gas flow. Finally, the material passes out through a ceramic nozzle and hits the substrate. It is good to mention that nitrogen gas is used as a supply gas for carriers and sheath gas in the AJP process. This sheath gas enables a consistent deposition and higher printing stand-off at 1–5 mm. The sheath gas also functions as a coaxial curtain for the droplets stream to prevent the droplets from adhering to the inside nozzle wall, which can cause clogging to the nozzle tip. Figure 5.3 shows a schematic diagram of the pneumatic AJP system, and Fig. 5.4 shows a photo of the actual AJP setup.

5.2.2 Selection of Polymer Optical Material for Aerosol Jet Printing Process

There are three main factors to be considered while choosing optical waveguide materials: optical loss, mechanical strength and manufacturability of the material with existing waveguide production processes. For waveguides with low losses

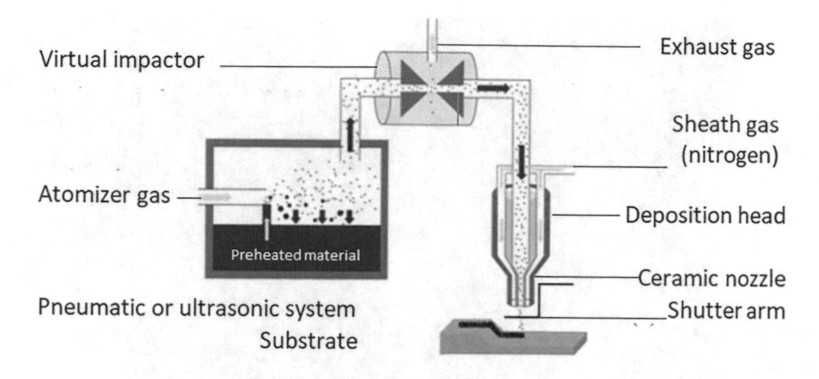

Fig. 5.3 Schematic diagram of the pneumatic AJP printing

and high strength, silica-glass is preferred compared to polymer material. Polymer (e.g., PMMA) has a different refractive index compared to silica-glass fiber. The index difference between the core and cladding determines the transverse modal structure of the waveguide [21].

In the OPTAVER research, polymer-based materials are used. The selection of the polymer optical material and its compatibility to be used in AJP is the essential factor toward the success of the research. Moreover, the substrate material functions as a lower cladding of the printed core structure. Hence, mechanical, optical and liquid characteristics of the POW material such as substrates, conditioning, core and cladding are critical and need to be thoroughly determined. Figure 5.5 shows the interaction diagram of the POW materials.

Prior to printing the POW core material, conditioning lines are applied to the selected substrate material, which acts as barrier for the printed core material. In order to obtain the desired quality of the printed POW, it is necessary to identify the liquid properties of the conditioning material for the flexographic printing process, such as surface tension that can allow sufficient wetting of the substrate material in the liquid state. Furthermore, the substrates should have the lowest possible surface energy after curing process in order to optimize the contact angle of the core material. The process control needs to be carefully determined to ensure the lowest edge ripple in order that accurate printed dimensions can be achieved. Details on substrates preparation can be found in Chap. 4 of this book. Table 5.2 shows the list of materials that have been successfully investigated in this research. Results analysis of the investigation is explained in the following section of this chapter.

Substrates material: OrmoCore and J+S 390119 are printed directly on the three substrates: PMMA, polyvinyl chloride (PVC) and polyimide (PI) without the conditioning lines. From the results in Fig. 5.6, it is clearly seen that J+S 390119 indicated much better printing structures on all the investigated substrates as straight edges and continuous lines are obtained. AJ printed OrmoCore material resulting in non-continuous structure is formed except for the PVC substrates

Fig. 5.4 Actual AJP setup for POWs printing in OPTAVER work [20]

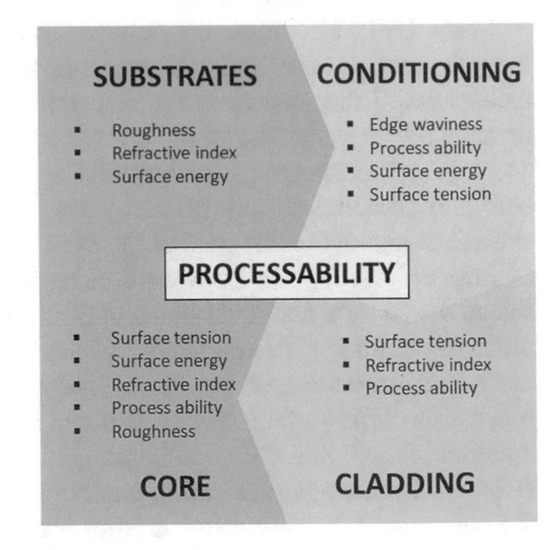

Fig. 5.5 Material interaction diagram of the AJ printed POW [20]

[22]. The POW core material structure requires a higher refractive index (RI) of the core than the cladding to guide the light adequately. This higher RI is essential to ensure total reflection and optical transmission can be achieved.

In the OPTAVER research group, the J+S 390119 material has been selected as POW core function. This material was AJ printed onto the PMMA substrate foil with preconditioned line structures made from Actega G8-372/NVK-S material, as shown in Fig. 5.7. The J+S 390119 material is a UV curing polymer that contains reactive binders. A solid chemical bond is formed as soon as it is excited by the

Table 5.2 List of materials investigated in the research work

Substrates	Core	Cladding
• Glass • Polyimide (PI) • Polyvinyl chloride (PVC) • Polycarbonate (PC) • Polymetymetacylate (PMMA)	• Norland NOA 61 • EpoCore 2,5,10,20 • Loctite 3105 • Loctite Ablelux OGR-146-TUV • Loctite Ablelux OGR-150-THTG • PolyTecVP 4641-1 • Actega G8-372/NVK-S • OrmoCore • UV Supraflex Varnish (J+S 390119)	• EpoClad 5

initiator. The initiator decomposes to radicals by UV irradiation, thus starting the polymerization chain reaction.

The following are the main factors of the J+S 390119 material suitability to be used throughout the OPTAVER research work:

- Suitable viscosity of 72.491 cP at 40 °C to be used in the pneumatic AJP.
- Resistance to sunlight.
- Long-term stability; material can be processed over several days.
- Process window parameters in a wide range to achieve desired printing result.
- Low shrinkage effect due to low solvent content (< 1.0% organic solvents).
- Fast curing process by using UV light without additional temperature control.
- No solvent residues on the optical waveguide surface.
- No cracking and detachment effect after UV curing (ductile properties).
- Higher refractive index of 1.52 at 850 nm compared to the PMMA substrates film of 1.49 at 850 nm.

5.2.3 Effect of the Material Temperature on the Mass Flow Output

To further investigate the polymer material behavior under specific temperatures, mass flow analysis was conducted. The polymer materials are J+S 390119 and EpoCore, Loctite Ablelux OGR-150THTG, OrmoCore. For measuring the mass flow, the sleeve temperature was set at 45 °C and a weighted aluminum foil was placed under the AJP nozzle to collect the aerosol output for five minutes. The aluminum foil with the deposited polymer was weighed again, and the total mass of polymer material deposited per minute can be calculated. Results analysis of the investigation is shown in Fig. 5.8. It can be seen that all four materials show the same trend. The mass flow outputs increased with increasing atomizer flow rate to a maximum yield before drop significantly.

Further investigation on the AJP shows that the mass flow dropped due to the massive droplet build-up at the nozzle tip, which leads to less material yield.

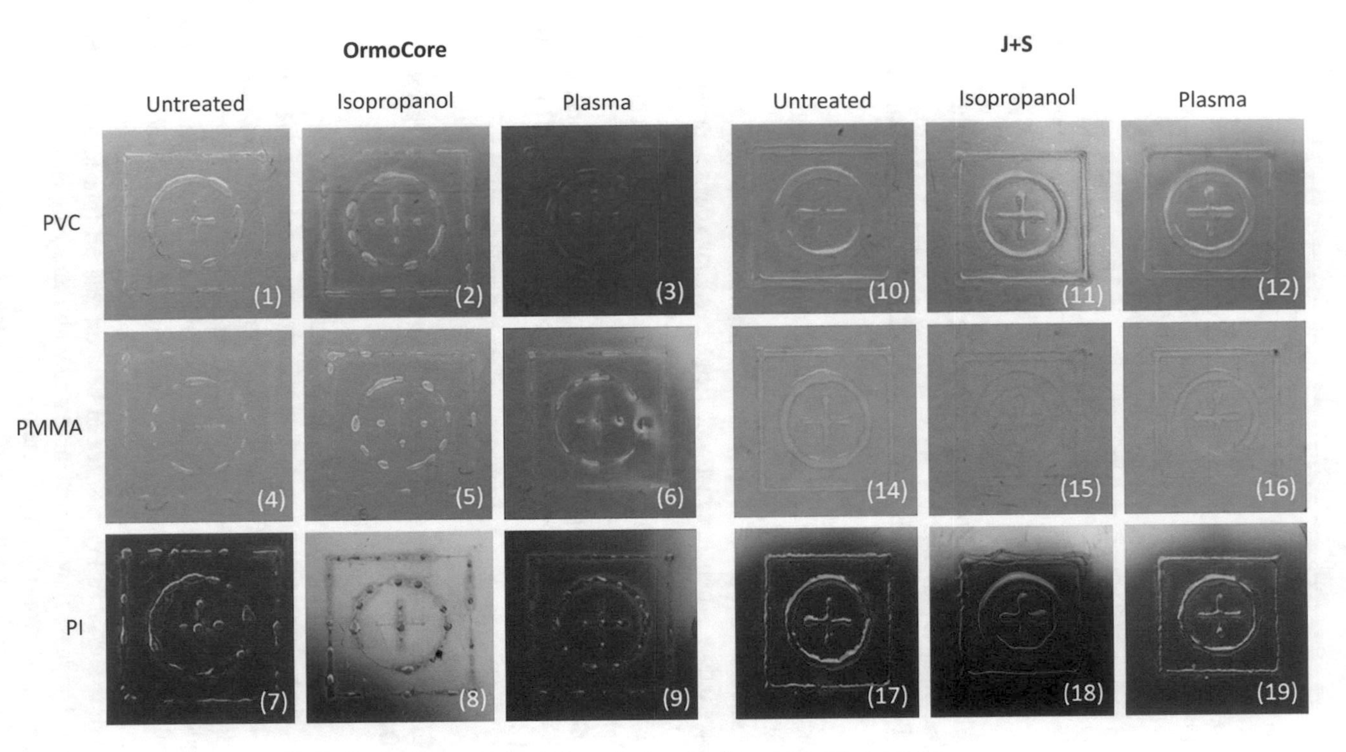

Fig. 5.6 AJ printed OrmoCore and J+S 390119 core material on selected substrate after pretreatment process

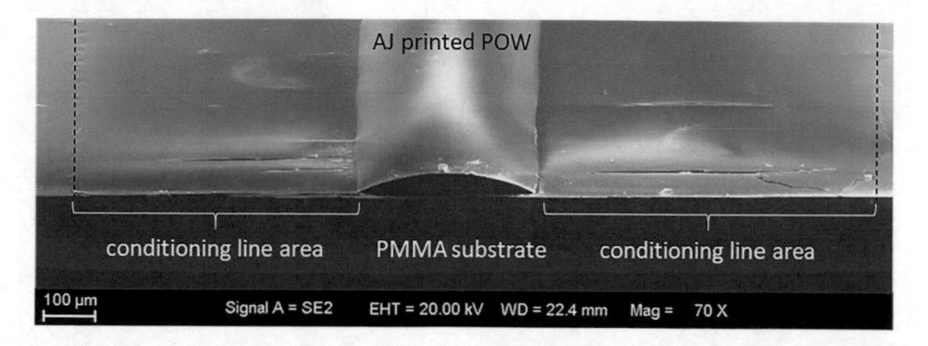

Fig. 5.7 Scanning electron microscope (SEM) image of AJ printed POW at optimum process parameter [23]

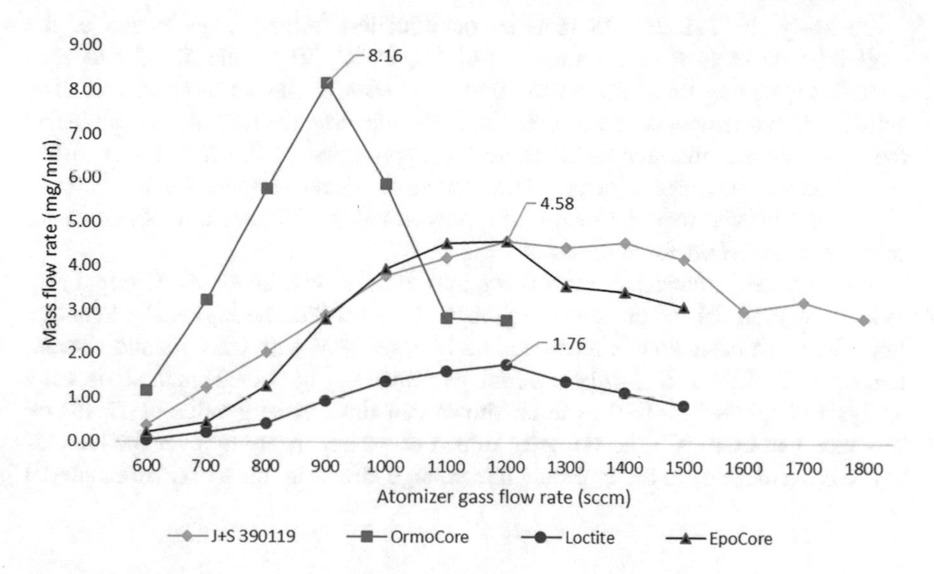

Fig. 5.8 Mass flow measurement results of J+S 390119, OrmoCore, Loctite Ablelux OGR-150THTG and EpoCore, preheated at 45 °C, for different atomizer gas settings. Mark value represents maximum yield mass flow before the build-up droplet starts to occur. [22]

Figure 5.9 shows the build-up material formed at the nozzle tip after its maximum mass flow. Furthermore, J+S 390119 and EpoCore showed the same maximum mass flow of 4.58 mg/min at 1200 sccm. The highest yield of mass flow of 8.15 mg/min at 900 sccm can be obtained from OrmoCore material, and the Loctite Ablelux OGR-150THTG produced a maximum mass flow of 1.76 mg/min at 1200 sccm.

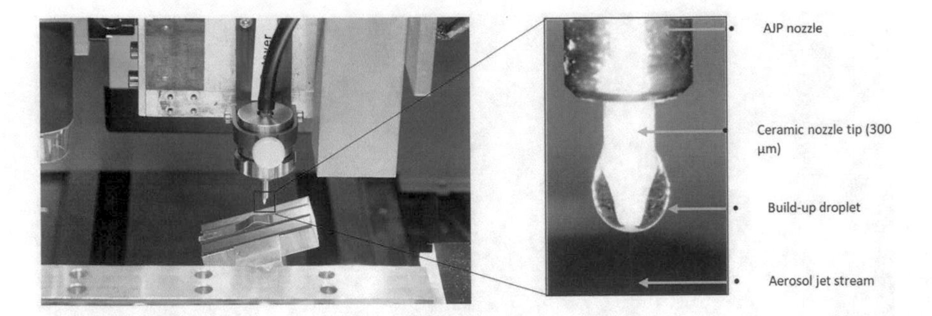

Fig. 5.9 Excessive polymer droplet build-up at the nozzle tip after mass flow reached its maximum yield

To study the J+S 390119 material qualification further, investigation of the material flow at sleeve temperature at 40 °C, 45 °C, 50 °C and 55 °C was conducted. Preheating time was set to 30 min for each of the set temperatures. The actual material temperature was measured by contacting the heated material inside the vial with the thermocouple. There are approximately 20–25% lower differences between the material temperatures to the set sleeve temperature, as shown in Fig. 5.10. The vial material made from polypropylene (PP) material significantly reduces the heat transfer to the material.

Furthermore, atomizer gas that flows into the vial with about 20 °C might also reduce the material temperatures. Figure 5.11 shows that the higher the temperature, the more mass flow is recorded and in agreement with the dynamic viscosities analysis of the J+S 390119 material, as shown in Fig. 5.12. Dynamic viscosity analysis of the J+S 390119 material shows that the viscosity value of 72.491 cP was recorded at 40 °C. The viscosity value decays exponentially over the increasing temperature due to the molecule interchange effect. In this work, core material

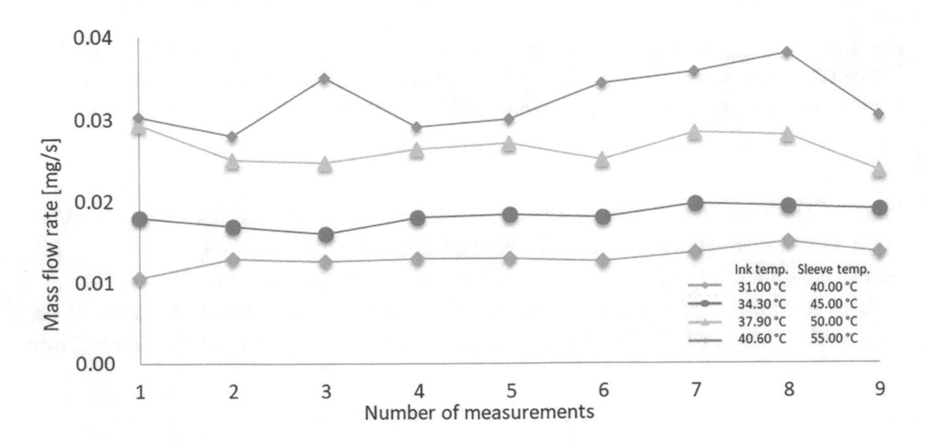

Fig. 5.10 J+S 390119 material mass flow at four selected sleeve temperature

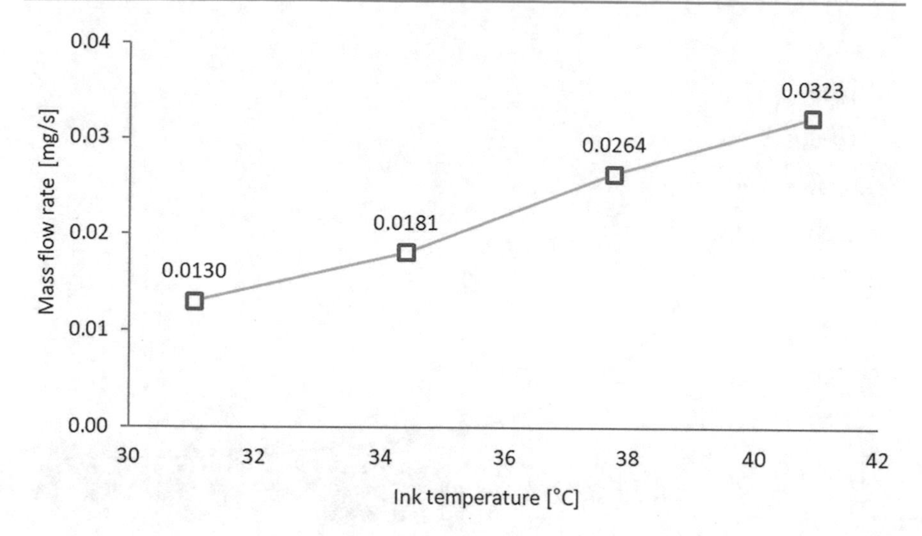

Fig. 5.11 Mean value mass flow versus actual temperature of the J+S 390119 material

from J+S 390119 yielded the best result compared to the other investigated material.

5.3 Fabrication Process Steps

In this work, the AJP system is mounted onto a 5-axes kinematic system from NeotechAMT. The kinematic system has the ability to route onto the desired 2D or 3D substrate surfaces based on the created numerical control (NC) code. Figure 5.13 shows the overview of the process steps in the fabrication of the POW. The process begins with the 2D or 3D model and the desired POW layout creation. Next, NC code is generated and transferred to the kinematic system. The kinematic system interprets the code for the execution of the printing process. Prior to the execution of the printing, optimized AJP process parameter values are selected. This is to ensure the best quality of the printed POW is achieved. It is good to mention that in the OPTAVER printing system, CADCAM, Remote-NC and AJP systems are not automatically linked together. Hence, each of the process steps needs to be set after one to another. Details of the process steps are explained in the next section.

5.3.1 Motion3D: CADCAM Design and NC Code Generation

For designing the printed POW layout on the desired substrate geometries in 2D or 3D, the CADCAM software (Motion 3D) from Neotech AMT GmbH was used. Figure 5.14 shows the user interface of the Motion 3D CADCAM software. The

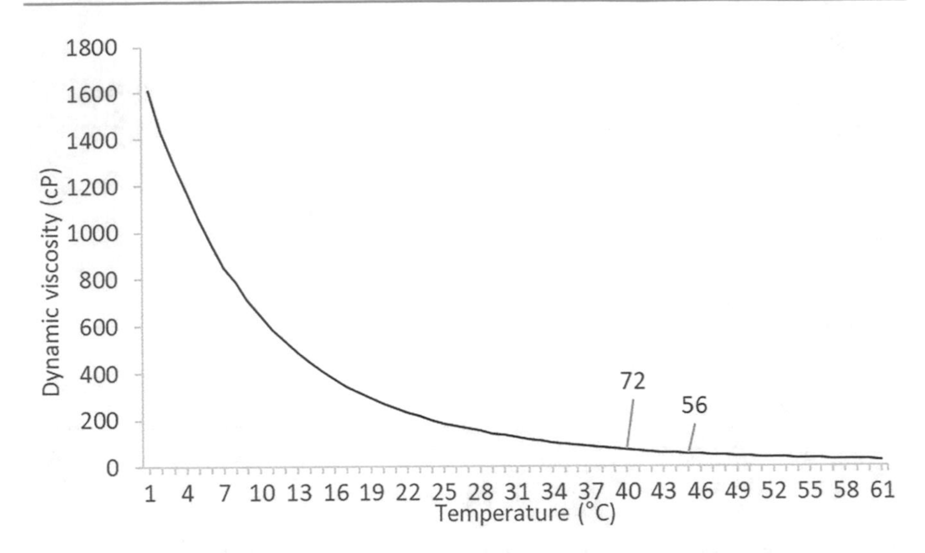

Fig. 5.12 Dynamic viscosity analysis of the J+S 390119 material over elevated temperature [20]

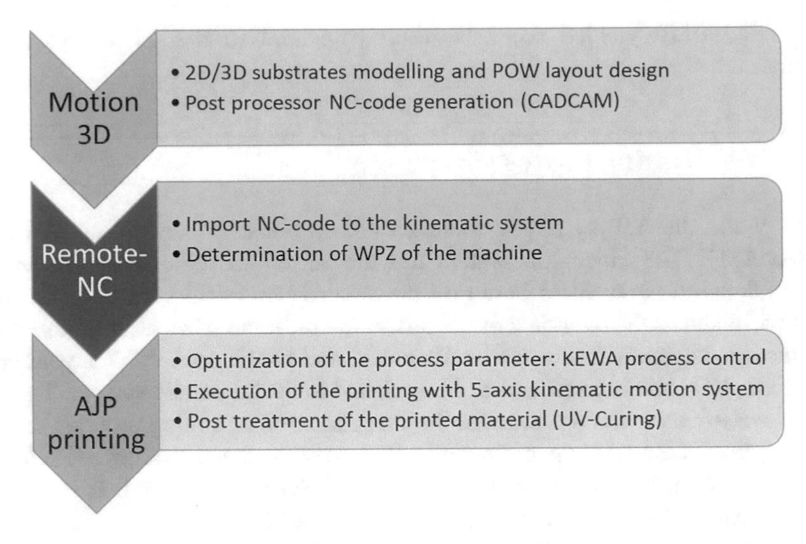

Fig. 5.13 Process steps of the aerosol jet printed POW of the research work

CADCAM software is configured for 2D/3D surface printing, which allows the printing nozzle to route across the geometrical surface. The process starts by creating the POWs layout on the CAD model. The CAD model can be drawn directly in the Motion 3D or by importing the (.step) file format from other CAD programs. Next, the CAM block with specific tool properties and the workpiece zero (WPZ) of the model design is determined and is automatically integrated into the NC code. WPZ functions as an origin reference for the printing program. Finally,

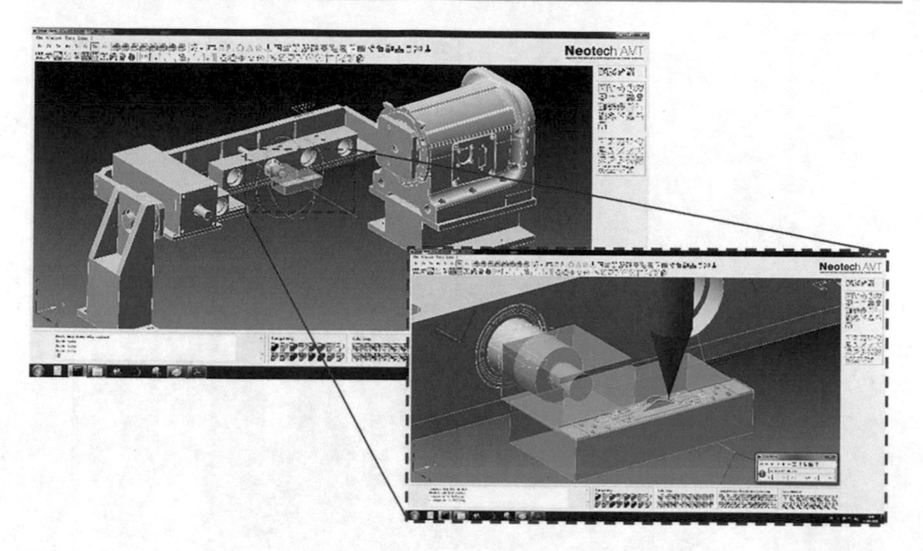

Fig. 5.14 Motion 3D CADCAM user interface for layout modeling and NC code generation

the NC code with fully functioned commands for axes motion and shutter function for the printing process execution is generated.

5.3.2 Remote-NC: 5-Axes Kinematic System

Remote-NC software is an interpreter program for controlling the kinematic motion between nozzle and substrate. Here, the generated NC code is imported into the Remote-NC program. Prior to the NC code execution, the WPZ of the machine in actual coordination is determined. This value has to be in the same position as in the WPZ-CADCAM. Besides, the desired stand-off distance from the nozzle to substrates surfaces is also selected. These steps are essential to ensure the machine operation works precisely according to the CADCAM design and collision of the printing head can be eliminated.

In the OPTAVER process, liquid polymer needs to be precisely printed in between two conditioning lines. MS-Excel tool for positioning the camera position relative to the printhead position was developed to precisely place substrates on the printing platform, as shown in Fig. 5.15. This MS-Excel tool is used to calculate the printing new position and ensure the printing process is in between the desired position. First, the substrate is fixed on the printing platform. Then, the camera is moved to the desired starting point of the printing. Coordinates in the actual machine position are entered into the MS-Excel tool, followed by selecting a new reference point. Next, the camera is moved to the end printing point and the machine coordinate is confirmed. The offset coordination between the camera and nozzle tip is determined based on the printing confirmed start and endpoint position. Finally, the new WPZ is set to the machine.

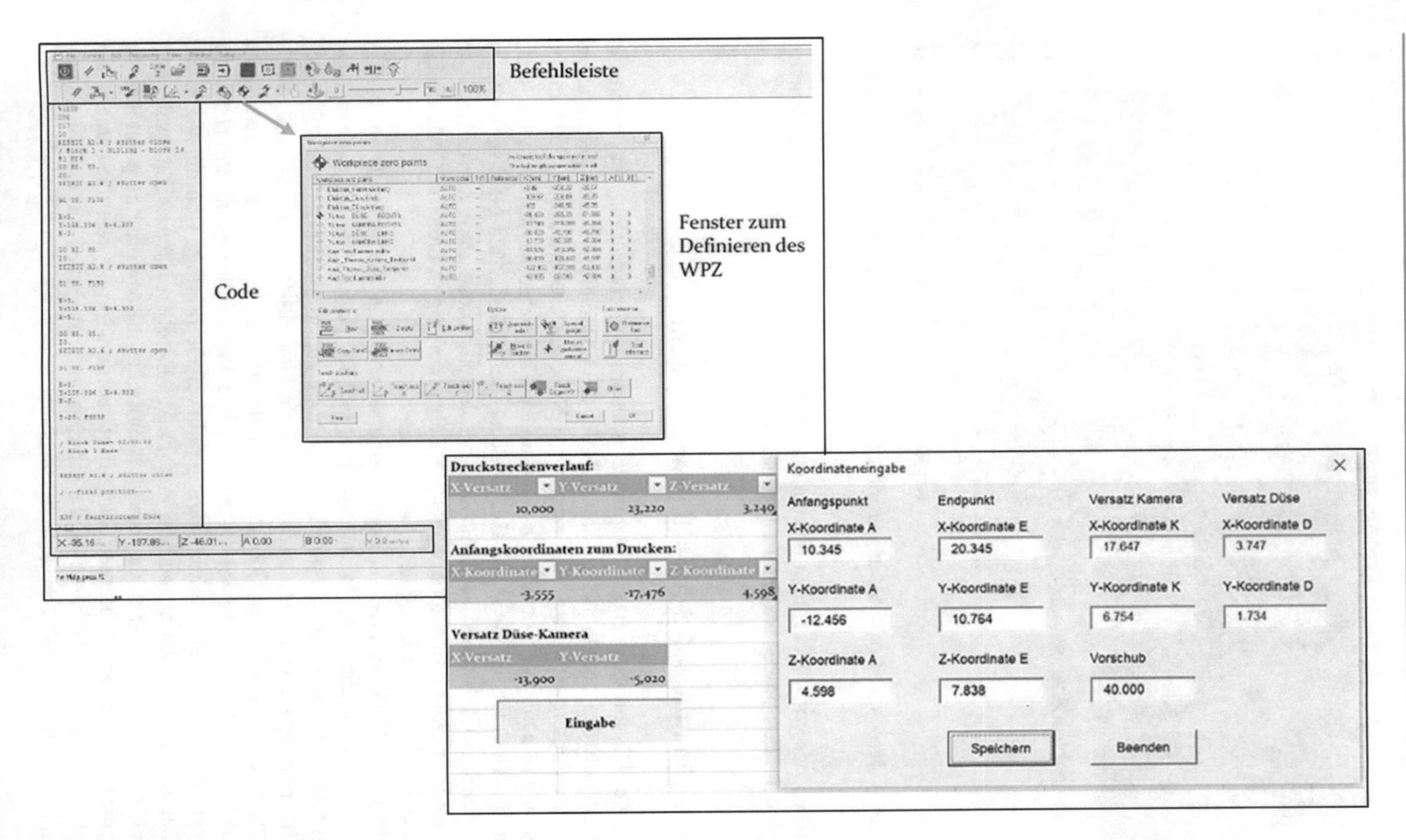

Fig. 5.15 Remote-NC control user interface and MS-Excel tool for printing positioning

5.3.3 AJP Process Parameter

Printing POW using the AJP-pneumatic system is promising and challenging as many parameters are involved during the printing process. There are three main parameters in the AJP process: sheath gas, carrier gas and exhaust gas. Those parameters are controlled by using OPTOMEC KEWA™ Process Control software. The KEWA software functions to start and stop the process and monitor the process parameter during the printing process.

To operate the AJP system, it is essential to switch on the gas flow in the recommended sequence. First, the sheath gas, and next, the exhaust gas and atomizer gas at last. Each gas needs to flow for approximately two to three minutes before switching on the other gas. This is to ensure every gas sufficiently flows to the system. Switching on in the wrong sequence will inevitably lead to the system clogging issue on printing units. Furthermore, there is a risk of damage to the nozzle tip if the gas flow is set to high pressure (> 2000sccm). A reverse-order sequence is required to shut down the system [24].

An overview diagram is created to comprehensively visualize the related AJP parameters in the system, as shown in Fig. 5.16. Here, the designation of the parameters in the AJP-pneumatic system is represented in abbreviations form, namely atomizer gas flow (AtmG), exhaust gas flow (ExhG), dense atomizer gas flow (DAtmG), sheath gas flow (ShG) and net output gas flow (OG). The gas flow rate unit is in standard cubic centimeters per minute (sccm) unless mentioned otherwise. The AJP process starts with AtmG entering the system and then passing through the impactor unit. In the impactor unit, the low-dense droplet is vacuumed out (ExhG) from the system resulting DatmG exiting the impactor unit. This DAtmG then passes through the printing head. Here, ShG enters the system to curtain and combine with DatmG, resulting in the final OG exiting the system through the nozzle. This diagram is further used in the later printing process in this chapter.

The process window of a material is defined by the minimal and maximum nozzle pressure of the aerosol jet system. The nozzle pressure should not be above one pound per square inch (psi), corresponding to 68.94 mbar. Exceeding this limit can lead to damage AJP system. Therefore, the amount of flow to reach pressure limits is critical and needs to be identified. For this purpose, all flow

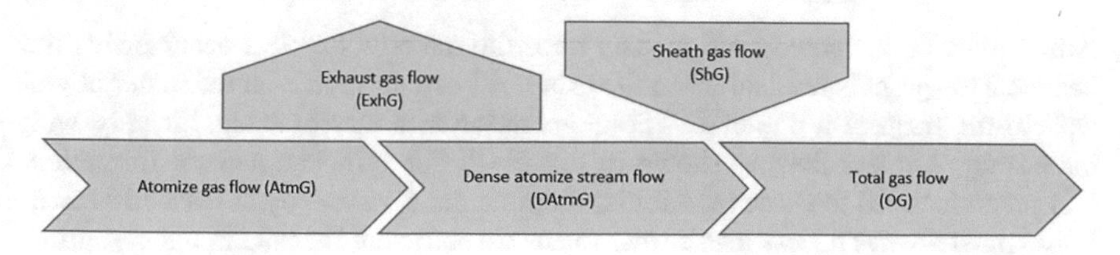

Fig. 5.16 Schematic representation of the gas flows in the aerosol jet printing with the abbreviations used in each case

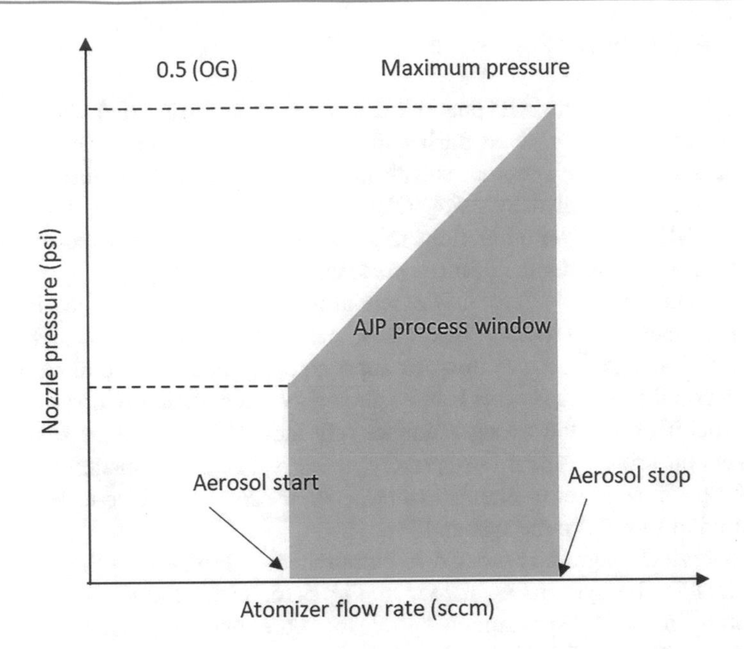

Fig. 5.17 Graphical representation of the general process window of processable material of the aerosol jet printing process

regulators in the KEWA control software are set to zero and gradually increase ShG value until the nozzle pressure sensor indicates about one psi. The relationship concerning the gas flows is shown in Eq. 5.1

$$\text{OG} = \text{ShG} + \text{DAtmG with ShG} \geq \text{DAtmG} \tag{5.1}$$

During the printing process, ShG and DAtmG add up and form the total flow through the printhead. Higher DAtmG than ShG can cause adhesion of aerosol material to the nozzle wall, which results in printhead blockage. This condition results in a maximum limit for DAtmG of 0.5 OG as shown in Fig. 5.17. The lower limit of the process window is described when AtmG sufficiently generates an aerosol. For pneumatic atomization, DAtmG is calculated from the difference between AtmG and ExhG, as shown in Eq. 5.2

$$\text{DAtmG} = \text{AtmG - ExhG with ExhG} < \text{AtmG} \tag{5.2}$$

AtmG must be increased continuously until the material droplets occur inside the beaker. This point is reached when the droplets become visible on the material vial wall in the form of a fine mist. The atomization rate increases significantly with increasing atomizer flow, as shown in Fig. 5.18. For sufficient aerosol formation, it is recommended to exceed the initial atomization flow rate by at least 150 sccm. ExhG must always be less than AtmG to ensure sufficient DAtmG in the system.

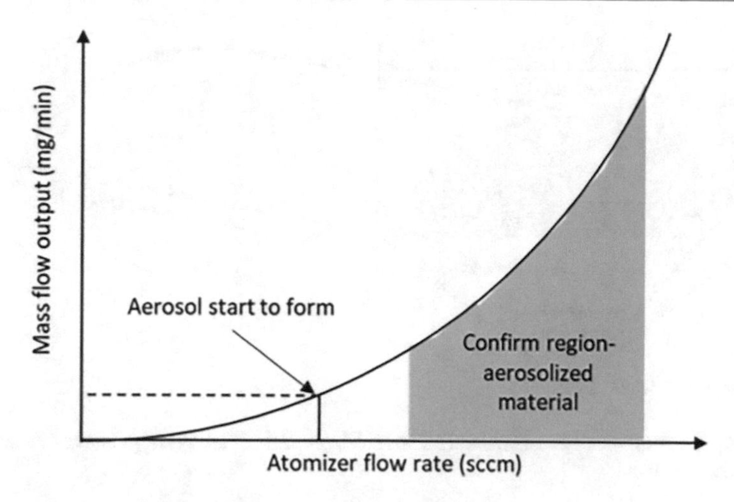

Fig. 5.18 Mass flow rate as a function of atomizer flow rate

Hence, the following points need to be considered when selecting suitable AJP parameters of POWs fabrication.

- The material mass flow increases proportionally to the temperature and significantly drops after reaching its maximum yield due to the excessive material starting to build up at the nozzle tip.
- ShG should always be greater than or equal to DatmG in order to prevent printhead blockage. Decreasing ExhG at the constant AtmG and ShG leads to increasing material flow at OG.
- The atomizer gas flow should be set at least 200 sccm lower than the maximum value of the material mass flow rate to ensure stable processes.
- ShG and DAtmG should be set at 80 sccm and 60 sccm, respectively. POW printing is conducted after five minutes, the parameter is set, and sleeve temperature at 45 °C is selected throughout the printing process.

5.4 Theoretical AJ Printed Polymer Optical Waveguide Geometry

In the OPTOVER research, AJ printed POW must be able to reach the desired dimension of a circular segment, as shown in Fig. 5.19. A direct relationship can be identified between the printed waveguide cross-sectional area and the ideal geometry of a circular segment.

The printed POW circular segment is derived from the POWs width (s) and the contact angle (Θ). This calculation is essential to prevent printed material from flooding out of the conditioning lines. Moreover, printing speed can be determined

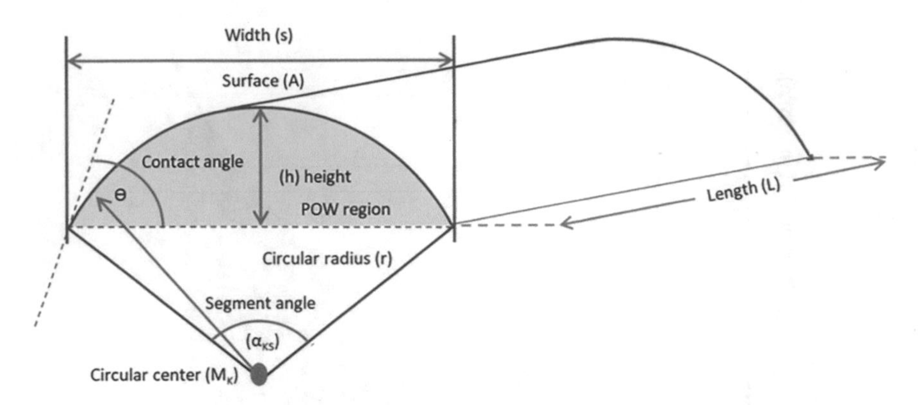

Fig. 5.19 Derivation of the mathematical model from the actual POW geometry dimension in the circular segment

after knowing the amount of material flow in the function of time. Furthermore, the calculation is necessary to avoid a bulging issue on the printed POW due to excessive material and full utilization of the POW material can be achieved.

The mathematical model presented here is based on the optimum contact measurement of $\Theta = 61.4°$ between the core and conditioning lines [25]. The printed POW geometrical dimension is calculated with the given contact angle and the distance between the two conditioning lines, as shown in Fig. 5.20 [26].

First, β is determined according to Eq. 5.3

$$\beta = 90° - \Theta \tag{5.3}$$

The angle sum in the triangle is 180°. Subtracting the 90° of the right angle and β gives $\alpha/2$ according to Eq. 5.4. Consequently, $\alpha/2 = \Theta$.

$$\alpha/2 = 90° - \beta \tag{5.4}$$

Here, the radius r of the circle can be calculated via Eq. 5.5

$$r = \frac{0{,}5 \cdot s}{\cos(\beta)} \tag{5.5}$$

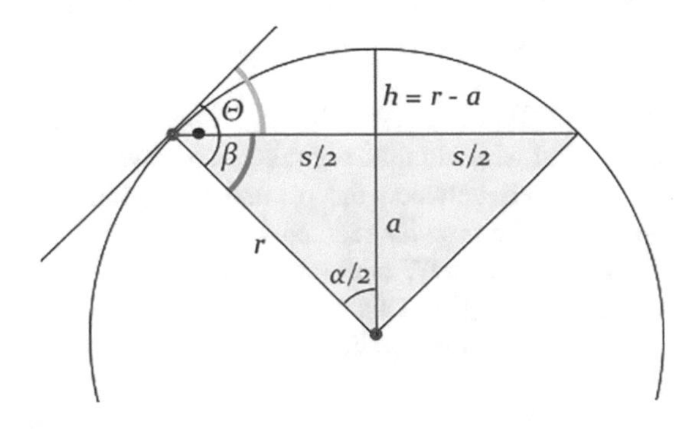

Fig. 5.20 Geometric principles for calculating the optical waveguide height as a function of the contact angle (Θ) and the distance between the conditioning lines (s)

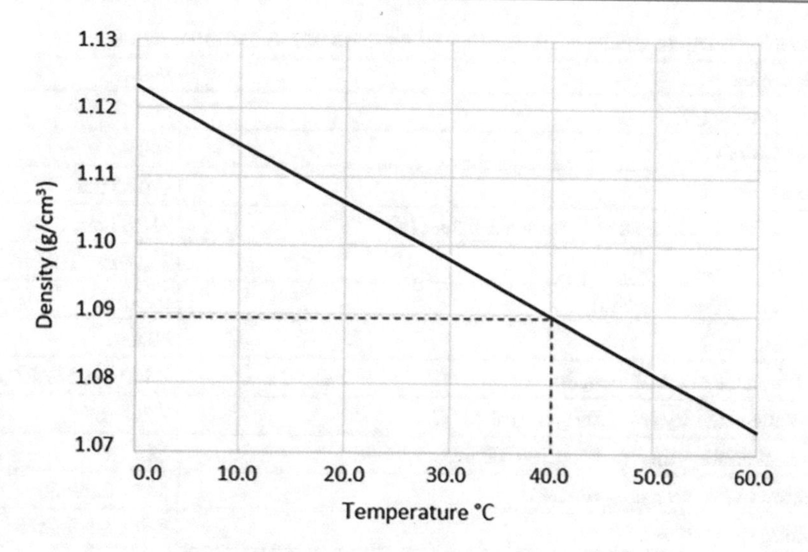

Fig. 5.21 Density versus temperature of J+S 390119 material

Via r and sin (β) a can be determined with the help of Eq. 5.6

$$a = r \cdot \sin(\beta) \tag{5.6}$$

The theoretical maximum height (h) can be calculated with Eq. 5.7

$$h = r - a \tag{5.7}$$

The volume (V) of the material can be determined by subtracting the volume of the circle segment at the angle (α) to the triangle volume (red), as shown in Fig. 5.20. Theoretical volume calculation is shown in Eq. 5.8.

$$V = L \cdot \left(\pi \cdot r^2 \cdot \frac{\alpha}{360^\circ} - \frac{s \cdot a}{2}\right) \tag{5.8}$$

The total mass (m) of the volume can be determined via Eq. 5.9 using the density (ρ) of the material. The material density value is shown in Fig. 5.21.

$$m = V \cdot \rho \tag{5.9}$$

The material printing time t_M can be determined according to Eq. 5.10

$$t_M = \frac{m}{q_M} \tag{5.10}$$

where q_M is the mass flow, i.e. flowing mass per time.

The printing speed (v_V) to be set on the machine can be calculated via the length of the optical waveguide L and the material application time according to Eq. 5.11.

$$v_V = \frac{L}{t_M} \tag{5.11}$$

Table 5.3 Theoretical calculation of AJP printing speed of J+S 390119 material

Description	Value
Contact angle (Θ)	61.4°
Printed width (s)	300 μm
Circle radius (r)	170.85 μm
Length from circle center to waveguide datum (a)	81.78 μm
POW height (h)	89.06 μm
Aspect ratio (height:width)	0.297
POW length	60 mm
POW volume at 60 mm length	1,140,818,562.87 μm^3
POW mass density ($\rho = 1.09$ g/cm^3) at 40 °C	1.24 mg
Material printing time (t_M) at $q_M = 0.02$ mg/s	62.17 s
Printing speed (vV) at L = 60 mm	57.91 mm/min
Printing speed for ten layers	579.1 mm/min

An example of volume and printing speed calculation for J+S 390119 material is shown in Table 5.3

5.5 Strategies in Aerosol Jet Printed Polymer Optical Waveguide

Two printing strategies have been studied to fabricate the POW via AJP, namely single-layer printing and multi-layer printing. In the OPTAVER process, UV light is used to cure the POW material and place it next to the AJP print head. Figure 5.22 shows the orientation of the platform for 2D and 3D substrates in the OPTAVER research work.

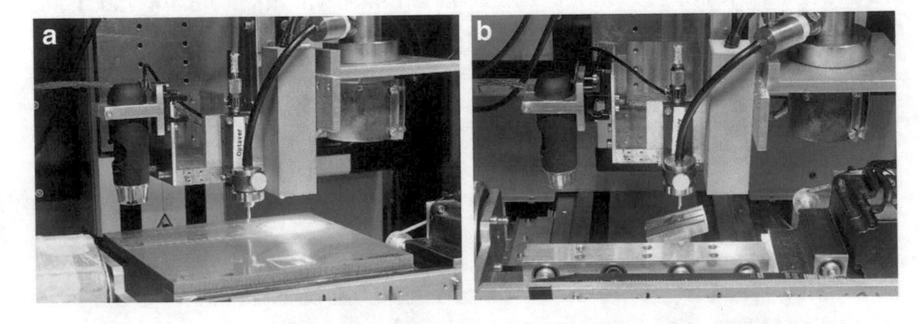

Fig. 5.22 Substrates orientation for printing POW onto substrates **a** 2D and **b** 3D

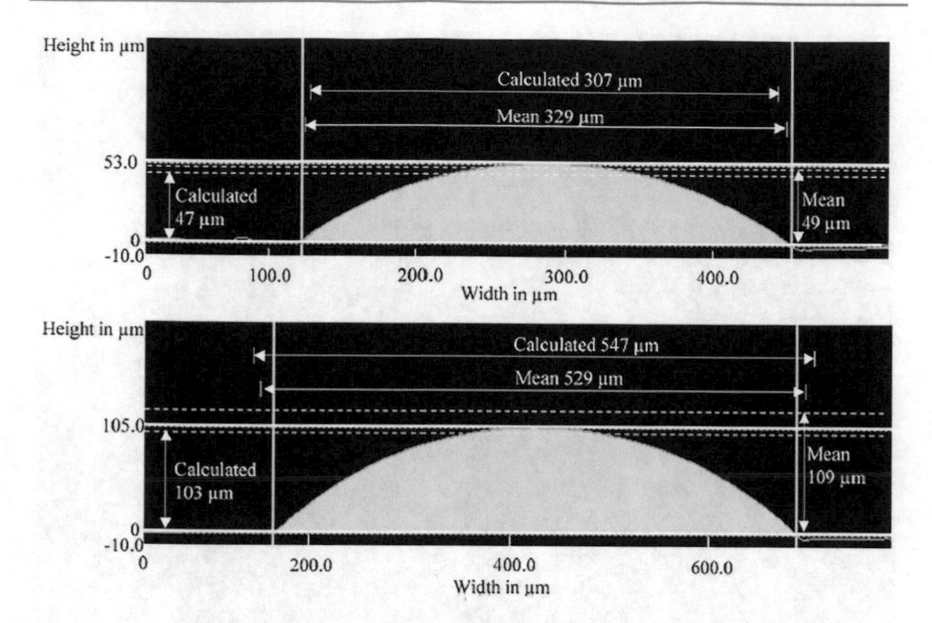

Fig. 5.23 Mean height and width of the POW with single-layer printing strategy

5.5.1 Single-layer Printing: Immediate UV Curing Process

Here, the POW material is printed at a low velocity. Next, printed materials are directly exposed to UV curing after printing. It is essential that the printed material sufficiently fills the area between the conditioning lines to ensure the desired dimension of the printed POW can be obtained. Our previous work shows that this strategy can reach height and width at 49 µm and 329 µm, respectively. It can be seen in Fig. 5.23, the POW aspect ratio (height: width) of 49:329 and 109:529 was obtained [27].

5.5.2 Multi-layer Printing: Immediate UV Curing Process

POW layers are printed on top of the other layer, followed by intermediate UV curing process as shown in Fig. 5.24. It can be seen that each layer formation can be distinguished and significant surface roughness is observed. This pyramidal structure and a massive overspray on the side with significant roughness lead to the negative quality of the POW. Further analysis of the cross-sectional view of the printed POW indicates that the line interlayers are observed. Moreover, multiple UV curing also significantly changes the color of the printed POW material. The early printed layer is darker than the outer printed layer.

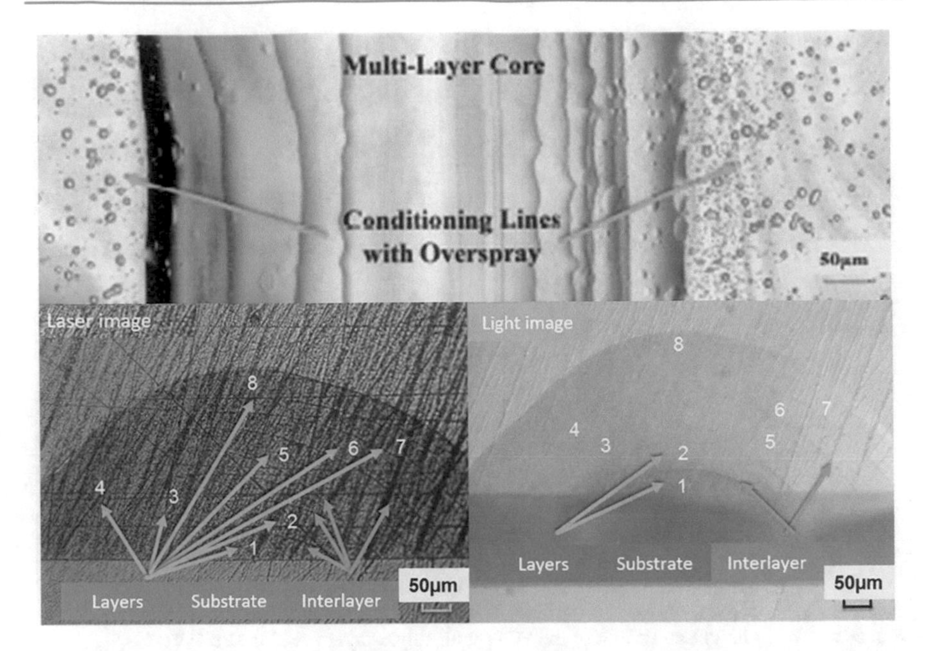

Fig. 5.24 Microscope image of AJ printed POW with intermediate curing process of each layer **a** top view **b** cross-sectional laser image **c** cross-sectional light image

5.5.3 Multi-layer Printing: UV Curing After the Complete Layer Printing

This printing strategy is the same as previously mentioned multi-layer printing, except the UV curing process is taking place once all the layers have been successfully printed. Based on our research work, this strategy can produce quality and desired POW aspect ratio. Figure 5.25 shows the dimension analysis of the printed POW by using this strategy.

5.6 Qualification and Characterization of the Printed Polymer Optical Waveguide

Based on the experimental work, higher quality of AJ printed POW can be obtained by printing multi-layer of six and ten. The lower number of printed layers with lower printing speed leads to the occurrence of defect patterns which will be discussed later in this chapter. Figure 5.26 shows the results of several layers of AJ printed POW on PMMA substrates with gap distance and width of conditioning line of 300 μm and 500 μm, respectively. AJP process parameter was set as follows: ShG = 80 sccm, AtmG = 800 sccm, ExhG = 740 sccm), nozzle diameter = 300 μm, sleeve temperature = 45 °C and preheating time = 30 min.

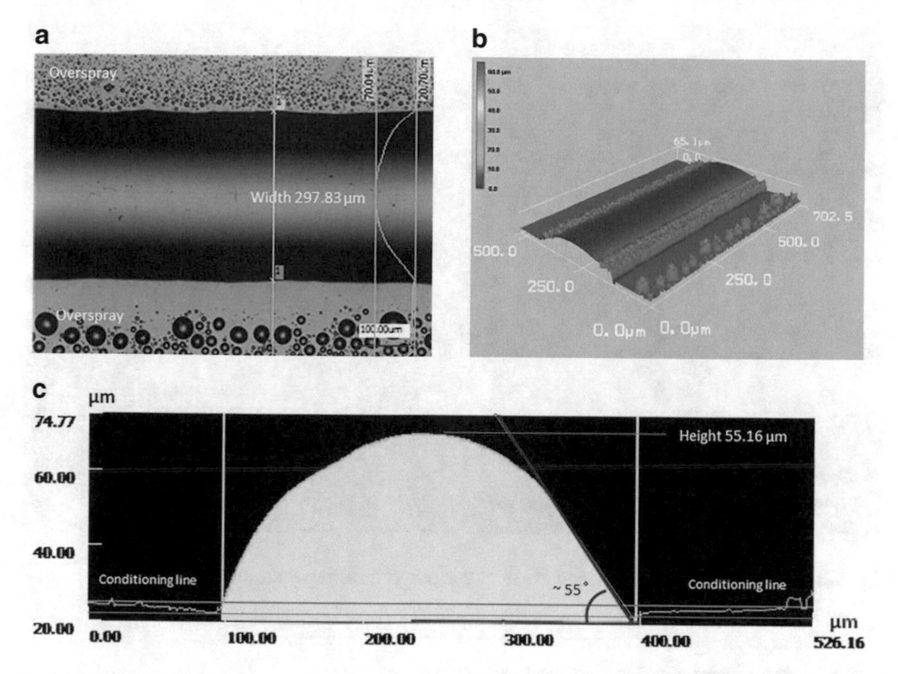

Fig. 5.25 Microscope image of AJ printed POW with multi-layer without intermediate curing

It is clearly seen that the excess and aggregated material occurs when printing with one to three layers. Printing with six and ten layers shows better results as a nearly symmetrical geometrical structure is obtained. Droplet formation at the start and end of the conditioning lines is observed. This is due to the open structure at both ends of the conditioning lines, which lead printed material to flow out to the area with higher surface energy.

Printed POWs are characterized thoroughly to determine the quality of the POW based on the required geometrical, mechanical and optical properties. For geometrical characterization, printed POWs undergo morphology inspections using a digital light microscope, laser microscope and scanning electron microscope (SEM). The microscope inspection analysis is used to define the desired geometry toward the quality of the printed POWs.

Moreover, the defects that occur during the process are analyzed to understand the critical parameter in the fabrication process. For testing the mechanical performance of the printed POW, hot-pin-pull tests are conducted. This test is to ensure their robustness to survive in harsh conditions. Finally, the printed POWs are tested for their optical quality for light data transmission. These tests determine the optical losses, energy distribution (near-field) and waveguides performance at high frequencies.

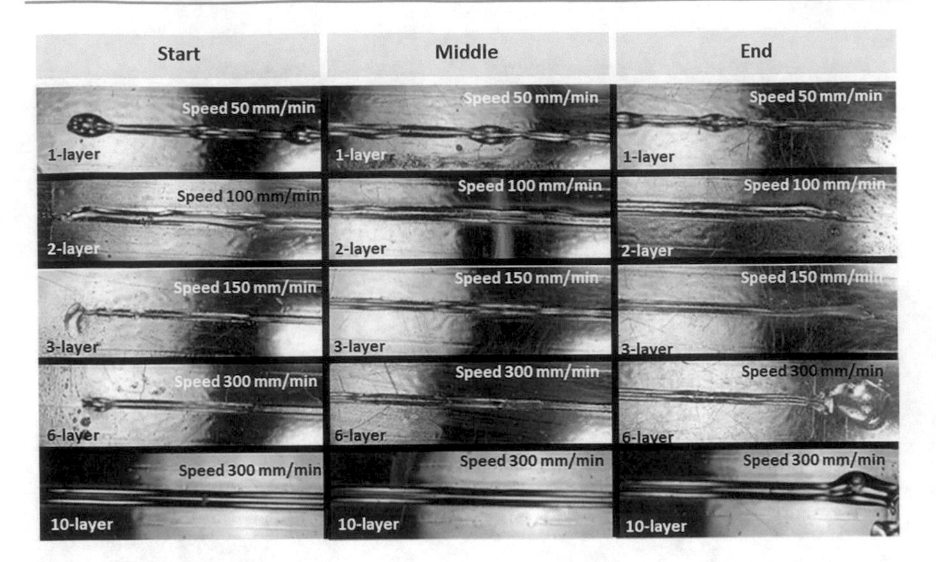

Fig. 5.26 Camera images of printed POW at various printing speeds and layers

5.6.1 Geometrical Properties

The printed POWs quality criteria are determined by measuring the width, height, waviness, contact angle and aspect ratio parameters. 3D laser microscope VK-9700 from Keyence is used to capture images of printed POWs with high magnification and measurement analysis, as depicted in Fig. 5.27. With the VK analyzer software, dimension profiles of the printed POWs are characterized. The POWs width is determined as the distance between two conditioning lines. The height is measured from the PMMA substrate surface up to the tip of the POWs.

5.6.2 Profile Dimension

The minimum cross-sectional requirement for multimode transmission is 50 μm × 50 μm. In this work, the contact angle at the range of ~30° to ~60° was successfully achieved. Printed POW should not exceed the contact angle of ~60° to keep the material in between conditioning lines. As shown in Fig. 5.27, the printed POW width and height are measured approximately at 328 μm and 66.3 μm, respectively. Therefore, the aspect ratio (height: width) at 0.2 was achieved.

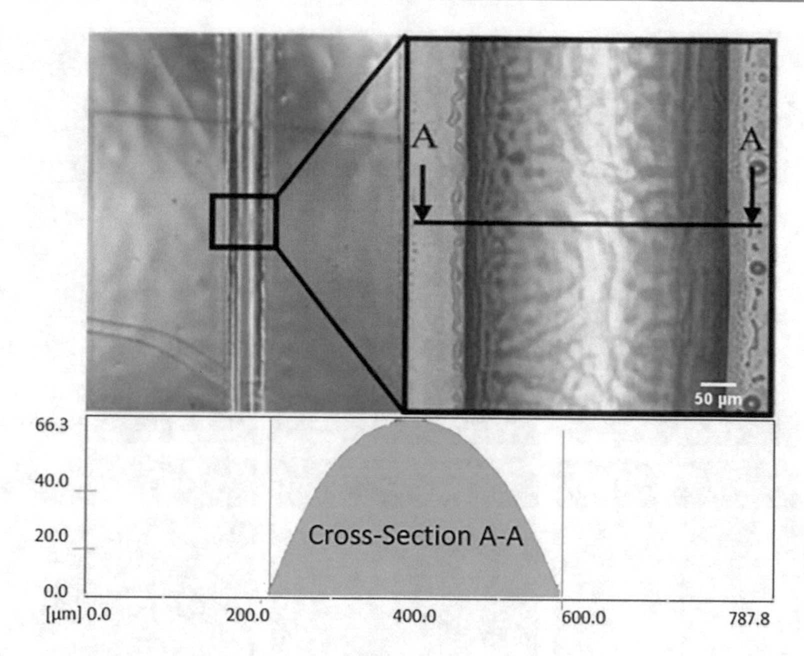

Fig. 5.27 Microscope examination of the printed POWs [22]

5.6.3 Effect of the Surface Roughness and Waviness

Another vital criterion in POW characterization is surface roughness (Ra). Surface roughness significantly affects POW quality due to the optical loss [28, 29]. Studies have shown that the desired surface roughness of optical waveguides should be below 30 nm and roughness values more than 50 nm are considered rough [28]. A Ra of 30 nm can be reached for freshly printed POW samples without impurities. The roughness and cleanness of the printed POW significantly affected the optical losses value [30, 31]. Analysis of the sample in Fig. 5.27 shows that the Ra value was measured at 0.06 μm [22]. Even though the value is slightly high, this work remains significant as the printing process is made not in the cleanroom environment.

5.7 Morphology Analysis

Further examination of the microscope images can determine defects that occur on the printed POWs. Overspray, edge waviness, bubble and impurities are among the defects seen on the printed POW samples, as shown in Fig. 5.28.

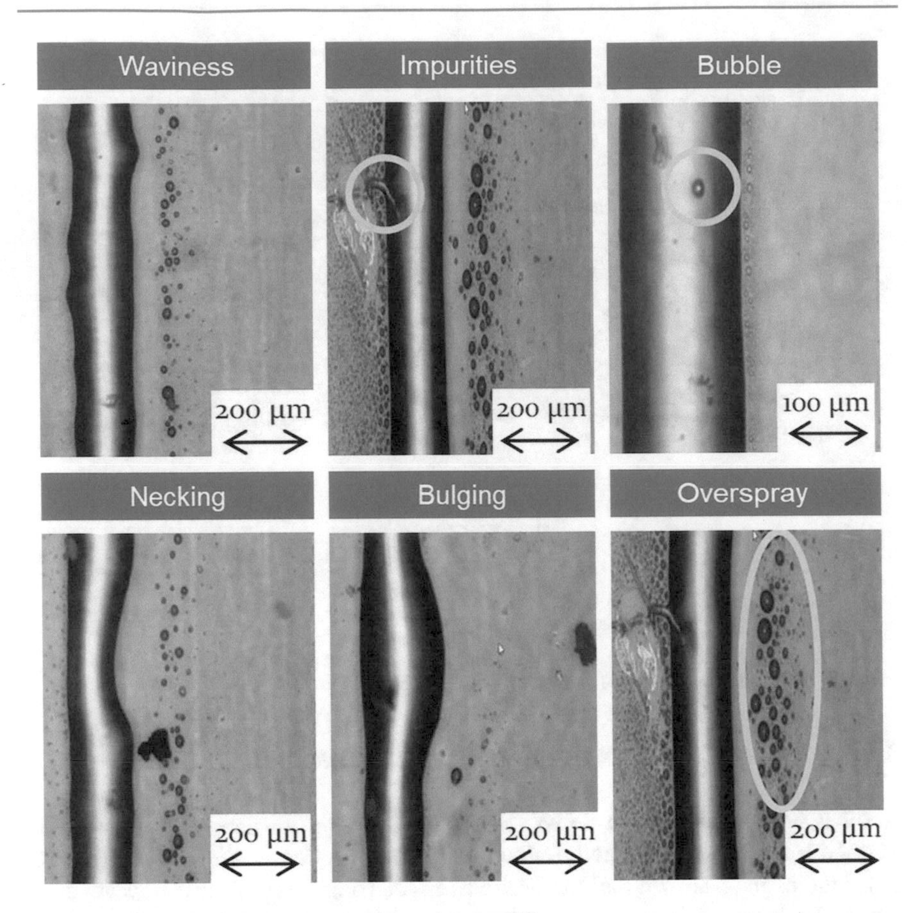

Fig. 5.28 Type of defects occurred on the printed POW

5.7.1 Overspray

Overspray is an inevitable issue in the AJP process. The overspray is caused by the impurities inside the nozzle wall, leading to an imbalance of the aerodynamic droplet flow, printing material and the process parameters selection. Studies have shown that the overspray occurrence can be reduced using solvent-rich material [26] and by optimizing the process parameters [32]. A suitable parameter setting was successfully achieved by knowing the process window range based on the desired printing strategies. To overcome the overspray issue is by ensuring the nozzle cleanliness prior to printing. Despite that, not every overspray occurred, causing a direct threat to the printed POW performance. The overspray deposited on top or beyond the conditioning lines and not directly in contact with the printed core POW does not influence the POW quality. This is because light can propagate inside the printed POWs main structure without having any disruption.

5.7.2 Waviness

Waviness of the printed POWs also plays a significant role in obtaining the desired quality. It shows a similar effect as the rough surfaces, which can cause optical loss, possibly in a more significant factor. Through our studies, the waviness of the printed POWs occurred due to the imperfect conditioning lines (cracks). Imperfect conditioning lines can cause the printed core material to flow through the cracks (capillary effect) and reduce homogeneity of the profile.

Hence, defect-free conditioning line edges are essential in ensuring high-quality printed POW can be obtained. Besides, POW material also needs to be deposited with the right volume. A higher deposition volume than anticipated has caused the conditioning lines being unable to contain the POW material. The deposited material then overflowed on top of the conditioning lines and lost its function as barrier.

5.7.3 Impurities

Impurities defect also can be observed in the printed POWs. Since the process was conducted in a non-controlled environment system, foreign particle most probably adheres to the substrate during the printing or curing process. Hence, to overcome that issue, the substrates were blown with nitrogen gas before printing the POWs. It can be concluded that keeping the entire printing system and working station clean before, during and after printing is crucial in producing quality POWs.

5.7.4 Bulging

Bulging formation is another defect that occurred on the printed POWs. It occurs at the point where the width is more significant compared to the symmetrical profile of the printed POWs. Several factors caused bulging in this work, 1) improper preparation (pre-defect) of the conditioning lines. The defect on the conditioning lines (crack, discontinuous, or gap) plays a vital role in the bulging formation. During POW printing, deposited material tends to accumulate at these points. 2) An excessive amount of the deposited material also contributes to the bulging formation. As mentioned previously, depositing the correct volume is among the critical factors in producing quality POWs. 3) The longer the material stays in the liquid phase after deposition, the higher chances for the bulging to occur.

This is due to the low surface energy of the substrates, which hardly holds deposited material in the position for the required duration of time. Thus, a short time interval between the printing and curing process is desired.

5.7.5 Bubble or Air Pocket

This defect is due to the air entrapped during the material coalesce to the substrates. Delaying in curing the material after the deposition is one of the solutions as the entrapped air will have more time to escape and thus making the dynamic of the material more stable. However, this will lead to the bulging issue mentioned above. One of the options to overcome this issue is using higher solvent material, but it can lead to shrinkage issue of the printed POW after curing process.

5.8 Mechanical and Optical Properties of the Printed Polymer Optical Waveguide

To identify the mechanical properties of the waveguides, hot-pin-pull and shear forces testing are conducted. These data are important to quantify the robustness quality of the printed POW at a certain force condition.

5.8.1 Hot-Pin-Pull Test

For adhesion measurement of the J+S 390119 (core material) on the PMMA substrate, DAGE 4000PLUS multifunction tester is used. The measurement starts with printed POWs is clamped on the equipment platform. Next, the test pin is inserted into the load at specified height. A specific amount of adhesive is applied to the printed J+S 390119. The tip is then lowered to make contact with the adhesive. The tip is then heated up to 175 °C to melt the adhesive and bond to the printed J+S 390119 material. To ensure the test tip and printed J+S 390119 material are cured perfectly, the bonding area is blown with compressed air. During the

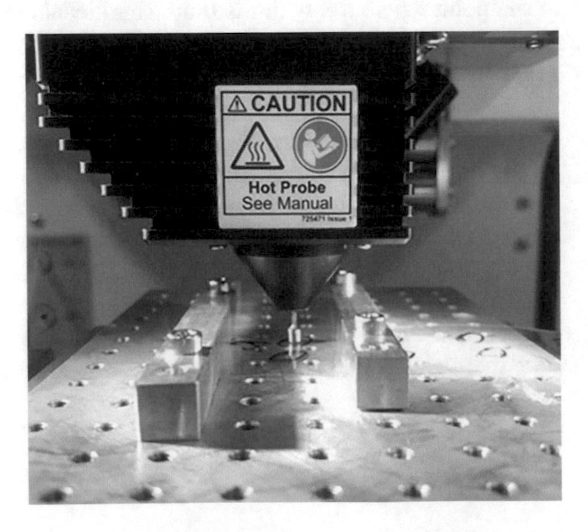

Fig. 5.29 Hot-pin-pull test of the printed POWs

test, the test tip is moved vertically upwards at a speed of 0.5 mm/s and forced displacement data are generated. Figure 5.29 depicts the hot-pin-pull test conducted on the printed J+S 390119.

The analysis of five printed J+S 390119 materials in Table 5.4 shows that average bond strength of 1.35 MPa is achieved. One mega Pascal corresponds to one Newton per square millimeter. In this case, the printed POW dimension is 140 mm long and 0.3 mm in width. Thus, a vertically upwards force of 56.7 N is theoretically required for the complete detachment of a polymer optical waveguide with 42 mm^2.

5.8.2 Shear Force Test

Shear force test is conducted using XYZTEC CONDOR 150-3 equipment with a chisel size of 0.6 mm. Shear value is calculated by fractioning the applied parallel force (F_{max}) to the sectional area (S_o) as shown in Eq. 5.12.

$$\tau_{aB} = \frac{Fmax}{s_0} \tag{5.12}$$

$$S_O = \text{Chissel size} \times \text{POW width} \tag{5.13}$$

The sectional shear area is measured at 0.6 mm × 0.35 mm, whereas 0.6 mm is the chisel size and 0.35 mm is the conditioning lines gap, as shown in Fig. 5.30. It can be seen in Table 5.5, some printed POW samples in groups 10L and 6L that have not gone through the climate chamber test tend to be in plastic deformation. On the other hand, most printed samples are prone to have brittle properties after the climate chamber test.

5.8.3 Lifetime Performance Test

In order to determine qualification of the printed POWs in a specific application such as the aerospace sector, climate testing is conducted. The test on the samples is based on four standards as follows:

Table 5.4 Hot-Pin-Pull test results of five printed J+S 390119

Printed sample	Area [mm^2]	Maximum force [N]	Adhesive strength [MPa]
JS1	3.9953	3.1521	0.7890
JS2	3.8349	7.7366	2.0174
JS3	2.3275	3.5772	1.5369
JS4	2.8685	3.5236	1.2284
JS5	2.4357	2.9465	1.2097
Average	**3.0924**	**4.1872**	**1.3540**

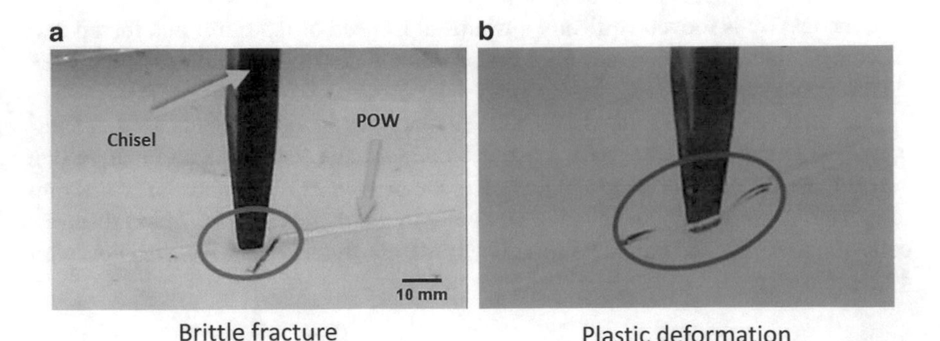

Fig. 5.30 Fracture types of the printed POWs samples **a** brittle fracture **b** plastic deformation

Table 5.5 Shear test results of the printed POWs after the climate chamber test

POWs group designation	Max. path [μm]	Min. displacement [μm]	Max. force [cN]	Min. force [cN]	Total samples tested	Samples in plastic deformation
10L	109	104	200	162	4	1
6L	104	76	145	110	4	2
10L HT	240	208	145	130	4	0
6L HT	176	80	220	120	3	0
10L FL	512	508	185	160	3	0
6L FL	174	128	255	150	3	0
10L Z	288	275	375	280	2	0
6L Z	192	144	240	155	2	0

*L = printed layer, HT = high temperature, FL = air-humidity, Z = temperature cycles

- Standard 1: DIN EN 61300-2-18: High-temperature endurance test, 96 h at 60 °C
- Standard 2: DIN EN 61300-2-19: Humid air, 96 h at 93% relative humidity and a temperature of 40 °C
- Standard 3: DIN EN 61300-2-22: (Category C) Temperature cycling, five cycles in the temperature range from -10 °C to 60 °C
- Standard 4: DIN EN 61300-2-22: (Category E) Thermal cycling, 12 cycles in the temperature range from −40 °C to 85 °C

For samples tested under Standard 3 and Standard 4 conditions, the POWs samples are heated at maximum or minimum temperature for an hour. The transition chamber from cold to hot or vice versa is set at 1°/minute (± 0.2 °K) tolerance. In total, seventeen POW samples were tested under conditions Standard 1, Standard

Table 5.6 Data of the shrink percentage of the printed POW core after climate testing

Sample no.	Height (μm)		Shrinkage factor (%)
	Before test	**After test**	
1	54.52	51.09	−6.29
2	30.64	29.87	−2.51
3	34.96	35.38	1.20
4	38.23	34.69	−9.26
5	51.48	48.41	−5.96
6	44.05	43.64	−0.93
7	31.38	30.43	−3.03
8	48.41	47.46	−1.96
9	39.61	38.64	−2.45
10	31.87	30.98	−2.79
11	47.05	44.46	−5.50
12	35.6	32.56	−8.54

2 and Standard 3. The result shows no significant changes in the sample's geometry, surface roughness and waviness. However, some foreign particles that can be seen probably come from the residual particles inside the chamber test.

Thirteen printed POWs were tested under condition Standard 4. Each sample consists of ten printed layers and is cured immediately after the last printed layer. Upon completion of the climate chamber test, the samples are reexamined for their geometrical characterization as shown in Table 5.6.

Overall, it can be seen that the height values are decreased on average shrinkage by about 4%. Sample 3 shows an increase in the height of 0.426 μm. However, this value is within the measurement tolerance. Sample 4 shows a maximum shrinkage amounts to 3.54 μm for Sample 4 (9.3% decreased). The shrinkage factor is due to the residual air being escaped from the printed POWs and the outgassing solvents that are still present in the waveguide or the post-crosslinking of the polymer.

5.8.4 Optical Quality

After the geometrical and mechanical analysis of the printed waveguides, the optical properties need to be evaluated. For that reason, this sub-chapter includes measurements of the optical losses, the energy distribution at the waveguides end facet and the performance at high frequencies, which were first described in [33]. Since the core material that is used is not a typical waveguide material, the first step is to investigate the transmittance. For this purpose, the varnish is applied between two glasses, resulting in a film thickness of about 1 mm. This is followed by testing with a spectrometer according to ISO 15368. The result is depicted in Fig. 5.31.

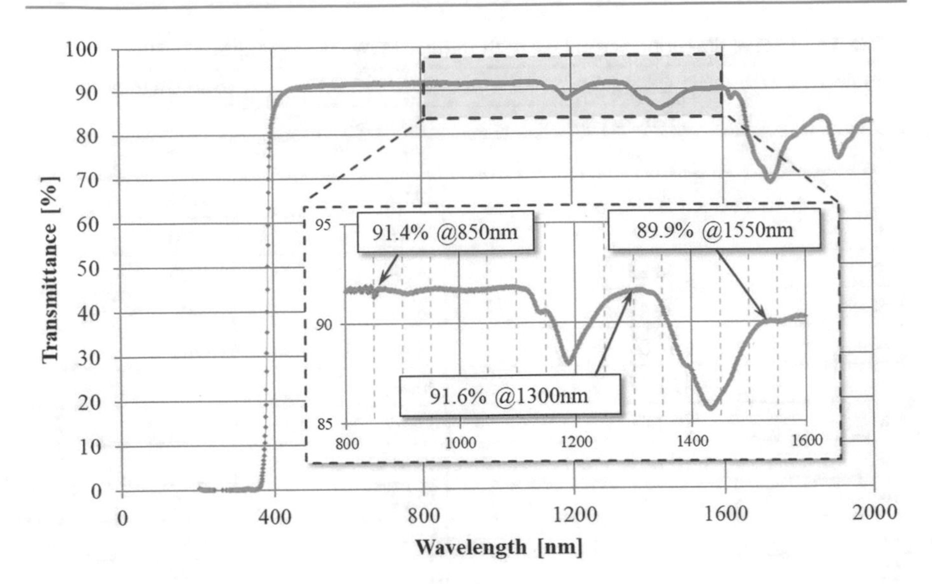

Fig. 5.31 Transmittance of the waveguides core material J+S 390119 related to the wavelength

The material shows a consistently high transmittance (> 91%) in the range between 400 and 1100 nm, which includes the first optical window for Datacom at 850 nm. After a small local minimum, values above 91% are again measured at the second optical window (1300 nm). Values close to 90% are also found for the third important wavelength window at 1550 nm. Only after that, at about 1600 nm, there is a significant decrease of the transmittance. Thus, the material can be attested a high suitability for waveguide fabrication.

The next step is to investigate the optical attenuation of the printed waveguides. Ten different samples are measured for this purpose. First, the end faces of the waveguides have to be prepared for the measurement. Wolfer et al. [34] have developed a suitable method for this. By means of cleaving, extremely smooth end facets are possible without polishing. Thus, reflections at the end face can be neglected.

A fiber-coupled, thermoelectrically stabilized Fabry Perot laser is used as the source for the attenuation measurement. The measurement was performed at 850 nm. A 10 μm (NA 0.1) fiber is used to launch the waveguide sample to minimize coupling losses. At the end of the POW, a 400 μm (NA 0.39) diameter detection fiber connects the sample to an integrating sphere. The alignment of the fibers to the POW is done actively, by manipulating the input and output fibers while measuring the optical power. This allows for the optimal coupling. The measured optical power is now compared to the reference value, which results from the direct butt-coupling of the launching and detecting fiber. As a consequence, the result also includes the reflection losses at the air gap between sample and detection fiber, which, however, can be neglected compared to the attenuation of the waveguides itself.

Table 5.7 shows the results of the attenuation measurement. While the best value of 0.20 dB/cm is a good result for any Datacom application, the measurement reveals a high deviation. This shows how difficult it is to control the AJP process. Mainly material inclusions and an unsteady waviness at the interface to the conditioning lines are the reasons for higher attenuation values. For a reasonable use in industry applications, the yield has to be improved, e.g., by controlling the material throughput directly at the printing nozzle. Nevertheless, the measurement shows the high potential of the additively manufactured waveguides.

Another special characteristic of the printed waveguides is their unique cross section. For the connection to peripheral components or other waveguides, the large width combined with lower height could cause a problem. For this reason, the energy distribution at the output of the printed waveguide needs to be investigated in more detail. A calibrated CCD camera with a magnifying lens is used for this purpose. This lens is focused on the end face of an illuminated waveguide with a length of 12.8 cm. Thus, the energy distribution under different coupling conditions can be investigated. For that, a single-mode fiber (SMF) with a mode field diameter of 5 μm and a 50 μm multimode fiber (MMF) are used for launching, each at three different positions (center, ± 50 μm horizontal misalignment). Figure 5.32 and Fig. 5.33 show the results of the measurement, which reveals that the energy is not uniformly distributed within the waveguide core. In fact, 80% of the overall energy is guided in the center of the waveguide, which is calculated by the evaluation software of the camera and illustrated with the dashed line. Furthermore, there is no significant difference whether the POW is launched by a SMF or a MMF. This indicates that the mode equilibrium in the waveguide is already reached after a short distance. In addition, the comparison between the overall output power shows no significant difference. With the MMF 91% total

Table 5.7 Measured attenuations of ten different aerosol jet printed waveguides

Sample number	Length [mm]	Overall attenuation [dB]	Attenuation per length [dB/cm]
1	60	3.36	0.56
2	62	3.53	0.57
3	62	3.91	0.63
4	63	5.42	0.86
5	51	4.79	0.94
6	63	3.47	0.55
7	62	1.92	0.31
8	63	3.97	0.63
9	63	1.58	0.25
10	63	1.26	0.20
	Min. /Average/Max 0.20/0.55/0.94		

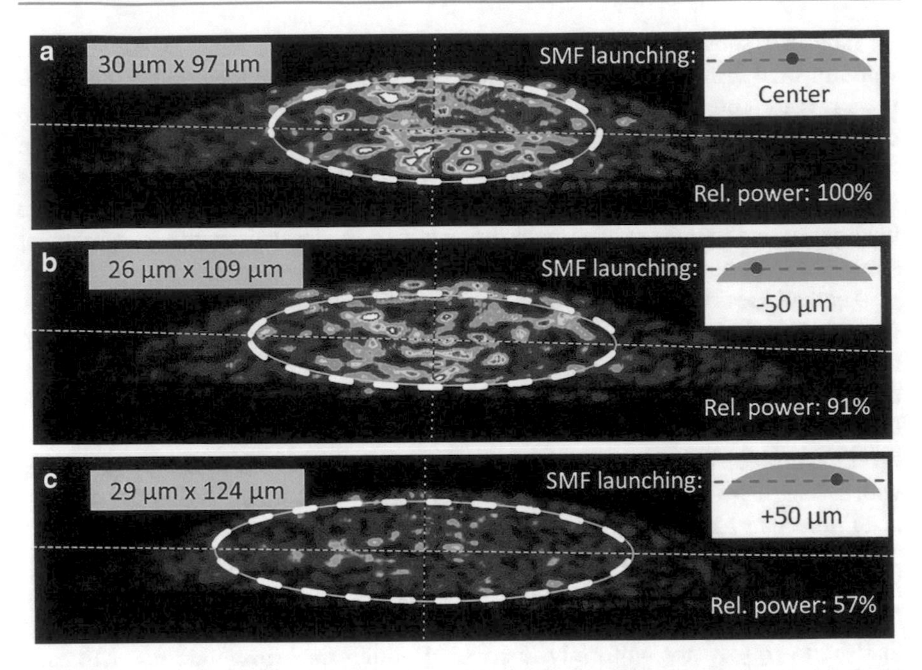

Fig. 5.32 Energy distribution at the end facet of an aerosol jet printed waveguide, launched with a single-mode fiber (SMF)

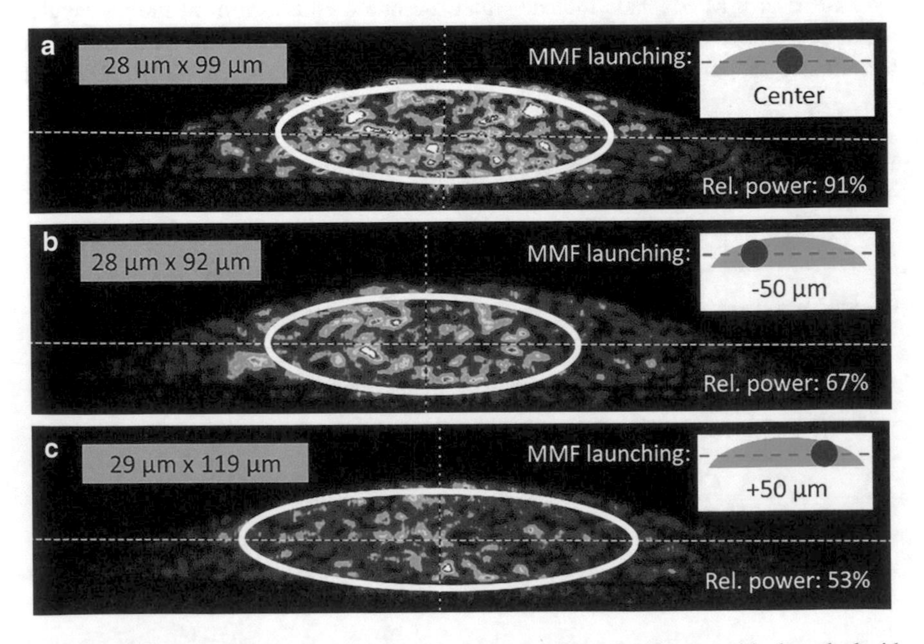

Fig. 5.33 Energy distribution at the end facet of an aerosol jet printed waveguide, launched with a multimode fiber (MMF)

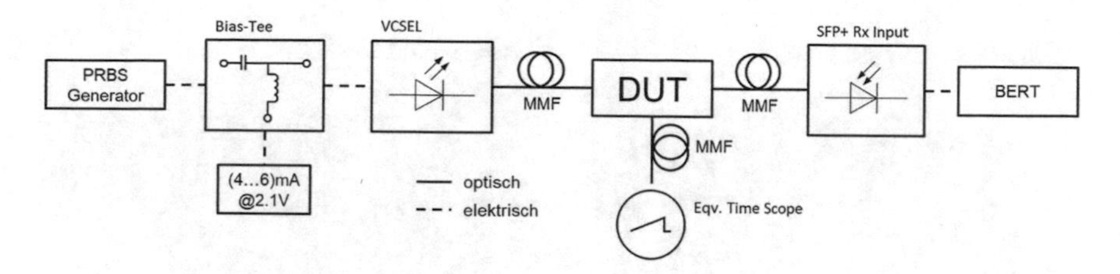

Fig. 5.34 Measurement setup for BER test and eye diagram

power are achieved, compared to the SMF (100% reference). Moreover, the fact that the light is mainly guided in the center of the waveguide is independent of the horizontal position of the input fiber. Nevertheless, the total output power is reduced by a misalignment of the launching fiber. Nevertheless, because the light is primarily guided in the center, low losses are expected when connecting to other multimode components in the diameter range between 50 μm and 100 μm.

The most important parameter to evaluate the suitability of aerosol jet printed waveguides for short-range communication is the possible data transmission rate. For this purpose, the eye diagram and the bit error rate are measured. Figure 5.34 shows a schematic diagram of the measurement setup used for this purpose. A pseudorandom binary sequence (PRBS) generator provides a signal at 12.5 Gbit/s, which is used for the direct modulation of a vertical emitting laser (VCSEL) via a bias-tee. The signal is coupled into the POW via a 50 μm MMF. At the output of the sample, a 50 μm fiber is then used to detect the signal and pass it either directly to the optical input of the equivalent time oscilloscope or to an SFP + module, which opto-electrically converts the signal and passes it to the bit error rate tester (BERT).

The results for the eye diagram, which are depicted in Fig. 5.35, show a satisfying opening for an 82 mm long aerosol jet printed waveguide, compared to the back-to-back measurement. No effects of dispersion or nonlinearities could be seen in the eye diagrams. Furthermore, the measured bit error rates in Fig. 5.36 allow for error-free data transmission, especially considering that no forward error correction is used in the measurement. In summary, the aerosol jet printed waveguides are suitable for short-range data transmission at high frequencies.

5.9 Modeling and Simulation Analysis of the Aerosol Jet Printhead

Computational fluid dynamics (CFD) is a numerical tool that helps to deliver quantitative prediction of the aerodynamics interaction of the AJP process. In order to optimize the outcome of the printed POW, the AJP-printhead is modeled and CFD simulation analysis of the velocity flows inside the nozzle is studied.

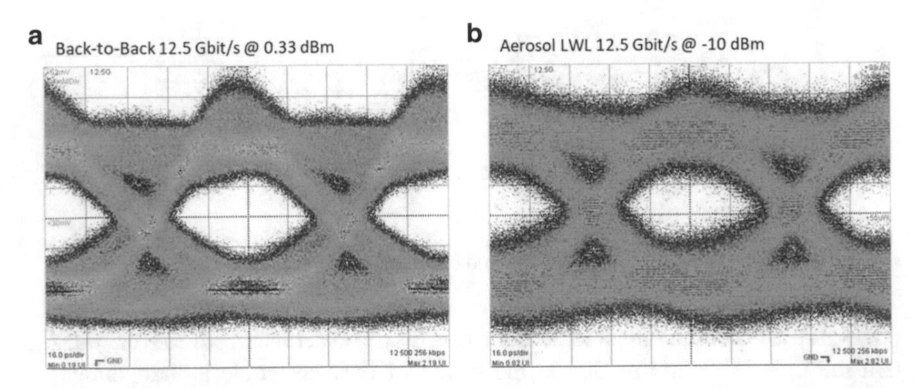

Fig. 5.35 Eye diagram **a** of a back-to-back transmission and **b** of transmission via a 82 mm long aerosol jet printed waveguide both at 12.5 Gbit/s

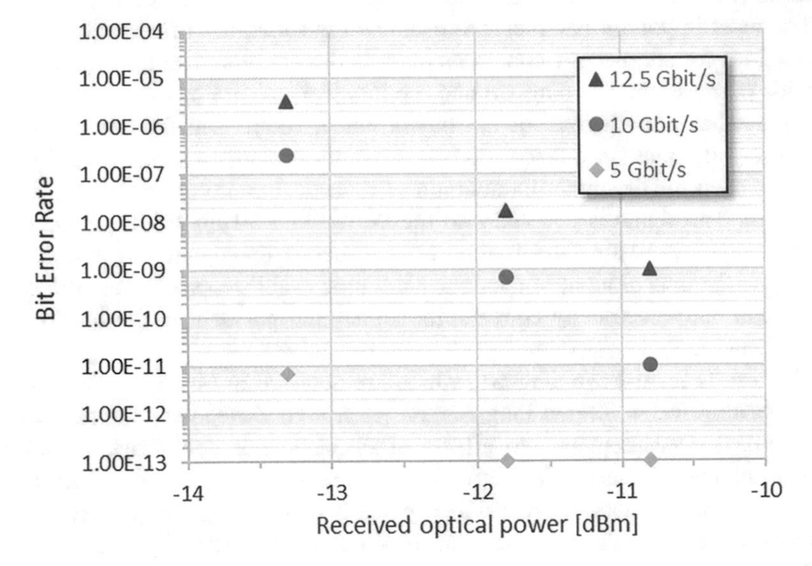

Fig. 5.36 Bit error rate test of a 82 mm long aerosol jet printed waveguide at three different data rates and received powers

5.9.1 Modeling and Operational Principles

AJP-printhead is an assembly of five components. All the parts secured with the O-rings can function without any foreign air interference during the printing process. Figure 5.37 shows the AJP-printhead parts and assembly cross section of the CAD model in SpaceClaim software tool used in this work.

The operational principles of the AJP-printhead during the deposition process start with the aerosol stream that brings along aerosolized/nebulized material entering the aerosol inlet. The stream flows then collimate with the sheath gas.

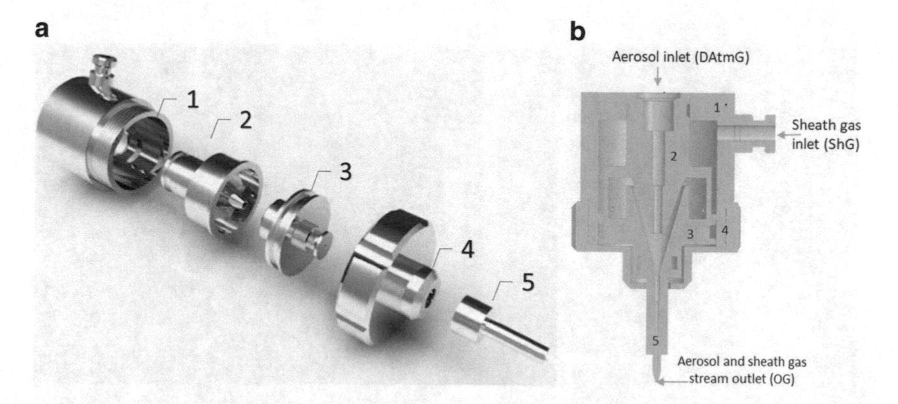

Fig. 5.37 AJP-printhead model **a** disassemble parts **b** cross-sectional model

Concurrently, the sheath gas entered and filled the upper plenum chamber (UPC). It then flows through eight holes with the approximate diameter of 1 mm to enter the lower plenum chamber (LPC) and form a cylindrically symmetric distribution of the sheath gas flow. Next, an annular jet forms an inner aerosol jet which is surrounded with sheath gas is channeled to the extended nozzle before exiting the nozzle orifice. The sheath gas not only functions to focus the annular flow but also prevents the nozzle from clogging. The gas exiting the nozzle orifice is steady as the sheath gas remains to curtain the flow up to certain distances (~ 5 to 10 mm).

Boundary conditions are set in the CFX-Pre software. The inlet gas is set to be sheath gas, atomizer gas and outlet at the nozzle tip. The sheath gas velocities are set at 0.085 m/s (80 sccm) and aerosol gas at 0.0097 m/s (60 sccm). The outlet relative static pressure and ambient pressure are set to zero Pascal, and nitrogen gas is set as fluid. Due to the low Reynolds number of 747 within the nozzle tip, the SST model, including the gamma–theta model, is selected. The solver settings are set to high resolution to generate a high-quality result, the execution is set to double-precision, and the target residual is set to 10^{-5}. Differential equations are performed in the CFX Solver Manager, and a total mesh node number of 450,000 are selected.

5.9.2 Simulation Results Analysis

The numerical analysis of the simulation is performed with CFD. It can be seen in Fig. 5.38a velocities as low as 2 m/s are presented inside the printhead system. The velocity increased dramatically at the nozzle output, which maximum velocity is reached up to 22.938 m/s. Besides, the simulation shows uniform gas distribution occurs across the entire nozzle after the sheath gas entering the system. The velocity contour at the nozzle tip is shown in Fig. 5.38b. It can be concluded that sheath gas flow is evenly distributed inside the printhead and completely envelope the aerosol gas, which prevents the print head system from clogging. Moreover, DAtmG velocity as low as 97 m/s and up to 22.938 m/s is successfully simulated.

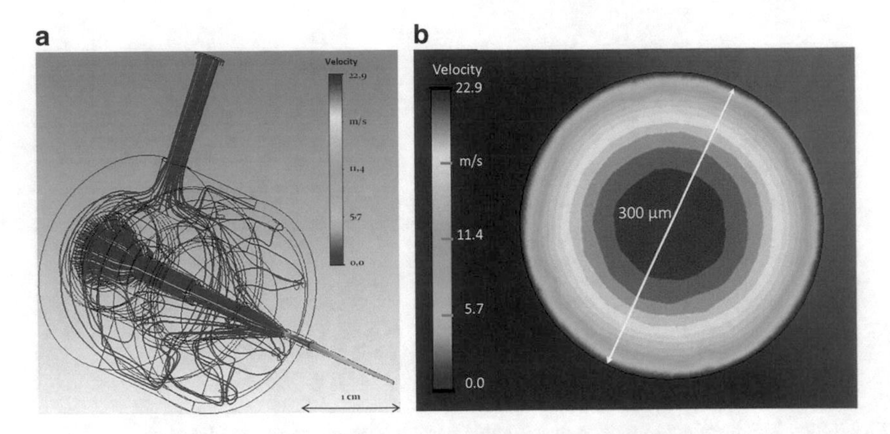

Fig. 5.38 Simulation of the aerosol gas velocity inside the AJP-printhead

Table 5.8 Requirement criteria of the AJ printed POW in this work

Description(s)	Recommended requirements in this work
Substrates material	PMMA
Conditioning distance/gap	~300–350 µm
Printed core material	J+S 390119
Printed core geometry	Semi-circle cross section
Width	~350 µm as per conditioning lines distance
Height	~half of the width (minimum)
Aspect ratio (H:W)	~1:5
Contact angle	~45° to 60 ^{O}
Printing strategies	Multi-layer printing without intermediate curing
Printing layer	6–10 layers
Curing	UV light
Atomizer gas	800 sccm
Exhaust gas	740 sccm
Sheath gas	80 sccm
Sleeve temperature	45 °C
Preheating temperature	30 min

5.10 Summary

This chapter has successfully presented cutting-edge POW fabrication by using AJP technology. Table 5.8 shows a requirement overview of the material, process parameters and strategies in producing quality POW via AJP process technology.

Acknowledgement This research work is funded by the German Research Foundation (DFG) of the research group "Optical integrated circuit packaging for module-integrated bus systems (OPTAVER)" (FOR 1660).

References

1. IPC International, Inc.: 2015 IPC International Technology Roadmap for Electronic Interconnections: Part D – Interconnections and Substrates. Section 6 - Electrical and Optical Performance (2015)
2. Lorenz, L., Ott, L., Nieweglowski, K., Bock, K.: Influence of Temperature Cycling on Asymmetric Optical Bus Couplers. In: IEEE Electronics System-Integration Technology Conference (ESTC)
3. Eldada, L., Zhu, N., Ruberto, M.N., Levy, M., Scarmozzino, R., Osgood, R.M.: Rapid direct fabrication of active electro-optic modulators in GaAs. J. Lightwave Technol. **12**(9), 1588–1596 (1994). https://doi.org/10.1109/50.320941
4. Chandross, E.A., Pryde, C.A., Tomlinson, W.J., Weber, H.P.: Photolocking-A new technique for fabricating optical waveguide circuits. Appl. Phys. Lett. **24**(2), 72–74 (1974). https://doi.org/10.1063/1.1655099
5. Tian, L., et al.: Polymer/silica hybrid waveguide bragg grating fabricated by UV-photobleaching. IEEE Photon. Technol. Lett. **30**(7), 603–606 (2018). https://doi.org/10.1109/LPT.2018.2805843
6. Verschuren, C.A., Harmsma, P.J., Oei, Y.S., Leys, M.R., Vonk, H., Wolter, J.H.: Butt-coupling loss of 0.1dB/interface in InP/InGaAs MQW waveguide-waveguide structures grown by selective area chemical beam epitaxy. J. Cryst. Growth **188**(1–4), 288–294 (1998). https://doi.org/10.1016/S0022-0248(98)00068-2
7. Ou, H.: Reactive ion etching in silica-on-silicon planar waveguide technology. In: Proc. ECIO (2003)
8. Vu, K.T., Madden, S.J.: Reactive ion etching of tellurite and chalcogenide waveguides using hydrogen, methane, and argon. J. Vac. Sci. Technol., A: Vac., Surf. Films **29**(1), 11023 (2011). https://doi.org/10.1116/1.3528248
9. van Steenberge, G., Hendrickx, N., Bosman, E., van Erps, J., Thienpont, H., van Daele, P.: Laser ablation of parallel optical interconnect waveguides. IEEE Photon. Technol. Lett. **18**(9), 1106–1108 (2006). https://doi.org/10.1109/LPT.2006.873357
10. Nseowo Udofia, E., Zhou, W.: 3D printed optics with a soft and stretchable optical material. Addit. Manuf. **31**, 100912 (2020) https://doi.org/10.1016/j.addma.2019.100912
11. Saito, Y., Fukagata, K., Ishigure, T.: Fabrication for low-loss polymer optical waveguide with graded-index perfect circular core using the Mosquito method. In: 2016 IEEE CPMT Symposium Japan (ICSJ), Kyoto, 07-Nov-16–09-Nov-16, pp. 147–148
12. Ishigure, T.: Multimode/single-mode polymer optical waveguide circuit for high-bandwidth-density on-board interconnects. In: Optical Interconnects XV, p. 936802. California, United States, San Francisco (2015)
13. Ishigure, T., Suganuma, D., Soma, K.: Three-dimensional high density channel integration of polymer optical waveguide using the mosquito method. In: 2014 IEEE 64th Electronic Components and Technology Conference (ECTC), Orlando, FL, USA, 27-May-14–30-May-14, pp. 1042–1047
14. Takahashi, A., Ishigure, T.: Fabrication for low-loss polymer optical waveguides with 90° bending using the Mosquito method. In: IEEE CPMT Symposium Japan 2014, Kyoto, Japan, 04-Nov-14–06-Nov-14, pp. 162–165
15. Kinoshita, R., Suganuma, D., Ishigure, T.: Accurate interchannel pitch control in graded-index circular-core polymer parallel optical waveguide using the mosquito method. Opt. Express **22**(7), 8426–8437 (2014). https://doi.org/10.1364/OE.22.008426

16. Soma, K., Ishigure, T.: Fabrication of a graded-index circular-core polymer parallel optical waveguide using a microdispenser for a high-density optical printed circuit board. IEEE J. Select. Topics Quantum Electron. **19**(2), 3600310 (2013). https://doi.org/10.1109/JSTQE.2012.2227688
17. Parekh, D.P., Cormier, D., Dickey, M.D.: Chapter 8: Multifunctional printing: Incorporating electronics into 3D parts made by additive manufacturing. In: Additive manufacturing: Multifunctional Printing: Incorporating Electronics into 3D Parts Made by Additive Manufacturing, A. Bandyopadhyay and S. Bose, Eds., pp. 215–258.
18. Chappell, J., Hutt, D.A., Conway, P.P.: Variation in the line stability of an inkjet printed optical waveguide-applicable material. In: 2008 2nd Electronics Systemintegration Technology Conference, Greenwich, 01-Sep-08–04-Sep-08, pp. 1267–1272
19. Klestova, A., Cheplagin, N., Keller, K., Slabov, V., Zaretskaya, G., Vinogradov, A.V.: Inkjet printing of optical waveguides for single-mode operation. Advanced Optical Materials **7**(2), 1801113 (2019). https://doi.org/10.1002/adom.201801113
20. Reitberger, T.: Additive Fertigung polymerer optischer Wellenleiter im Aerosol-Jet-Verfahren. PhD, Lehrstuhl für Fertigungsautomatisierung und Produktionssystematik (FAPS), Friedrich-Alexander-Universität Erlangen-Nürnberg, Erlangen (2020)
21. Ma, H., Jen, A.K.-Y., Dalton, L.R.: Polymer-based optical waveguides: materials, processing, and devices. Adv. Mater. **14**(19), 1339–1365 (2002). https://doi.org/10.1002/1521-4095(20021002)14:19<1339::AID-ADMA1339>3.0.CO;2-O
22. Reitberger, T., Hoffmann, G.-A., Wolfer, T., Overmeyer, L., Franke, J.: Printing polymer optical waveguides on conditioned transparent flexible foils by using the aerosol jet technology In: Printed Memory and Circuits II, San Diego, California, United States, 99450G (2016)
23. Lorenz, L., et al.: Additive waveguide manufacturing for optical bus couplers by aerosol jet printing using conditioned flexible substrates. In: 2017 21st European Microelectronics and Packaging Conference (EMPC) & Exhibition, Warsaw, 10-Sep-17–13-Sep-17, pp. 1–5
24. OPTOMEC, Aerosol Jet®: Print Engine. User Manual (2018)
25. Hoffmann, G.-A., Reitberger, T., Franke, J., Overmeyer, L.: Conditioning of surface energy and spray application of optical waveguides for integrated intelligent systems. Procedia Technol. **26**, 169–176 (2016). https://doi.org/10.1016/j.protcy.2016.08.023
26. Harris, J., Stöcker, H.: Handbook of mathematics and computational science. Springer, New York, London (1998)
27. Reitberger, T., Loosen, F., Schrauf, A., Lindlein, N., Franke, J.: Important parameters of printed polymer optical waveguides (POWs) in simulation and fabrication. In: Physics and simulation of optoelectronic devices XXV, p. 100981B. California, United States, San Francisco (2017)
28. Schröder, H., Ebling, F., Starke, E., Himmler, A.: Heißgeprägte Polymerwellenleiter für elektrisch-optische Schaltungsträger (EOCB)-Technologie und Charakterisierung. In: Proc. DVS/GMM-Conference, pp. 6–7 (2002)
29. Bierhoff, T., Sönmez, Y., Schrage, J., Himmler, A., Griese, E., Mrozynski, G.: Influence of the cross sectional shape of board-integrated optical waveguides on the propagation characteristics. In: 6th IEEE-SPI Workshop (2002)
30. Elson, J.M.: Propagation in planar waveguides and the effects of wall roughness. Opt. Express **9**(9), 461–475 (2001). https://doi.org/10.1364/OE.9.000461
31. Hamjah, M.K., et al.: Manufacturing of polymer optical waveguides for 3D-Opto-MID: Review of the OPTAVER process. In: 14th International Congress MID (2021)
32. Chen, G., Gu, Y., Tsang, H., Hines, D.R., Das, S.: The effect of droplet sizes on overspray in aerosol-jet printing. Adv. Eng. Mater. **20**(8), 1701084 (2018). https://doi.org/10.1002/adem.201701084
33. Lorenz, L., et al.: Aerosol jet printed optical waveguides for short range communication. J. Lightwave Technol. **38**(13), 3478–3484 (2020). https://doi.org/10.1109/JLT.2020.2983792
34. Wolfer, T., Bollgruen, P., Mager, D., Overmeyer, L., Korvink, J.G.: Printing and preparation of integrated optical waveguides for optronic sensor networks. Mechatronics **34**, 119–127 (2016). https://doi.org/10.1016/j.mechatronics.2015.05.004

3D-Opto-MID Coupling Concept Using Printed Waveguides

6

Lukas Lorenz, Carsten Backhaus, Karlheinz Bock and Norbert Lindlein

The aim of this chapter is the development of a 3D-Opto-MID packaging technology for the interruption-free optical waveguide coupling in large bus networks with a high number of connected modules. The aim is not to develop an application-specific integration concept, but a general coupling technology that can be used in different scenarios depending on the design. For that, this chapter answers the following question:

> *How is it possible to couple two multimode waveguides without interruption and achieve direction-dependent coupling rates in a compact 3D-Opto-MID package?*

L. Lorenz (✉) · K. Bock
Institut für Aufbau- und Verbindungstechnik der Elektronik, Technische Universität Dresden, Dresden, Deutschland
e-mail: lukas.lorenz@tu-dresden.de

K. Bock
e-mail: karlheinz.bock@tu-dresden.de

C. Backhaus · N. Lindlein
Institut für Optik, Information und Photonik, Friedrich-Alexander-Universität Erlangen-Nürnberg, Erlangen, Deutschland
e-mail: carsten.backhaus@fau.de

N. Lindlein
e-mail: norbert.lindlein@fau.de

J. Franke et al. (eds.), *Optical Polymer Waveguides*,
https://doi.org/10.1007/978-3-030-92854-4_6

6.1 Coupling Strategies for Large-Scale Optical Networks

6.1.1 Demands for Optical Bus Coupling

In order to use an optical bus system in the device communication sector, numerous requirements have to be fulfilled [1]:

- Coupling without waveguide interruption
- Asymmetric coupling ratios
- Tunable coupling ratios
- No power loss in the bus with disconnected modules

By definition, the realization of a bus system requires **coupling without interruption of the waveguides;** i.e., all transmitted signals of the base side (Tx) as well as all received signals of the connected modules (Rx) need to be transmitted via the same waveguide. In contrast to this are systems in which the bus signal is decoupled, o/e converted, processed, e/o converted and coupled in again at each node. Such bypass systems [2] are logically also bus systems, but not physically, which means significant effort in implementation.

The so far unsolved but crucial requirement for a system with numerous modules, are **asymmetric coupling ratios** depending on the coupling direction. As previously described, the modules share a bus waveguide and thus a bus signal. If no filters are used, the power level in the bus wavcguide drops at each node. In order to connect a large number of modules to the bus, however, it is necessary that the coupling from the bus to the module is only moderate, so that at the end of the system the power level is high enough to allow for successful detection. However, for a symmetrical coupling this means that the coupling rate for the module signal is also very low. The example in Fig. 6.1 shows a bus system with three nodes, where only the first module sends a (green) signal. The coupling rates

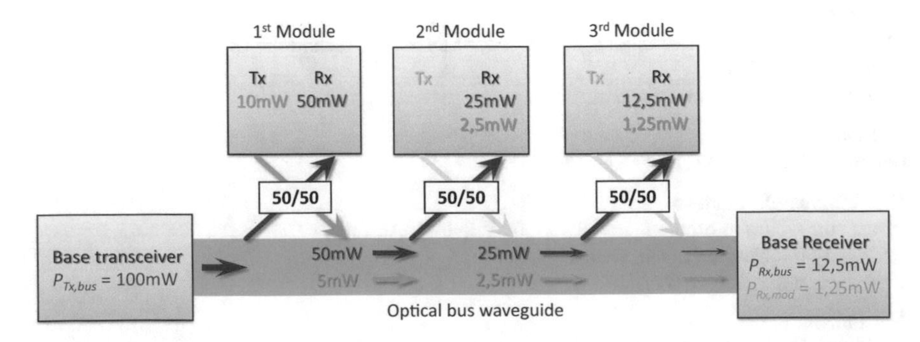

Fig. 6.1 Schematic of an optical bus system with three connected modules and a **symmetric** coupling, i.e., 50% for the bus-to-module coupling and 50% for the module-to-bus coupling [1]

at the nodes are assumed to be 50/50; i.e., 50% of the power of the bus couples into the module, and also 50% of the power of the module couples into the bus—this is called symmetric coupling. Thus, in the example, after three nodes 12.5% of the original power of the bus signal (red) remains in the waveguide. The received power $P_{Rx,bus}$, which is still available at the end of the bus, is calculated according to formula (6.1) with the quantities $P_{Tx,bus}$—the power coupled at the beginning, K_{bus}—the coupling rate between bus and module, and n—the number of connected modules.

$$P_{Rx,bus} = P_{Tx,bus} \cdot (1 - K_{bus})^n \tag{6.1}$$

$$P_{Rx,mod} = P_{Tx,mod} \cdot K_{mod} \cdot (1 - K_{bus})^{n-1} \tag{6.2}$$

The level of the Tx signal of the first module is also reduced at each coupling point according to formula (6.2). For large bus networks with more than three nodes, it can thus occur that the modules connected at the end of the bus no longer receive information from the bus, because the signal level has already dropped significantly. On the other hand, as the number of modules increases, it becomes more and more difficult to detect a module signal (green) from the beginning of the bus system at the end on the base side again. Hence, the coupling from thc bus to the module needs to be moderate, whereas the coupling from the module to the bus should be as high as possible. In this case, we speak of an asymmetric coupling, as shown in Fig. 6.2. The example shown there with a 10/60 coupling makes it possible to obtain almost six times the optical power of the bus signal at the end of the bus waveguide than it would have been possible with the previously shown symmetrical coupling. For systems with more than three nodes, this difference becomes even greater as the number of modules is included in potency according to formulas (6.1) and (6.2).

In a system with different modules, it can also be assumed that it is necessary to ensure **adjustable coupling ratios**. For example, if a module is connected to an Rx photodiode, which has a low signal-to-noise ratio compared to other connected modules, it is necessary to provide a higher power level of the bus signal; i.e., the coupling rate has to be higher.

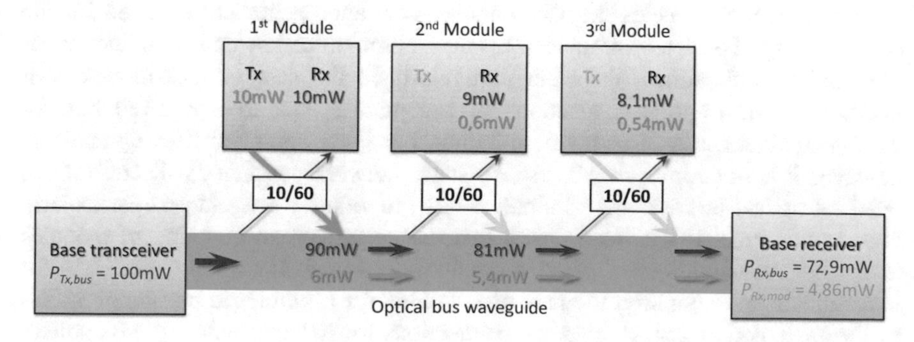

Fig. 6.2 Schematic of an optical bus system with three connected modules and an **asymmetric** coupling, i.e., 10% for the bus-to-module coupling and 60% for the module-to-bus coupling [1]

The last key requirement for bus coupling in device communication is that the **link can be disconnected without energy loss** for the system. For a flexible design of the network, it should be possible to vary the number and arrangement of the modules connected to the bus. This configurability helps to set up the network cost-effectively and allows for simple maintenance. For this purpose, it is necessary that modules can easily be connected and disconnected to the bus. In the case of a disconnected module, however, there must not be any loss of energy in the optical bus system.

Furthermore, there are several minor requirements. Costs play a significant role in the success of a technology. For this reason, the coupling should be adjusted in as few degrees of freedom as possible. This contributes significantly to keeping the packaging costs low. These are the majority costs for e/o assemblies besides the costs for testing [3].

A simple adjustment also creates the basis for field assembly; i.e., the coupling is established outside a manufacturing environment directly at the point of use. In contrast to the use of complex, predefined cable harnesses, the network can be configured on site according to the application and situation. The same reasons are behind the demand for coupling at arbitrary locations; i.e., it should be possible to attach a module anywhere on the bus waveguide.

To enable field assembly, the coupling mechanism has to be prepared in the component on the module side. This integration into the 3D-Opto-MID should make it possible to establish the bus coupling without extensive additional preparations, either in a field assembly process or by means of an assembly system (e.g., flip-chip bonder). The latter requires compatibility with electronics production. This means that the coupling principle has to allow the use of standardized manufacturing tools and assembly processes, which increases the cost efficiency as well.

The last requirement to be mentioned at this point is the possibility of transmitting high data rates via the coupling point. On the one hand, the amount of data to be transmitted—as already mentioned in Chap. 1—increases continuously, and on the other hand, a parallel transmission of several pieces of information may be necessary. Nevertheless, it can be assumed that, in contrast to applications in the data and computer communication sector, transmission rates in the lower Gbit/s range are sufficient.

Based on the demands described above, new approaches are required for the bus coupling. For telecommunication, data communication and computer communication applications (the latter particularly in the case of rack-to-rack connections), point-to-point connections are acceptable, since the space required for additional switches is often not a limitation. For (very compact) IoT applications, however, it is not appropriate to use extensive network peripherals. Therefore, the need for optical bus systems with only a few individual waveguides is particularly high in this area. It has been shown that the asymmetrical coupling of modules to the bus waveguide is of great importance. Although there are approaches for interruption-free coupling, the lack of solutions for asymmetric bus coupling has so far prevented optical short-range connections for IoT and Industry 4.0 applications from gaining acceptance.

6.1.2 State of the Art for Optical Bus Couplers

Directional Fiber Coupler

The longest known and commercially used type of interruption-free waveguide coupling is the directional fiber coupler. The basic principle is always the same. The fiber cladding is either completely removed or grinded off in the area of the coupling point [4]. The cores are brought into physical contact over a certain length. Depending on the length of the overlap, more or less energy is exchanged symmetrically [5]. Often the fibers are heated to fuse the two cores. It is also possible to produce a star coupler using several fibers [6]. While the fusion creates a permanent connection, there are also approaches for a (theoretically) detachable connection of both fibers. Here, the cores are twisted and joined with an optically adapted adhesive [7]. By using solvents, it is possible to detach the coupling later. However, the complete separation of the two elements is considerably more difficult because of the twisting. Furthermore, the integration of this coupling approach into 3D-Opto-MID assemblies is very difficult as well as the use of standard manufacturing processes, whereas field assembly is possible in principle.

Y-Cascade

One way to create a bus system on a circuit board is to use Y-splitters. This allows—as shown in Fig. 6.3—to design a network in which all the information sent and received is combined in one bus waveguide. By cascading, it is theoretically possible to connect an arbitrary number of modules to the bus [8]. Different coupling rates can be achieved by using different cross sections on the Y-splitters. Since the transmit and receive channels are separated in this network architecture, it is also possible to enable asymmetrical coupling for the Tx and Rx channels. By using imprint [9], photolithography [8] or ion exchange [10], a large number of standard manufacturing processes can be used. The major disadvantage of this type of bus coupling, however, is the rigid design, which is already predetermined and thus does not allow field installation and subsequent adjustment of the coupling rates. Furthermore, the connection can only be detached behind the actual bus coupling point, since the Y-coupler is permanently integrated in the layout. This has the

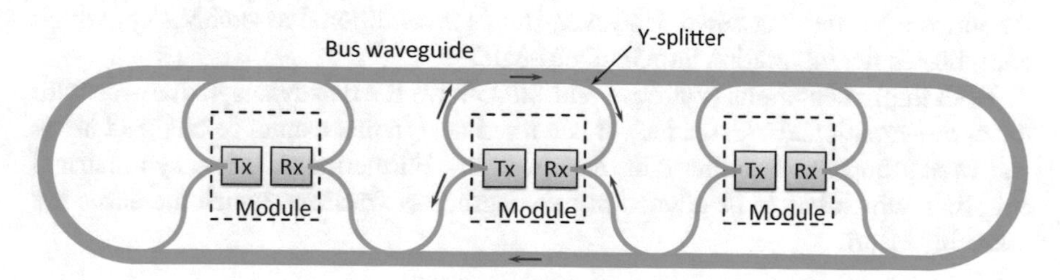

Fig. 6.3 Bus system realized by a Y-cascade (adapted from [9])

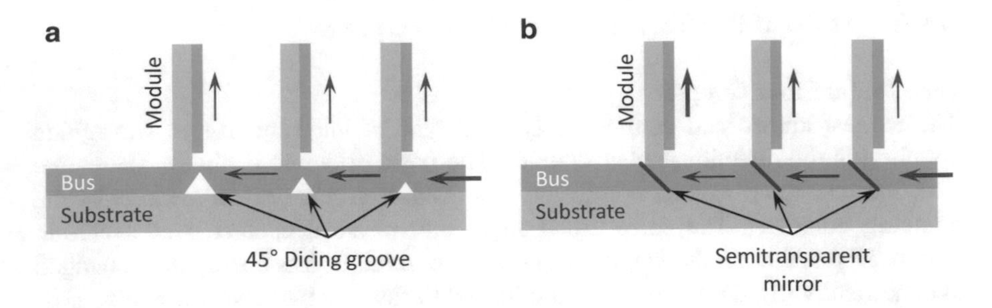

Fig. 6.4 Bus system on back planes realized by beam deflection with **a** diced v-grooves with different heights and **b** semitransparent mirrors [1]

disadvantage that it is not possible to couple at any point on the bus waveguide. In addition, the bus system also loses power at an unused coupling point.

Bus Systems with Beam Deflection

Bus systems can also be implemented using the principle of indirect coupling. In this case, a deflection optic is integrated into the waveguide, which allows for in- and out coupling of the optical signal at different angles [11, 12]. The mirrors are manufactured either by a selective removal of the waveguide (laser ablation [13], dicing [14]) or directly during the manufacturing of the waveguide (embossing process, UV angular exposure [14]). In both cases, a slanted mirror surface is created. The decisive factor here is that not all of the optical power is coupled out of the bus waveguide during beam deflection. To solve this, there are two main approaches, as shown in Fig. 6.4. In the first principle, the beam deflection is realized via diced v-grooves with different depths [15]. For that, a v-shaped dicing blade is used, resulting in a 90° light deflection. Due to the different depths, it is possible that a part of the light passes the mirror and remains in the bus waveguide. Furthermore, the use of the diced v-grooves as microfluidic channels allows for switching the coupling points on or off with a refractive index-adapted fluid [16]. The disadvantage, however, is that this system requires a very complex design.

The second possibility to achieve such a deflection is by using semitransparent mirrors, which are integrated into the bus waveguide at a 45° angle [17, 18]. For this purpose, a cut is made in the waveguide at the appropriate angle, into which the mirrors are then mounted. However, this is an additional assembly step, which complicates the integration into 3D-Opto-MID.

Both implementations (v-groove and filter) have the disadvantage that—as with the above-mentioned Y-cascades—it is a fixed design that cannot be changed afterward regarding the coupling rates and position. Furthermore, it is a symmetrical coupling, which has to be aligned in five degrees of freedom, which increases the assembly effort.

Directional Core–Core Coupling

The principle of directional mode couplers—which is finally presented—has been known for a long time [19]. Just like directional fiber couplers, it is based on physical contact of the waveguide cores, so that different modes cross-couple into the second core (see Fig. 6.5) [20, 21]. However, it is not possible to twist planar waveguides, so the contact needs to be made by pressure. This coupling principle offers many advantages. By adjusting the overlap length, it is possible to change the coupling rates even during operation. Furthermore, it offers the possibility to disconnect the modules without loss of power [22], as well as an implementation in 3D-Opto-MID and field assembly.

The major disadvantage of this principle, however, is the symmetrical coupling. Flores et al. [21] demonstrated coupling rates of up to 80%; i.e., 80% of the guided light couples from the module to the bus and vice versa. For a network with four connected modules, this means, according to formula (6.1), that the signal level in the bus waveguide drops below 0.16% after the fourth module, making it significantly more difficult to detect the signal. In the opposite direction for a module–bus coupling, this would mean that although a large amount of the transmitted power is coupled into the bus waveguide, 80% of it is already lost again at the next node. Hence, this symmetrical coupling method is unsuitable for the use in extensive optical networks.

Comparison of the State-Of-The-Art Solutions

The presentation of the state-of-the-art solutions shows that there are already technologies available, which allow interruption-free waveguide coupling. Hence, bus systems can be implemented with only one bus waveguide. However, the detailed analysis of the individual concepts shows significant disadvantages, so that none of the described approaches is able to meet all requirements. At this point, the main requirement of asymmetric coupling, which allows for optical networks with a large number of connected modules, must be especially highlighted. Only Y-cascades meet this requirement, which allow different coupling rates via two separate Tx and Rx channels. However, this already includes the major disadvantage that this coupling principle uses several coupling points. All other solutions only allow symmetrical coupling, whereas the directional core–core coupling has no further disadvantage and is most promising for an optical bus coupling system. All the advantages and disadvantages of the above concepts are summarized in Table 6.1.

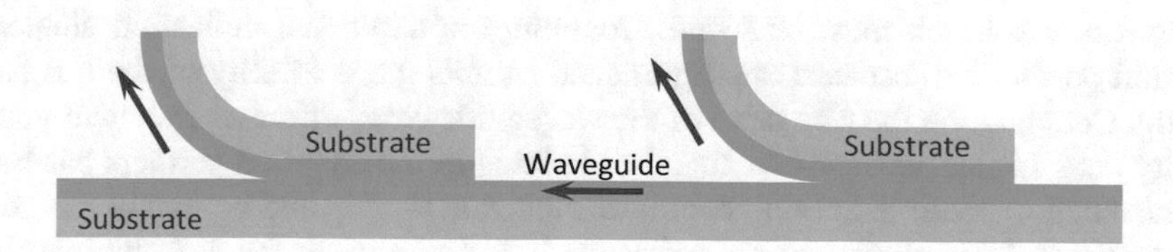

Fig. 6.5 Directional core–core coupler

Table 6.1 Advantages and disadvantages of the state-of-the-art approaches for interruption-free waveguide coupling for bus systems

	Directional fiber coupler	Y-cascade	Beam deflection	Directional core-core-coupler
Interruption-free waveguide coupling	Possible	Possible	Possible	Possible
Tunable coupling ratios	Depends on the twisting length	Only possible during design	Only possible during design	Depends on the overlap length
Asymmetric coupling ratios	Only symmetric coupling	Possible with separated Tx and Rx channels	Only symmetric coupling	Only Symmetric coupling
Detachable	Yes (with solvents at glued connections)	Yes (but only after the Y-splitter)	Yes	Yes
Energy loss at detached nodes	No	Yes	No (with opto-fluidic switches)	No
Degrees of freedom during coupling	1 Translation	Non (fixed coupler)	3 Translation 2 Rotation	1 Translation 1 Rotation
Suitable for field assembly	Yes	No (Fixed coupler)	Yes	Yes
Coupling at arbitrary position	Yes (only with a flexible cladding removal)	No	No	Yes (only with a flexible cladding removal)
Integration in 3D-Opto-MID	No	No	No	Yes
Data rates	n.a.	10 Gbit/s [9]	10 Gbit/s [16]	2,5 Gbit/s [20]
SMD compatibility	No	Yes	Yes	Yes

Requirement not fulfilled
Requirement can only be fulfilled with high effort
Requirement fulfilled

6.1.3 Theoretical Basis for the Asymmetric Optical Bus Coupler

With the requirements specified in Sect. 6.1.1, a solution for an optical bus coupler has to be found. As already mentioned, the asymmetric coupling behavior is crucial for that. At first, however, it is necessary to achieve interruption-free coupling. Hence, a coupling via the end face of the waveguides is not possible, which is why a new solution must be found. Coupling via additional deflection elements should be avoided, because this significantly limits the flexibility of the bus coupling. Coupling via the side faces of the waveguide cores offers an excellent possibility here. Under the condition that the cladding of the coupling partners has been removed, it is possible to bring both into contact at the top and bottom of the core, respectively. This eliminates the refractive index step to air (n_{air}) at the interface of the core, assuming both coupling partners have the same refractive index n_{core}. Thus, the coupling point behaves over the length of the contact like a single core with an increased diameter. Depending on the order of the modes and the length of the overlap, coupling occurs, as shown in Fig. 6.6 in the side view.

It can be seen that higher-order modes and thus—in simplified terms—beams with a larger angle to the optical axis have a higher probability of coupling into the second waveguide. In this case, it is therefore a mode coupler that meets the requirement of interruption-free waveguide coupling. Because higher-order modes tend to couple over more frequently in directional core–core coupling, it could be assumed that these modes will immediately return to the original waveguide at the next coupling point. For real waveguides with a rough core–cladding interface, however, this effect is neglectable, since the imperfections lead to fast mode coupling within the core [23], so that the initial condition of the mode distribution for each subsequent coupling point can be considered constant. Preliminary investigations have shown that in the case of the lithographically fabricated waveguides, the mode equilibrium is restored after only 20 mm [24].

With the described approach, it is possible to meet the demand for adjustable coupling ratios. By varying the size of the contact area A, it is possible to influence the coupling behavior of the individual modes. As shown in Fig. 6.6, a longer coupling length increases the probability that low-order modes will also couple over into the second waveguide. This allows for tunable coupling ratios. Such a variation of the coupling area can be realized by changing the contact force which presses the two cores together. Due to an elastic behavior of the two coupling partners, a larger contact area is created at higher forces. The coupling areas for two different waveguide cross sections are depicted in Fig. 6.7.

The approach also includes the ability to disconnect the link without power loss for the system. Since the two waveguide cores are not actively modified for the purpose of coupling, they can be restored to their original state when the connection is detached. This ensures that there is no undesired decoupling of light from the waveguide when the node is inactive. By restoring the refractive index difference at the core–air interface, total internal reflection is restored and the light is guided exclusively in the source waveguide.

The solution for an asymmetrical coupling is more difficult. A principle has to be developed in which only one of the two coupling paths is influenced without

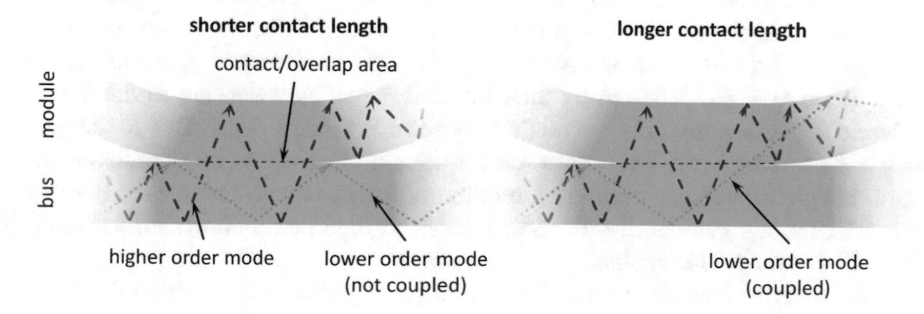

Fig. 6.6 Side view of two waveguide cores in physical contact and the simplified behavior of different modes at (left) a shorter contact length and (right) a longer contact length

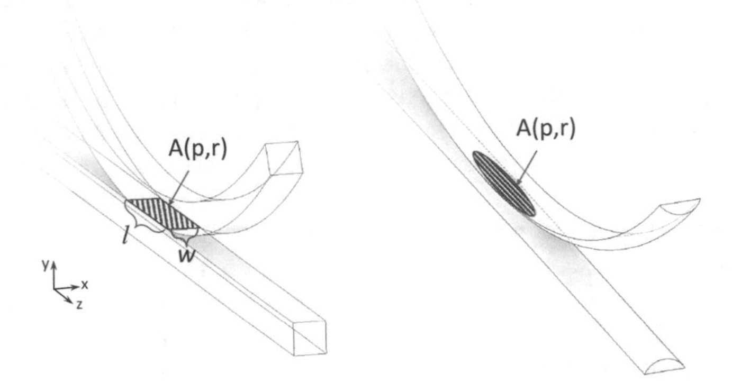

Fig. 6.7 Contact/overlap area A as a function of the pressure p and the waveguide bending radius r for (left) rectangular cores and (right) for cores with a circular segment cross section

changing the other. This suggests that a change needs to be made to one of the two coupling partners. In the context of this work, one of the actually disadvantageous properties of optical waveguides will be used: the losses due to bending. At radii of bending, there are transition and bending losses in the waveguide; i.e., the light is emitted into the environment [25–27]. The aim is to use this light specifically for coupling into a second waveguide. In this context, the bending must not affect the overall power budget of the system significantly.

Asymmetric bus coupling, as proposed in this work, is based on the fact that the changes in the optical conditions in the waveguide core under bending are utilized. While too much bending leads to high losses in the waveguide, an effect is used in bent waveguides where high losses do not yet occur. It is known that in bent multimode waveguides there is a shift of the intensity maximum to the outer radius of the bending [28–30]. Hence, it can be assumed that there is more power for coupling at the outer radius of the waveguide, which supports higher coupling rates at this point.

A simplified ray optics model can explain this phenomenon, as shown in Fig. 6.8. It is assumed that only rays parallel to the optical axis propagate in the core. These are characterized by a constant distance d_x to the inner core–cladding interface. If the core turns into a bending, the rays hit the outer side face of the core. There they are deflected by total internal reflection according to the law of reflection. At this point, it is crucial that no transition losses occur. Due to the geometrical conditions, the beam does not hit the side face of the inner radius afterward, but only approaches it up to a minimum distance d_x, which is already given at the beginning. Parallel rays therefore never hit the inner radius interface in the case of a bending. This applies to any core dimension and any bending radius.

This behavior leads to the formation of a propagation area in which the beam can be found. The red ray in the example has a propagation area from d_1 to d_{core}. For example, this means that in the area of the inner radius, a region is formed

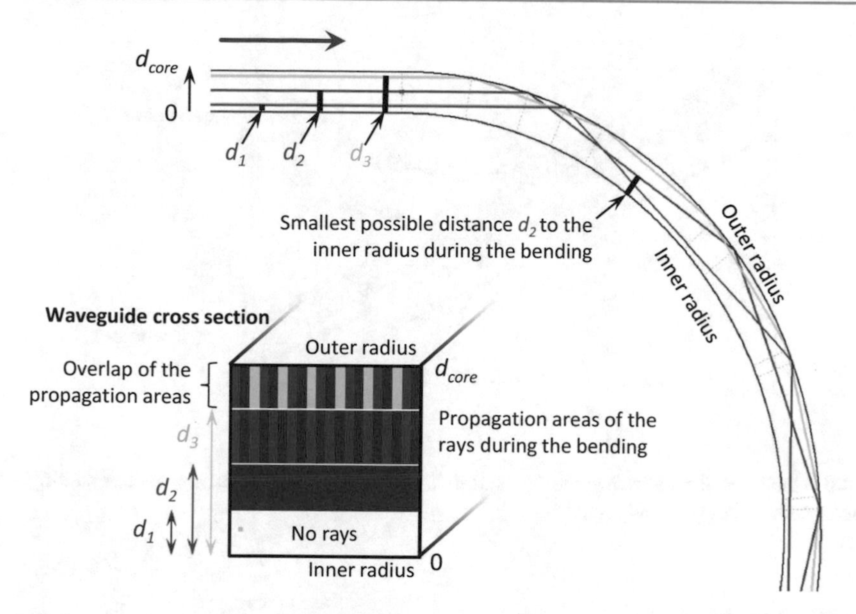

Fig. 6.8 Schematic of three rays propagating in a bent waveguide (side view) and the resulting propagation areas and their overlap in the cross section [1]

in which the ray never propagates (0 to d_1). At the opposite interface, on the other hand, an area of overlap is formed in which there is a probability to find all three rays at once. For a waveguide with several thousand rays, an increase in the number of guided rays occurs toward the outer radius. Since this observation refers only to rays parallel to the optical axis, the effect is strongly dependent on the angle at which the rays are guided, which depends essentially on the NA of the waveguide. The exact effects on the distribution in the core are examined in Sect. 3.2.3.

Based on the energy redistribution within the core, the bus coupling solution in this work uses a defined bending of one of the coupling partners, while the other one remains straight. Hence, more energy is provided at the coupling point when the light is coupled from the bent to the straight waveguide. In the opposite direction, the bent waveguide does not influence the coupling from the straight to the bent waveguide, as it is depicted in Fig. 6.9. Because of that, asymmetric coupling ratios (depending on the coupling direction) are achieved.

Crucial for optical coupling are the tolerances, since the costs increase enormously with increasing adjustment effort. It is therefore necessary to keep the degrees of freedom for the coupling to a minimum. For the selected coupling at the side faces, there are already only two relevant degrees of freedom, as shown in Fig. 6.10, without considering a specific realization. The remaining four possible movements of the waveguides relative to each other have no influence on the coupling, since they do not change the basic structure of the coupling point. This means a significantly reduced adjustment effort compared to the butt coupling of

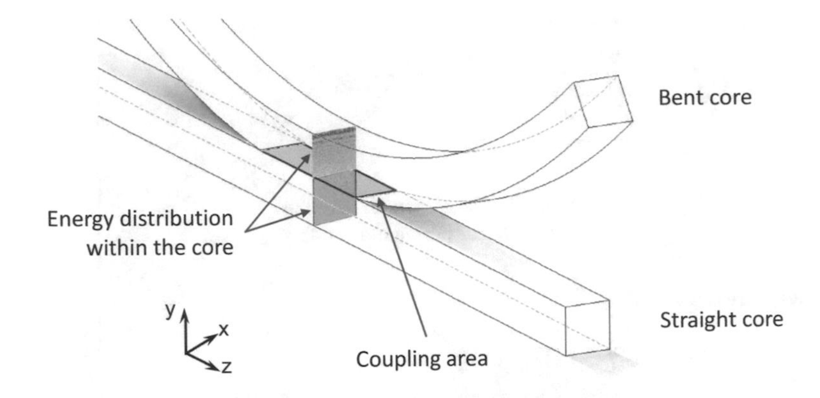

Fig. 6.9 Coupling of two waveguide cores at their side faces with the energy redistribution to the outer radius in the bent waveguide

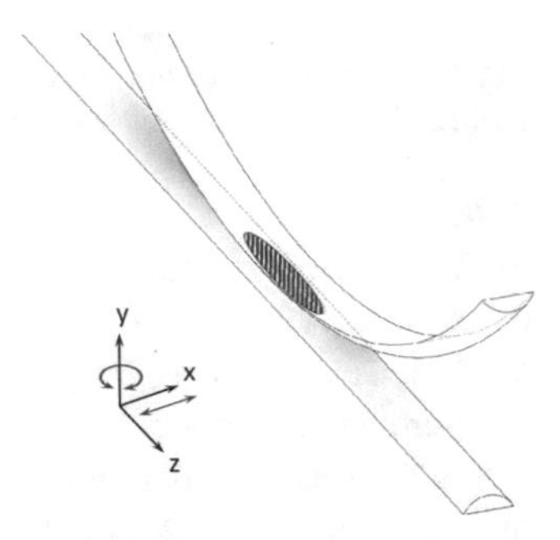

Fig. 6.10 Definition of the coordinate system for the coupling point with the two relevant tolerances for the coupling

two waveguides. The movement along the y-axis has a special importance. The translation in this direction is used on the one hand to establish or disconnect the coupling and on the other hand to adjust the coupling rate.

6.2 Simulation of the Asymmetric Optical Bus Coupler

6.2.1 Parameters for the Mathematical Model of the Bus Coupler

The optical simulation of the asymmetric optical bus coupler requires a suitable simulation algorithm. Therefore, waveguides can be integrated into a whole range of different categories, with different simulation methods for each category, as it is described in Chap. 3. Since interference effects at the coupling interface occur, raytracing simulations are no longer applicable and a wave optical approach has to be used. The most frequently used simulation methods for waveguides are based on a beam propagation method (BPM). But, here for the highly multimode polymer optical waveguides [31], the wave propagation method (WPM) [32] will be used since it is quite fast, as already mentioned in Sect. 3.3.

To represent the asymmetric optical bus coupler in the simulation algorithm, a mathematical model of its surfaces has to be derived (Fig. 6.11). In the case of a two-dimensional simulation, i.e., only one lateral coordinate y- and the z-coordinate in propagation direction, the straight waveguide can easily be described by its width w. However, a far more complex description is needed for the flexible waveguide, which also has the width of w. A distance d_{Ger} describes the length of the straight part of the flexible waveguide before and after the bended part. The bended part can be described by three circles with the radius r, which describes the bending radius. While the first circle describes the deformation downwards,

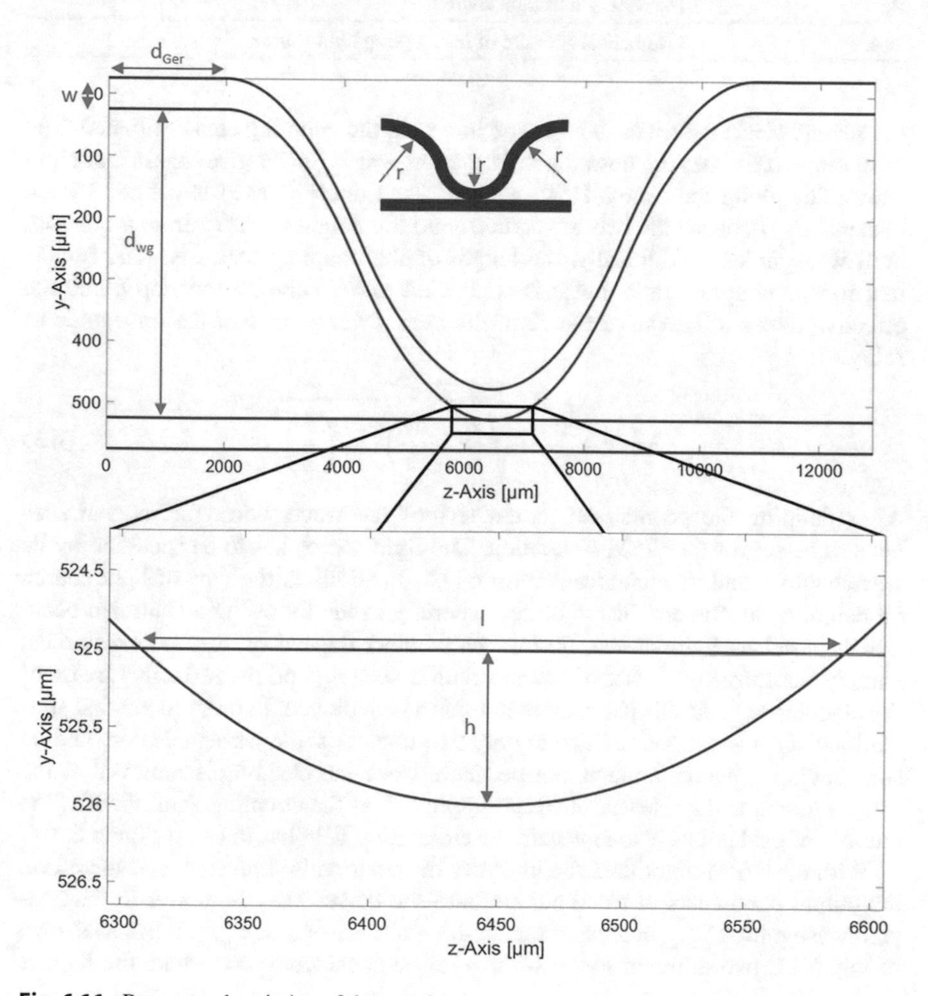

Fig. 6.11 Parameter description of the coupler

Table 6.2 List of all input variables for the simulation of an asymmetrical optical bus coupler

Variable	Description
h	Height of overlap
l	Length of coupling area
w	Width of waveguide
d_{Ger}	Length of straight waveguide at the start and end of flexible waveguide
d_{wg}	Distance between the two waveguides
r	Radius
n_{in}	Refractive index of waveguide
n_{out}	Refractive index of cladding
N	Number of grid points
λ	Wavelength of light source
NA	Numerical aperture of in-coupling light source

the second circle describes the u-shape in which the coupling area is situated. The third circle describes the upwards deformation which finally gives again a straight waveguide along the z-axis. How much of each circle is used is defined by the distance d_{wg} between the two waveguides and the height of the overlap h between both waveguides. Additionally, the length of the coupling area l is quite important for the coupling ratio, but it is connected to the height of overlap h and the effective radius of curvature $r+w/2$ of the bent lower surface of the waveguide as follows:

$$h = r + \frac{w}{2} - \sqrt{\left(r + \frac{w}{2}\right)^2 - \left(\frac{l}{2}\right)^2}. \tag{6.3}$$

After defining the geometrical parameters of the waveguides, further variables have to be set for the WPM simulation. The light source has to be specified by its wavelength λ and its numerical aperture NA. In addition, the type of light source (coherent or incoherent) has to be set, whereby in the following a Gaussian beam will be used as light source. Further parameters depend on the used materials, namely the refractive index of the waveguide core n_{in} and the refractive index of the cladding n_{out}. At this point, an assumption is deployed: In order to get fast simulations, the system is simplified to only two materials, the waveguide core and its surrounding. This assumption can be made since the cladding is removed at the coupling area and no second material is present at the coupling area. Finally, the number of grid points N to represent the simulation field has to be set (Table 6.2).

With the WPM-algorithm, the intensity distribution is simulated, and therefore, the output parameters of the simulation are the power loss, the power in the coupled waveguide P_{couple} and the power in the main waveguide P_{main}. Here, as shown in Fig. 6.12, two different scenarios have to be considered: one where the light is

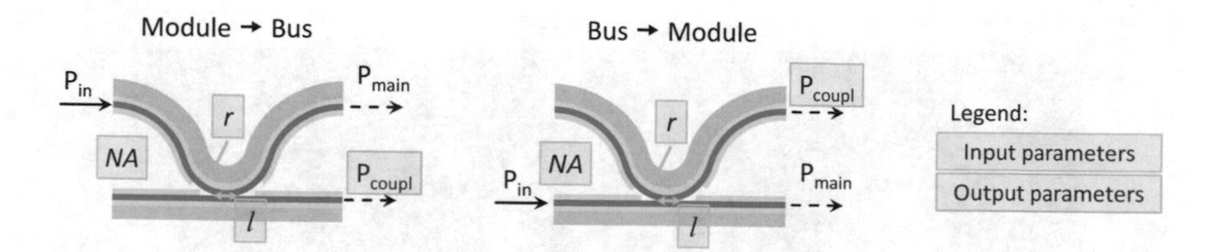

Fig. 6.12 Definition of the coupling scenarios and the resulting output powers for (left) module-to-bus coupling and (right) bus-to-module coupling considering different input parameters

introduced to the flexible waveguide (module-to-bus coupling) and one where the light is introduced to the rigid waveguide (bus-to-module coupling).

6.2.2 2D Simulation

After defining the relevant parameters of the coupler in Sect. 6.2.1, a two-dimensional simulation can be performed. In order to show the functionality of the simulation, we choose the parameters of a typical experimental setup [33–35]. To simulate the light source, we use a coherent Gaussian beam with a NA of 0.1 which is coupled into the center of the front face of the waveguide. The whole set of variables for the simulation can be found in Table 6.3.

First, a simulation of the coupler is carried out where the module couples into the bus. Consequently, the light is introduced into the flexible (here upper) waveguide. Figure 6.13 shows how the electrical field propagates with almost no loss in

Table 6.3 Values of the parameters for the two-dimensional simulation of the optical bus coupler

Variable	Value
l	282 μm
w	50 μm
d_{Ger}	500 μm
d_{wg}	499 μm
r	10,000 μm
h	0.99 μm
n_{in}	1.5455
n_{out}	1.525
Nx	4096
x-field	1080 μm
Nz	30,000
z-field	9896.9 μm
λ	0.85 μm
NA	0.1

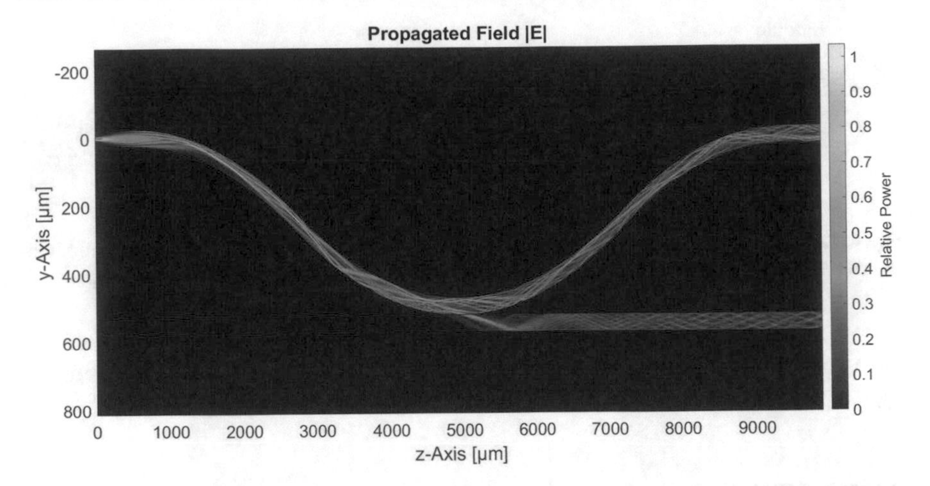

Fig. 6.13 Resulted E-field distribution for a coherent light source coupled into the flexible waveguide

the waveguide up to the point where the two waveguides touch. Upon this point, the rigid (here lower) waveguide carries light as well as the flexible waveguide. It is possible to already guess that more light is guided in the flexible waveguide than in the rigid one.

To quantify this guess, one can look in each slice of the simulation at the power distribution (= sum over the respective pixels of the squares of the electric field multiplied by the refractive index) and relate it to the spatial distribution of the rigid and flexible waveguide. This can be seen in Fig. 6.14. A quite small amount of intensity can be found outside of any waveguide and is therefore defined as loss. An increase of loss in the used geometry can always be found, whenever the waveguide undergoes a curvature. These losses are mainly errors of the simulation

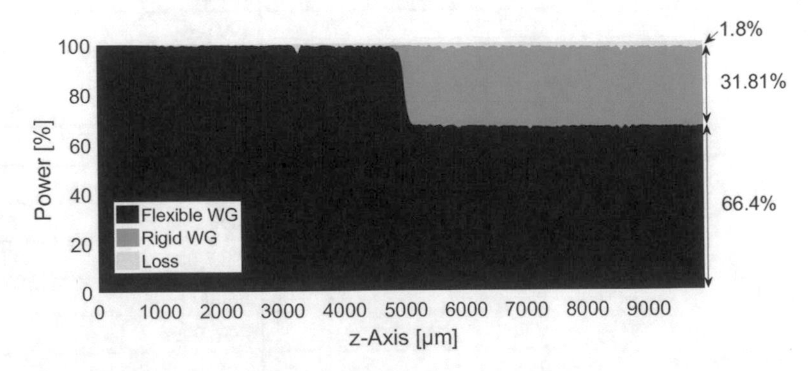

Fig. 6.14 Power distribution for the simulation field when light is coupled from the flexible waveguide to the rigid waveguide

method due to the discretization of the curved waveguide. They might be reduced by a higher number of grid points, but then by increasing the calculation time. But, since the losses are smaller than 2% even at the end of the waveguide, this can be tolerated.

The graph shows very clearly the coupling of light of the flexible to the rigid waveguide at $z \approx 5000$ µm. After this coupling, no further changes happen and the amount of light in each waveguide does not change. However, a slight increase of loss is observed which is correlated with the up-curve of the flexible waveguide and resulting in a decrease of the light guided in the flexible waveguide. At the end face of the coupler, the distribution of the power gives the expected result: 66.4% of the initial light stays in the flexible waveguide, while 31.81% couples into the rigid waveguide. A neglectable amount of 1.8% couples out of the waveguides and does not couple back into one of the waveguides.

After investigating the module-to-bus coupling, the vice versa setup is of high interest. In terms of simulation adjustments only the point of the light source has to change in regard to the previous simulation setup, while all parameters of Table 6.3 are still the same. The result of the simulation is shown in Fig. 6.15.

The distribution of the light between the waveguides can already be seen: Most of the light stays in the rigid waveguide and very little couples into the flexible waveguide. This is supported by Fig. 6.16 as it shows that 99.76% of the introduced light stays in the rigid waveguide while only 0.2% couple into the flexible waveguide. Further, an interesting result can be found while investigating the loss: Since most of the light is guided in a straight waveguide the loss is as low as 0.04%. However, one remark has to be made concerning the very small coupling into the flexible waveguide: As can be seen in Fig. 6.15 the Talbot effect (see Chap. 3), which appears for the 2D simulation, causes just quite near to the

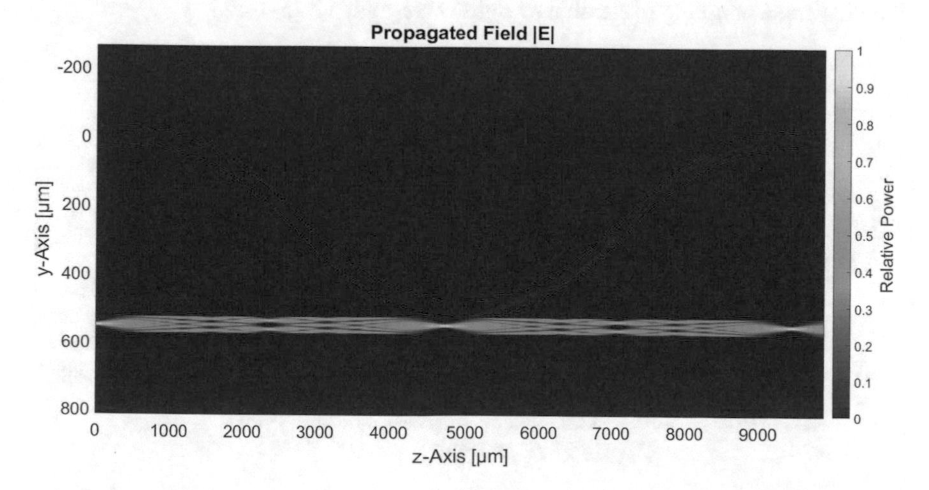

Fig. 6.15 Resulted E-field distribution for a coherent light source coupled into the rigid waveguide

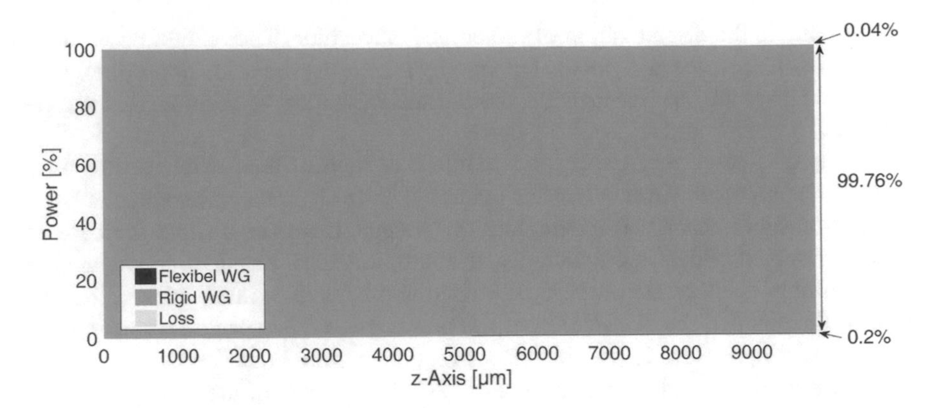

Fig. 6.16 Power distribution for the simulation field when light is coupled from the rigid waveguide to the flexible waveguide

coupling area a self-imaging of the light source, and so nearly no light couples into the flexible waveguide. For the later 3D simulation, the Talbot effect will not appear due to the circle segment cross section.

After simulating the bus-to-module and module-to-bus coupler, the asymmetry of the coupler can be observed. To fully understand this asymmetry, a systematic approach by varying the overlap length l and the usage of a radius r of 5 mm or 10 mm is deployed (see Fig. 6.17).

This analysis shows that a longer overlap length leads to stronger coupling. This result can be translated to the experiment in the way that more force on the flexible waveguide leads to a stronger coupling of light. Further, the asymmetric nature of the coupler can be verified: Independent of the radius and overlap length, a coupling from the rigid to the flexible waveguide is much lower (0–0.21) than the vice versa coupling of flexible to rigid waveguide (0.25–0.58).

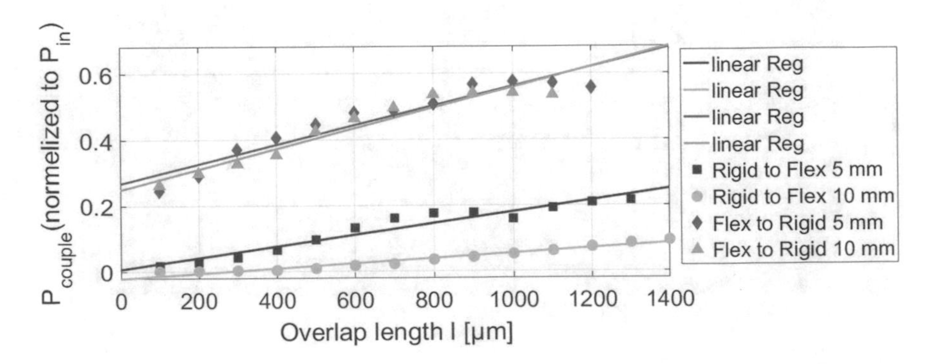

Fig. 6.17 Analysis of the power at the end of the coupled waveguide in dependence on the overlap area of the two waveguides

6.2.3 3D Parameter Extension

An upright Torus (Fig. 6.18) can be given in parametric form with the parameters (u,v) in the following:

$$\begin{aligned} x &= r \cdot \sin(v) + Mx \\ y &= (R + r \cdot \cos(v)) \cdot \sin(u) + My \\ z &= (R + r \cdot \cos(v)) \cdot \cos(u) + Mz \end{aligned} \tag{6.4}$$

with R the major radius and r the minor radius. The point $M = (Mx, My, Mz)$ is the center of the Torus, $v \in [0{,}360]$ the angle in degrees for the minor radius and $u \in [0{,}360]$ the angle in degrees for the major radius.

In order to get the POWs symmetry, the range of the possible angles has to be limited. To derive these limitations, one has to recall the definition of the round side of the cross section. First, the radius r_M of a waveguide with height h and width b can be calculated:

$$r_M = \frac{(4 \cdot h^2 + b^2)}{(8 \cdot h)} \tag{6.5}$$

Second, the cross section can be given by this function:

$$f(x) = \sqrt{r_M^2 - x^2} + h - r_M \qquad \forall x \in [-b/2, b/2] \tag{6.6}$$

This definition holds for a plane and straight waveguide. To translate this description for a Torus, the domain of Eq. (6.6) has to be obtained by the angle v. The angular range holds:

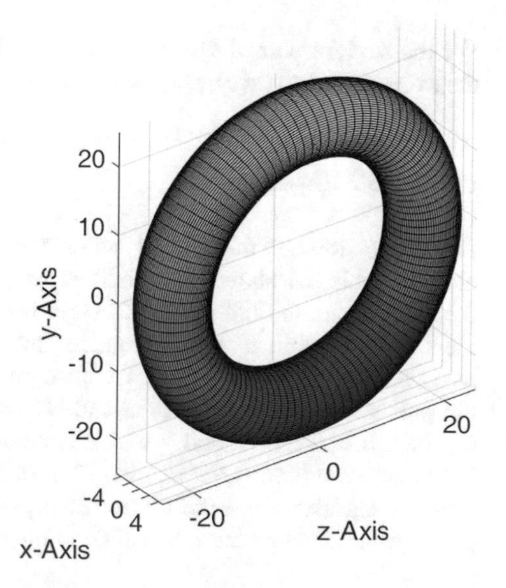

Fig. 6.18 Upright Torus defined in the coordinate system

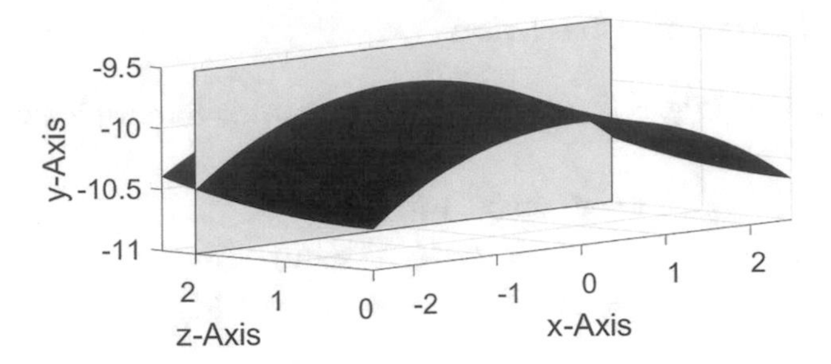

Fig. 6.19 Demonstration how the two-dimensional parametric curve lies in a plane with the surface normal in z-direction

$$\alpha = \cos^{-1}\left(\frac{r_M - \mathrm{h}}{r_M}\right) \tag{6.7}$$

After solving in Eq. (6.4) z toward u, one gets:

$$\mathrm{u(v)} = \cos^{-1}\left(\frac{\mathrm{z - Mz}}{\mathrm{R + r \cdot \cos(v)}}\right) \qquad \forall \mathrm{v} \in [-\alpha, +\alpha] \tag{6.8}$$

In order to simulate the coupler with the WPM, the three-dimensional space has to be divided into slices. Choosing carefully the coordinate system in the way, the surface normal of these slices are in the direction of the z-axis. For a plane in $z = z_1$, the two-dimensional parametric curve can be calculated (Fig. 6.19):

$$\begin{aligned} \mathrm{x(v)} &= \mathrm{r \cdot \sin(v) + Mx} \\ \mathrm{y(v)} &= \mathrm{(R + r \cdot \cos(v)) \cdot \sin(u(v)) + My} \\ \mathrm{z_1} &= \mathrm{(R + r \cdot \cos(v)) \cdot \cos(u(v)) + Mz} \end{aligned} \tag{6.9}$$

For the straight part of the cross section, a line can be found which is defined by the extreme-points (x(v),y(v)) with $\mathrm{v} \in \{-\alpha,\alpha\}$ of Eq. (6.8).

6.2.4 3D Simulation

In order to simulate the coupler with nearly the same parameters as in the 2D simulation, Table 6.4 shows the modified parameters, where now the width b along the x-direction and the width or height w along the y-direction of the circle segment cross section of the waveguide have to be defined. In order to get results in a decent simulation time but with a good resolution, the amount of the sample points was set to Nx = 2048 and Ny = 4096 which gives a sampling distance of about 0.2 μm for x and y which is below half of the wavelength of the wave in the core medium ($\lambda/(2^*n_{\mathrm{in}}) = 0.27$ μm). This is important for the simulation of bent waveguides since the discretization of the waveguide structure will lead to pixels which change from one slice to the next slice from core to cladding. If the

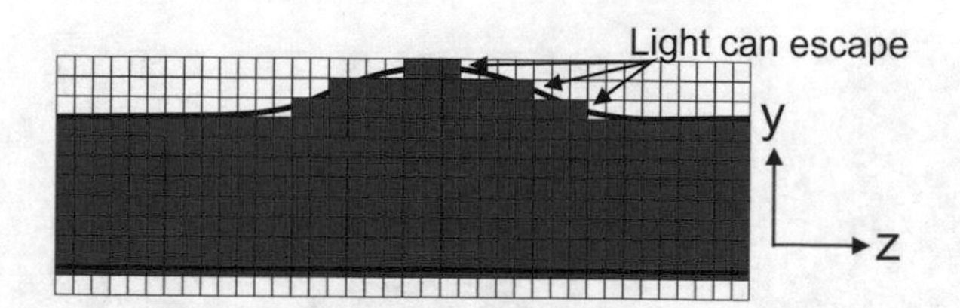

Fig. 6.20 Scheme of the discretization of a bent waveguide here shown for the case of a droplet on top of the waveguide. If the lateral pixel size along the y-direction is larger than half a wavelength in the medium, light can escape from the waveguide

lateral extension of these pixels is larger than half a wavelength in the medium, light can leave the waveguide although for a smooth continuous surface the angle of incidence would not be below the critical angle of total internal reflexion (see Fig. 6.20). For the z-direction, the sampling distance was increased to 0.5 μm, i.e., Nz = 25,200 sampling points for a propagation length/z-range of 12,600 μm (Table 6.4).

First, the coupling from the flexible waveguide to the rigid waveguide is investigated (Fig. 6.21). The result of the simulation does show the same characteristics as the 2D simulation. After the two waveguides had contact to each other, a clear accumulation of light can be seen in the rigid waveguide (see Fig. 6.21 at $z \geq 7.2$ mm). At the end of the simulation, light is guided in both waveguides, but is asymmetrically distributed.

Table 6.4 Changed simulation parameters for Sect. 6.2

Variable	Value
l	282 μm
b	300 μm
w	40 μm
d_{Ger}	2000 μm
d_{wg}	499 μm
R	10,000 μm
h	0.99 μm
n_{in}	1.5455
n_{out}	1.525
Nx	2048
x-field	420 μm
Ny	4096
y-field	800 μm
Nz	25,200
z-range	12,600 μm

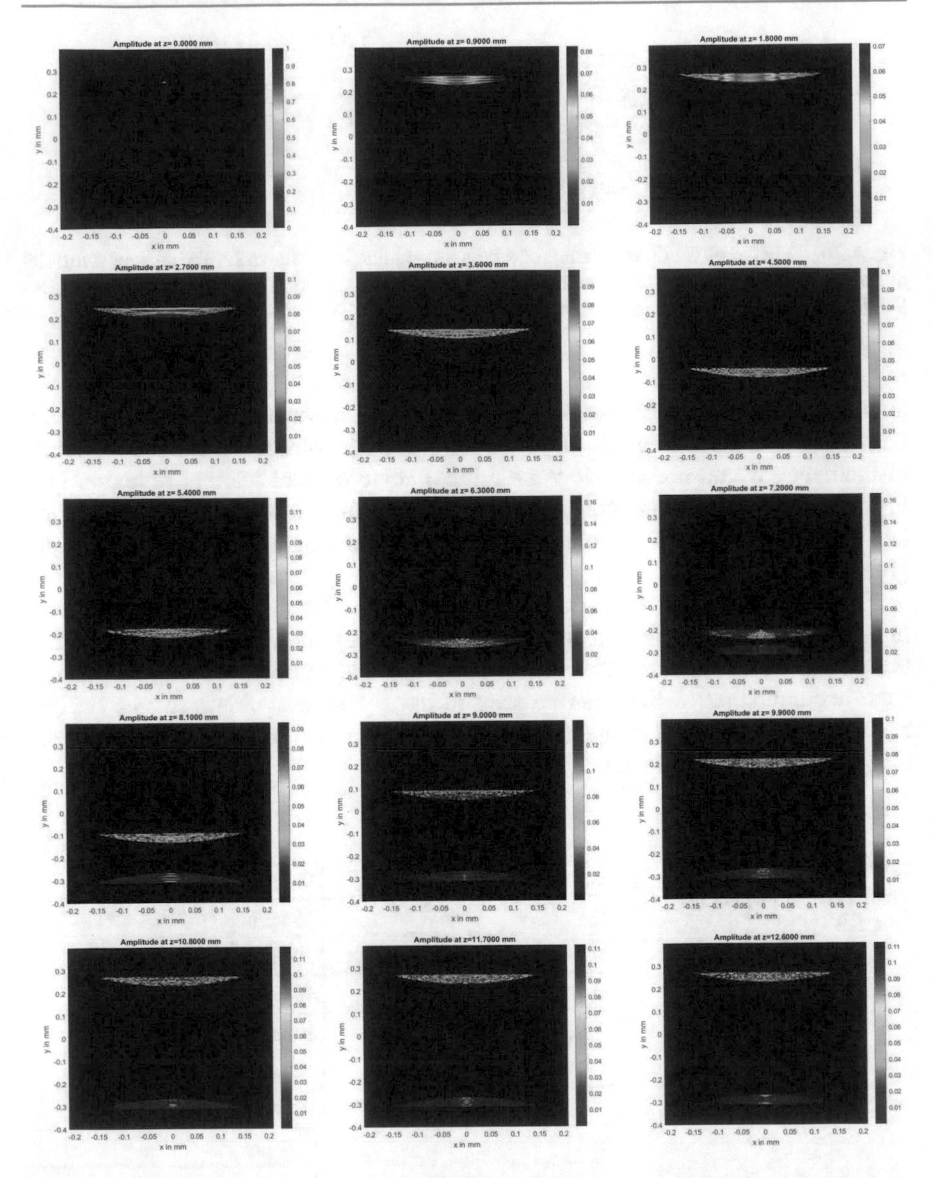

Fig. 6.21 Exemplary slices through the coupler displaying the amplitude distribution in the waveguides and their surroundings. The light was initially coupled into the flexible waveguide

The power distribution is again analyzed and is divided into three areas: the flexible waveguide, the rigid waveguide and the surrounding area, which is assumed as loss. At the end of the coupler, 10.5% of the light coupled into the rigid waveguide and 87.5% remained in the flexible waveguide. Only, about 2% of the light is lost by coupling out of the waveguides due to simulation errors. The asymmetric nature can clearly be demonstrated.

Fig. 6.22 Same simulation as shown in Fig. 6.21, here with the light coupled into the rigid waveguide

Since the 2D simulation suggested a weaker coupling ratio when coupling from the rigid to the flexible waveguide, this phenomenon is also investigated in the 3D simulation (see Fig. 6.22). The result of the simulation does match the expected field distribution for an again asymmetric coupler by displaying much less light in the coupled (flexible) waveguide compared to the vice versa case. This can be

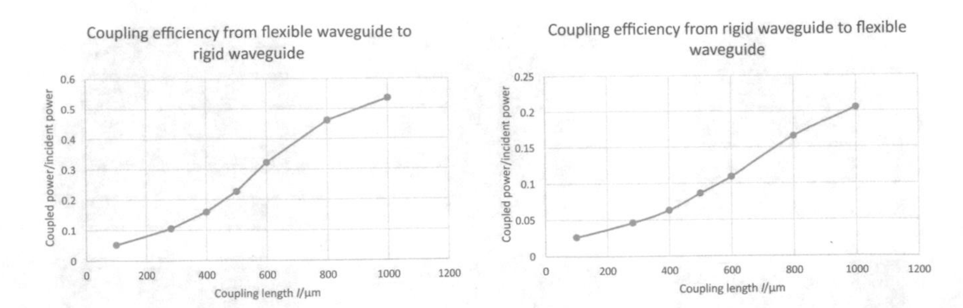

Fig. 6.23 Coupling efficiency as function of the coupling length l. Left: Amount of light which couples from the flexible waveguide to the rigid waveguide. Right: Amount of light which couples from the rigid to the flexible waveguide

explained by the behavior of light in the bent waveguide: The light in the flexible waveguide is always guided on the outside of the bend. At the coupling area, light is therefore accumulated on the border to the other waveguide and has in addition a general propagation direction into the rigid waveguide. In contrast, for light in the rigid waveguide, the light is accumulated in the whole waveguide and with the general propagation direction parallel to the z-axis of the waveguide. This results in a much weaker coupling into the flexible waveguide and very low losses. At the end of the coupler, only about 0.3% of the power is lost and 4.6% is coupled into the flexible waveguide, while 95.1% remain in the rigid waveguide.

Finally, several simulations with different values of the coupling length l were made in order to show the coupling efficiency from the flexible to the rigid waveguide and vice versa as function of the coupling length (see Fig. 6.23). For the coupling from the flexible to the rigid waveguide, the coupling efficiency first increases slowly, but then faster up to about $l = 600$ μm. Then, the curve increases again only slowly since there is already more than 50% coupled over for $l = 1000$ μm. The quite fast increase in the middle region can be explained by the circle segment cross section. This means that with increasing coupling length l also the lateral extension of the coupling area increases so that the coupling area itself increases with about the square of l. A similar relative increase can also be observed for the coupling from the rigid to the flexible waveguide, but the absolute value of the coupling efficiency remains always much smaller. The ratio of the coupling efficiencies from the flexible to the rigid waveguide and from the rigid to the flexible waveguide increases from 2.0 for $l = 100$ μm to a maximum of 3.0 for $l = 600$ μm and then decreases again slightly with an average value of about 2.5, which again shows the asymmetric behavior of the coupler.

6.3 Performance of the Asymmetric Optical Bus Coupler

6.3.1 Measurement Setup

The measurement of the AOBC is very complex. The setup used for this purpose is shown schematically in Fig. 6.24 and 6.25 in reality. It is necessary to couple in and out of the sample waveguide at four points simultaneously. To avoid coupling losses which could affect the actual measurement, it is recommended to use active coupling. This means that the coupling is varied via linear axes until an optimal coupling is achieved. In addition, a camera system monitors the overlap area to ensure constant coupling conditions. Furthermore, the module waveguide has to be fixed in its position so that it follows a predefined radius, which can be realized using corresponding radius templates.

With the help of the described measurement setup, it is now necessary to provide experimental evidence for the function of the AOBC. Using a red laser, the interruption-free coupling can already be visually verified for aerosol jet printed waveguides. Figure 6.26 shows an image of a successful module-to-bus coupling without the need to interrupt one of the two waveguides. For better visibility, a simple stamp replaces the radius template.

The visual experiment already proves that the interruption-free coupling works with printed waveguides. However, the repeatability to print waveguides with the exact same properties is quite challenging. Hence, standard $50 \times 50\ \mu m^2$ waveguides ($NA = 0.25$) fabricated by photolithography are used for the quantitative verification of the coupling principle to eliminate additional influences by varying waveguide properties.

The greatest challenge for the coupling experiments is the monitoring of the overlap area during coupling, which results from the applied coupling force. A specially developed and patented [36] visual real-time monitoring of the coupling point is used for this purpose. A camera is directed through a glass carrier (no. 1 in

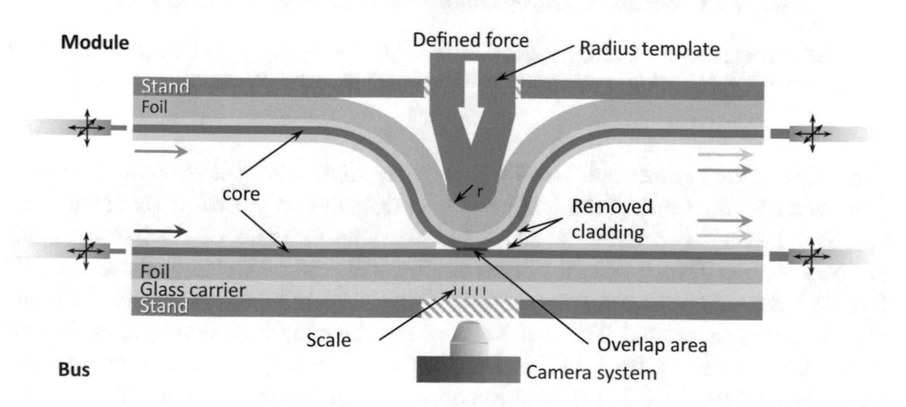

Fig. 6.24 Schematic of the experimental coupling setup

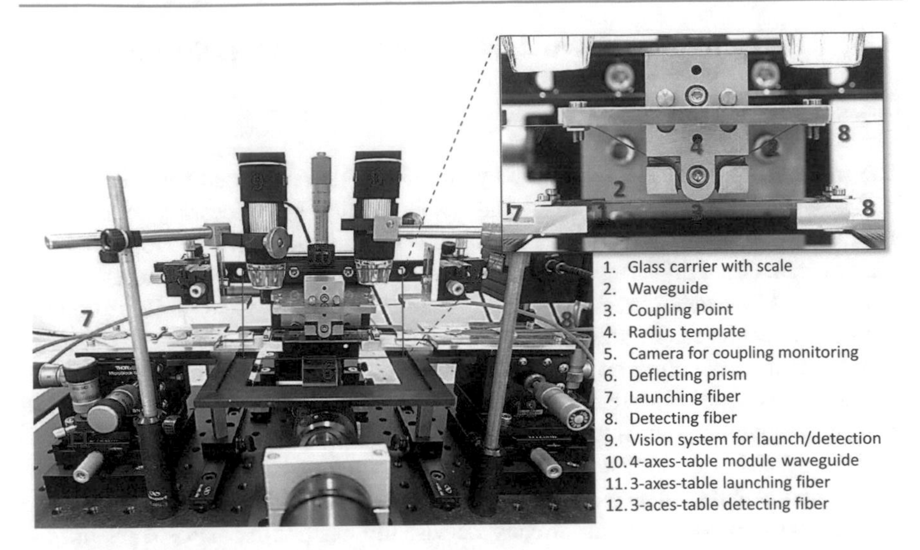

Fig. 6.25 Photograph of the experimental coupling setup

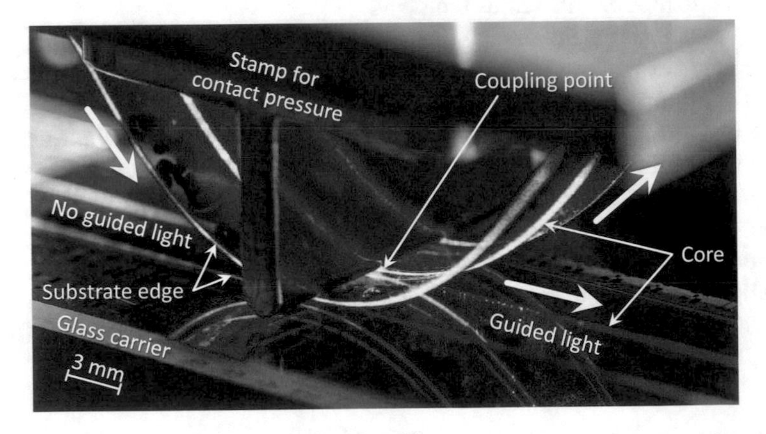

Fig. 6.26 Photograph of the interruption-free waveguide coupling of two aerosol jet printed waveguides at $\lambda = 635$ nm (a simple stamp replaces the radius template for better visibility)

Fig. 6.25) at the contact point of the two waveguides. Using a scale on the transparent carrier, the overlapping lengths can be determined precisely. By using a red auxiliary laser, it is possible to make the overlapping areas of the two coupling partners visible. The comparison between Fig. 6.27 and 6.28 clearly shows the difference between coupled and uncoupled AOBC, whereby in both images the module waveguide is excited. The scattered light at the coupling point in combination with the scale allows for a measurement of the overlapping area. A micrometer screw (no. 10 in Fig. 6.25) is used to adjust the contact pressure, which also varies the overlap area. It is revealed, that the width of the coupling area is constant,

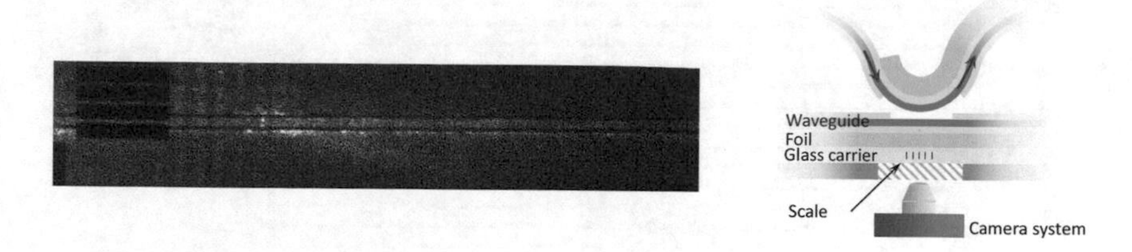

Fig. 6.27 Camera recording of the module waveguide through the glass carrier and the foil substrate **without** coupling

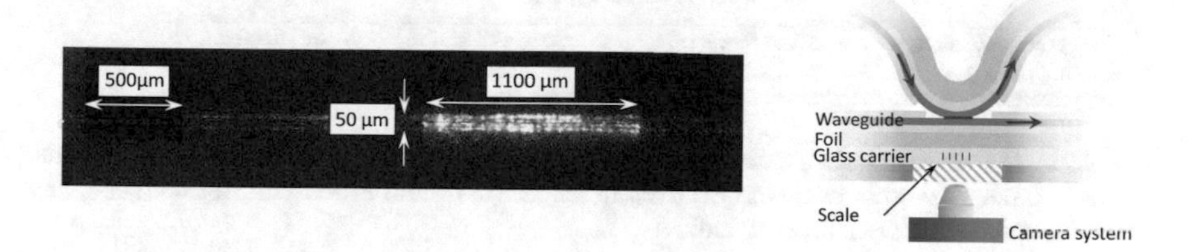

Fig. 6.28 Camera recording of the module waveguide through the glass carrier and the foil substrate **with** asymmetric optical bus coupling; Because of scattered light, the overlap area is visible

because of the rectangular cross section of the test waveguide. Hence, the following measurements are only related to the overlap/coupling length *l*. In addition, camera-based monitoring allows the two coupling partners to be precisely aligned to ensure constant measurement conditions when using different waveguide samples.

6.3.2 Coupling Results

The results of the coupling rate analysis at $\lambda = 850$ nm with relation to the overlap length are summarized in Fig. 6.29, while the parameters of the experiment are shown in Table 6.5. The diagram shows the coupling rates in relation to the coupling length for two different radii ($r = 5$ mm and $r = 10$ mm). For a better comparability to the simulation, the coupling rates are normalized to the sum of the output powers P_{coupl} and P_{main}, as they are defined in Fig. 6.12. Hence, losses are eliminated. The experiment shows the influence of the bent waveguide on the coupling and proves the asymmetric coupling behavior. As expected, the asymmetrical coupling depends on the radius for the module-to-bus coupling (blue and green graph), whereas the opposite coupling direction remains unaffected (red and orange graph). The asymmetric behavior benefits from a smaller bending radius of the module waveguide. However, increased bending losses need to be considered

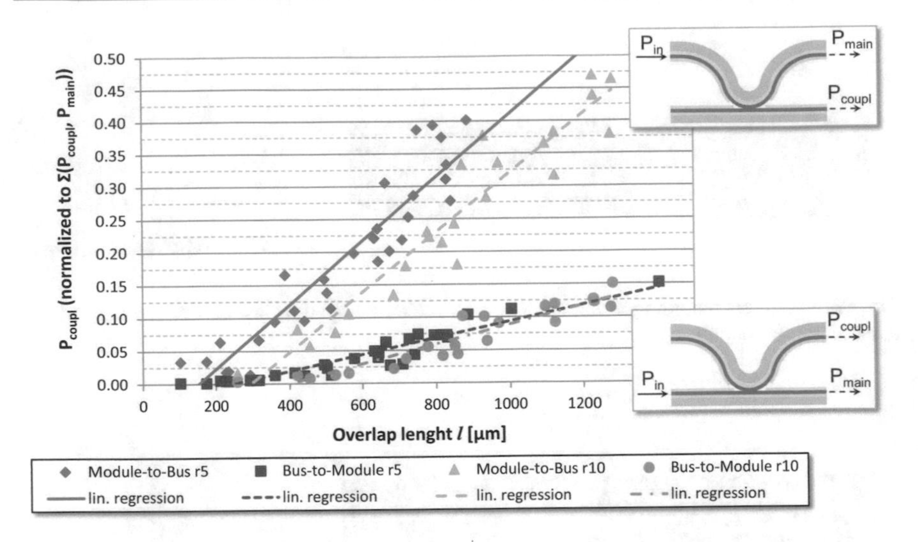

Fig. 6.29 Measured coupling rates and the corresponding linear regression normalized to the sum of the output powers related to the overlap length (the width is constant) for both coupling scenarios and two different bending radii

Table 6.5 Overview of the boundary conditions and parameters for the coupling experiment

Parameter	Value	Note
Wavelength	$\lambda = 850$ nm	Standard wavelength for optical short-range connections
Waveguide dimensions	$(50 \times 50)\ \mu m^2$	
Numerical aperture waveguide	$NA = 0.25$	Standard NA for optical short-range connections
Length of the waveguide samples	$l_w \approx 90$ mm	
Diameter source	$d_{source} = 10\ \mu m$	Fiber-coupled semiconductor laser
Numerical aperture source	$NA_{source} = 0.1$	Comparable to commercially available VCSELs
Diameter detector	$d_{detect} = 200\ \mu m$	Fiber-coupled integrating sphere
Numerical aperture detector	$NA_{detect} = 0.39$	
Bending radius	$r = \{5\ mm, 10\ mm\}$	
Coupling length	$150\ \mu m \leq l \leq 1400\ \mu m$	Less than 150 µm are not detectable. After 1400 µm, the coupling length cannot be further increased by additional pressure
Index matching	None	

for smaller bending radii. Furthermore, higher coupling rates are achieved for longer overlap lengths; i.e., the coupling rate is adjustable. It is revealed that the coupling rates increase linearly. The slope of the graph is steeper for the module-to-bus coupling, which means the asymmetry of the coupler increases with greater overlap areas.

Besides the consideration of the coupling rates themselves, the attenuation caused by the asymmetric bus coupling plays a significant role. This value is determined as follows: First, the output power of the system in the uncoupled state is measured. With inactive coupling, the attenuation of the waveguide under consideration (bus or module) is thus obtained. If light is now transmitted into the second waveguide via the asymmetrical coupling point, the output power of the system needs to be measured again. In this case, this is the sum of P_{coupl} and P_{main}. This gives the insertion loss of the system including the AOBC. By subtraction, the attenuation at the coupling point a_K can then be determined. This does not take into account the attenuation in the coupling waveguide after the coupling point. Due to the selected detector fiber, however, it can be assumed that this attenuation can be neglected.

The results, as depicted in Fig. 6.30, show that the bus coupling has little effect on the system's attenuation at short coupling lengths. If the coupling length is increased, the losses at the coupling point also increase. This is caused by increasing scattered light, which is emitted from the system in the area where the two waveguides overlap, as it can be seen in Fig. 6.28 when determining the coupling length. In this context, the asymmetrical coupling behavior leads to higher losses for module-to-bus coupling than in the opposite direction.

Since the attenuation at the coupling point depends on both the radius and the coupling length, it is necessary to consider the coupling rates under these aspects. For this purpose, the coupling power is shown again in Fig. 6.31 (as already

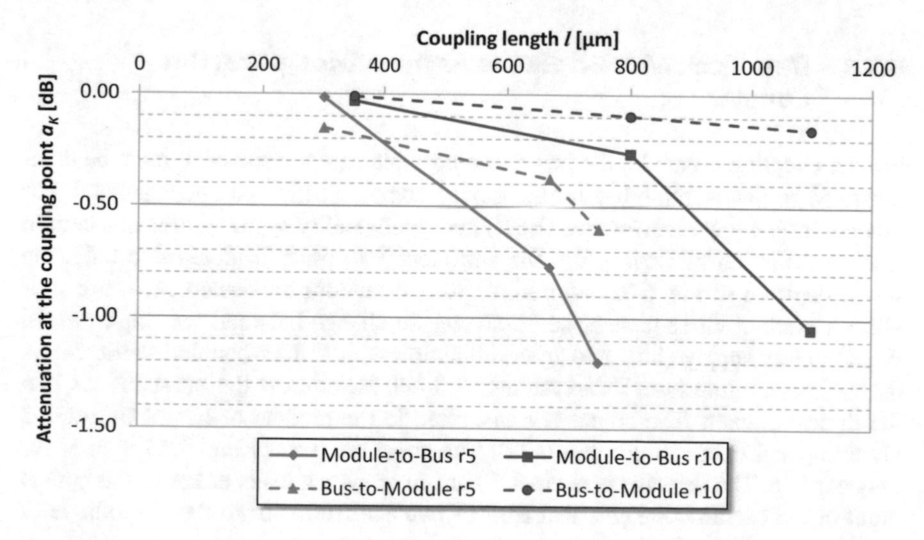

Fig. 6.30 Attenuation at the coupling point depending on the coupling length for different radii

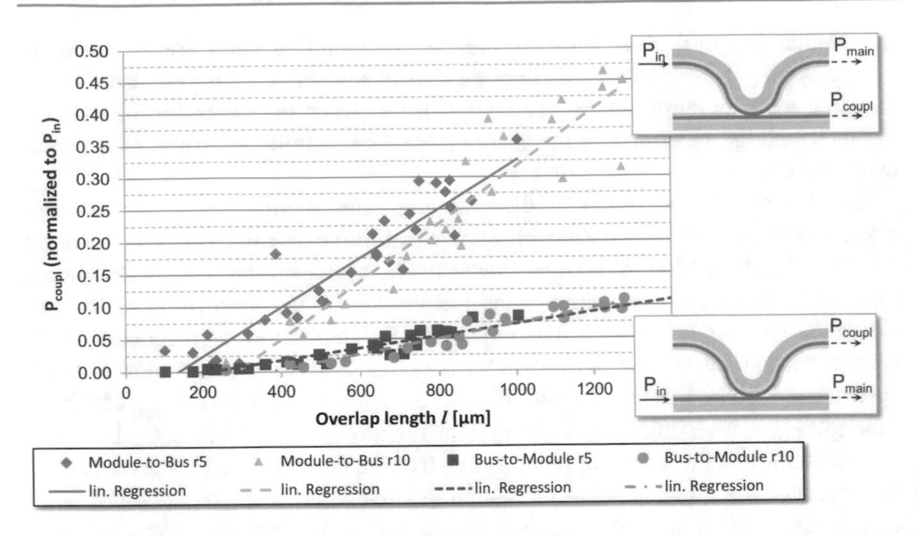

Fig. 6.31 Measured coupling rates and the corresponding linear regression **normalized to the input power** related to the overlap length (the width is constant) for both coupling scenarios and two different bending radii

shown in Fig. 6.29), but with the difference that this time it is normalized to the input power P_{in} (not the sum of P_{coupl} and P_{main}). It can be seen that, especially at longer overlap lengths, the coupling rates are lower because the attenuation at the coupling point becomes increasingly significant. Furthermore, it is revealed that higher coupling rates at smaller bend radii are more and more compensated by the losses at longer coupling lengths. Nevertheless, the asymmetric behavior is still very clearly visible.

6.3.3 Data Transmission via the Asymmetric Optical Bus Coupler

For the coupling principle, the proof of a possible data transmission must be demonstrated so that a suitability for the device communication can be evaluated. For this purpose, a low bit error rate (BER) and good quality eye diagrams at a certain data transmission rate can proof the sufficiency. In order to determine both, the setup, depicted in Fig. 6.32, is used. A pseudo-random bit-stream generator provides a random bit sequence, i.e., a stochastic change between two logic levels. A DC power supply (bias-tee) in combination with a fiber-coupled vertical-cavity surface-emitting laser (VI-Systems V40-850 M) convert the bit sequence into the optical domain. The signal is transmitted to the module or bus of the AOBC via an optical fiber (50 µm, NA 0.22). The same setup as in Fig. 6.25 is used for the coupling. The detector is again a 50 µm fiber, which leads either to the optical input of the oscilloscope (Anritsu BERTWave MP2100B) or to the photodiode of

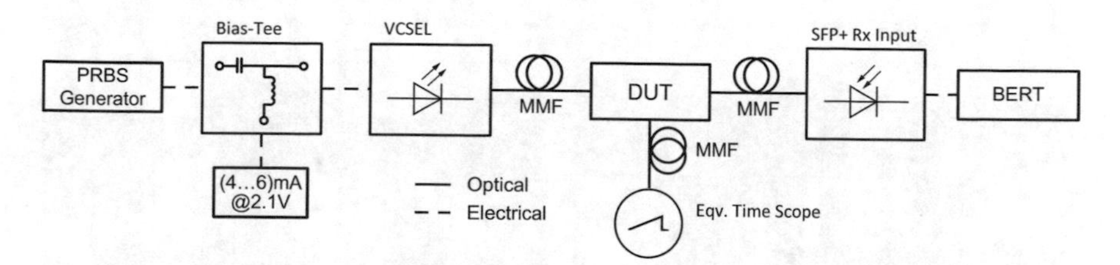

Fig. 6.32 Measurement setup for the eye diagram and bit error rate analysis of the AOBC

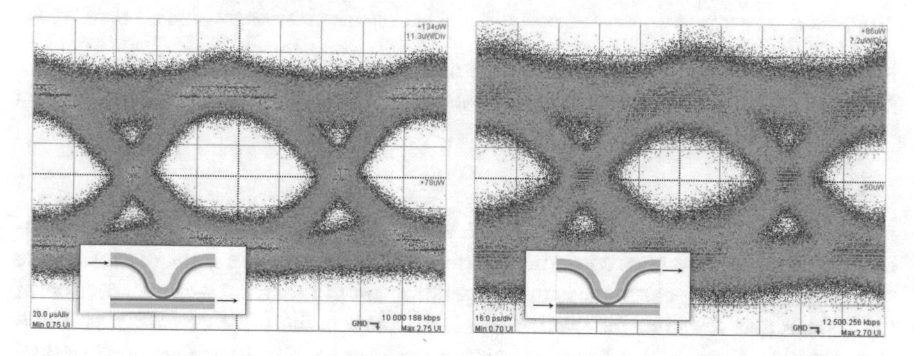

Fig. 6.33 Eye diagrams at 12.5 Gbit/s (left) for the module-to-bus coupling at −9.1 dBm and (right) for the bus-to-module coupling at − 10.8 dBm (each at l = 1400 μm, r = 10 mm)

an SFP+module, which converts the optical signal back to the electrical domain to use the electrical input of the BER tester (Anritsu BERTWave MP2100B).

The results at the output of the transmission path are shown in Figs. 6.33 and 6.34. The former shows the two possible coupling paths, whereas the latter image presents the signal remaining in the bus. The remaining signal in the module waveguide is not relevant for a later application and can be omitted. It is revealed that the AOBC is in principle suitable for data transmission at high data rates (min. 12.5 Gbit/s). However, the received signal for the bus-to-module coupling is of lower quality. This can be explained by the modes that couple into the second waveguide. While only higher-order modes are coupled from the straight bus waveguide to the module, also lower-order modes are coupled in the opposite direction due to the bent waveguide. Hence, the signal quality is better for the module-to-bus coupling. Nevertheless, both signals are still clearly detectable by the receiver. In addition, the remaining signal in the bus waveguide, which is used for further coupling, is of good quality. Compared to the back-to-back signal of the bus waveguide without coupling (Fig. 6.34 left), the remaining signal shows no degradation. Hence, the coupling of further modules to the bus is not negatively affected.

Besides the signal quality, the BER is highly relevant for data transmission paths. The results related to the received optical power, depicted in Fig. 6.35,

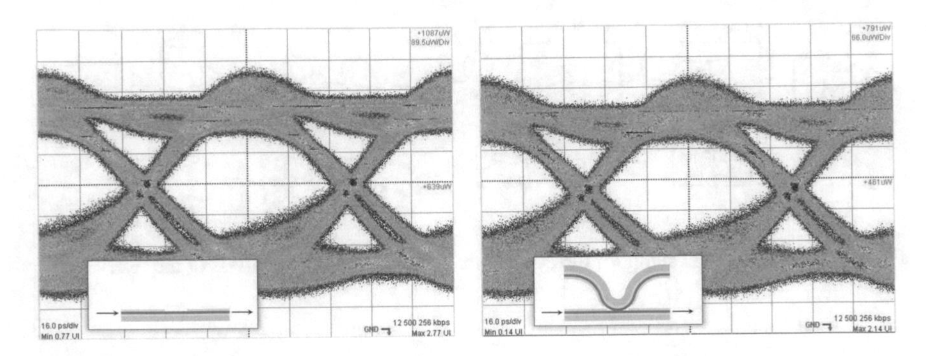

Fig. 6.34 Eye diagram at 12.5 Gbit/s (left) for the back-to-back measurement for the bus waveguide and (right) at − 1.5 dBm for the remaining signal in the bus after coupling (same coupling sample as in Fig. 6.33)

reveal an error rate between 10^{-3} in the worst case and 10^{-12}, which is the minimum of the measurement equipment. This represents the pure transmission of the data without any signal processing. Depending on the forward error correction of the transmitted signal or the evaluation electronics on the receiver side, it is therefore possible to receive the transmission error-free even at lower received optical power. Furthermore, it is revealed that the BER behavior is independent of the coupling direction, which underlines the bidirectional character of the AOBC.

6.3.4 Long-Term Stability of the AOBC

After successfully testing the AOBC in terms of coupling and data transmission, the next step is the analysis of the long-term stability. In a first experiment, the coupling is monitored during seven days. Figure 6.36 shows the results for two different sets of waveguide samples manufactured by photolithography and aerosol jet printing. The results are normalized to the initial starting value, which is marked by a red line in the diagram. In both experiments, the coupling remains stable until the end of the measurement, except for some minor fluctuation, which is probably caused by small temperature changes. Hence, the long-term stability of the AOBC under constant conditions is not influenced by the manufacturing method of the waveguides.

The measurement setup is now modified, in order to test the AOBC under changing temperature conditions. For that, heating elements are added to the metal radius template, which carries the module waveguide, and to the glass carrier, on which the bus waveguide is applied. Figure 6.37 shows two images of the extended setup. Furthermore, important for the measurement is the mechanical decoupling of the samples fixing points. Since the radius template and its guidance are made of metal, it expands during heating. Hence, this would affect the coupling between waveguide sample and launching/detecting fiber (no. 7/8

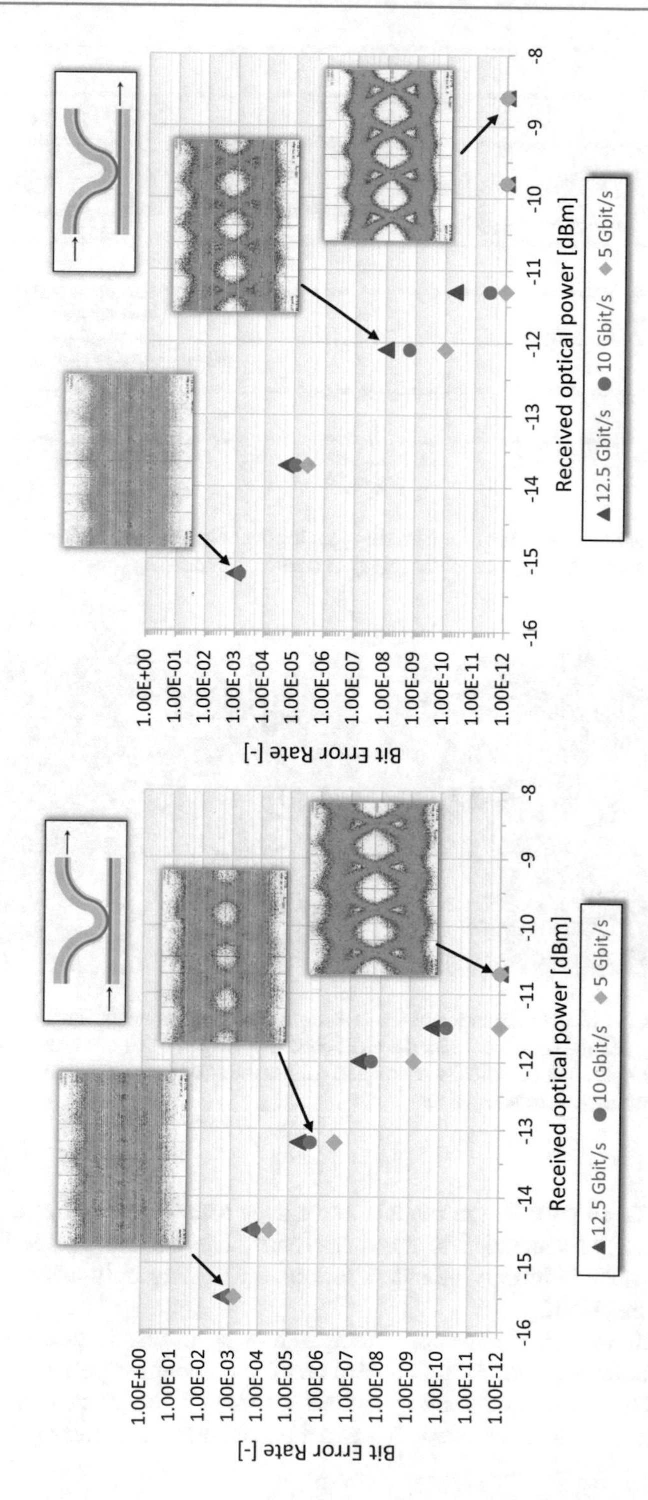

Fig. 6.35 Bit error rate related to the received optical power for the two coupling scenarios @ l = 1400 µm

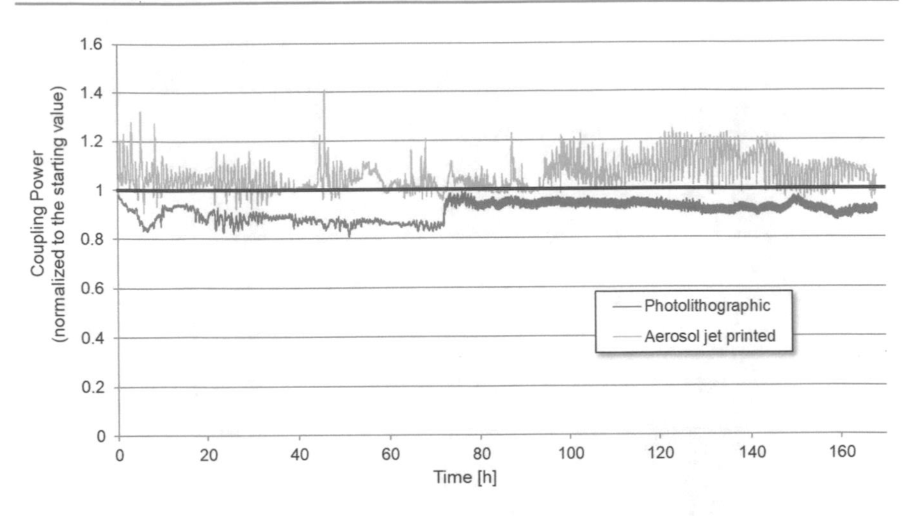

Fig. 6.36 Long-term measurement for a bus-to-module coupling with two different set of waveguide samples, manufactured by photolithography and aerosol jet printing

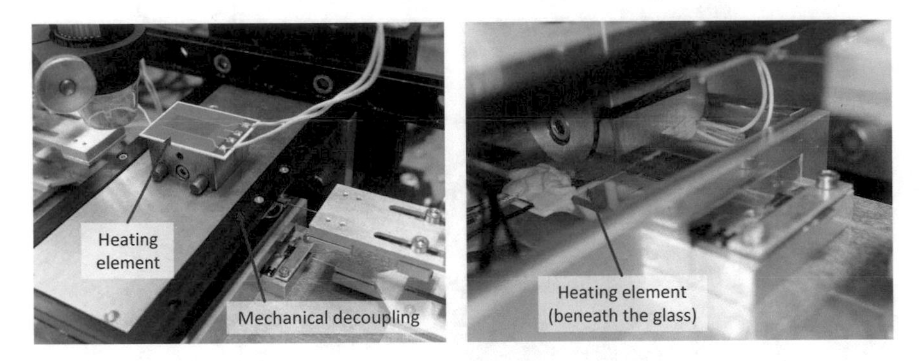

Fig. 6.37 Extension of the measurement setup with two heating elements (left) on top of the metal radius template and (right) beneath the glass carrier of the bus waveguide; the mounting of the waveguide samples is mechanically decoupled from the rest of the setup to avoid undesired losses at the launching and detecting fibers

in Fig. 6.25). To avoid this, the module waveguide sample is fixed at a polymer frame, which is not connected to the metal parts. The bus waveguide lies on a glass carrier, which is loosely mounted. Hence, the heating only affects the coupling point of the AOBC.

With the aforementioned setup, a temperature cycle test is performed. The high and low point is set to 60 °C (30 min) and 22 °C room temperature (60 min), respectively. The temperatures are measured at three points (module waveguide, bus waveguide and room temperature) and implemented in a control loop. The

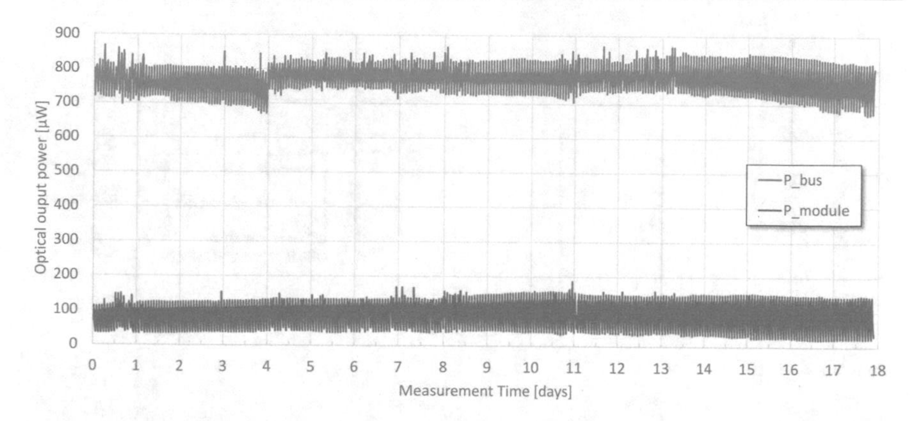

Fig. 6.38 Long-term stability of the bus-to-module coupling during a 60 °C/22 °C temperature cycling test with photolithographic manufactured waveguides

AOBC is measured with photolithographic waveguides with rectangular cross sections first. Figure 6.38 shows the measured optical output powers of both waveguides for a bus-to-module coupling. The experiment reveals two main aspects. First, both optical powers remain stable over 18 days, proving successfully the long-term stability. Second, the optical power alternates during the whole measurement. By taking a closer look onto two temperature cycles in Fig. 6.39, it is revealed that with increasing temperature, the coupled power in the module waveguide increases as well. On the other hand, the remaining power in the bus waveguide decreases accordingly. As the temperature decreases, the power levels return to their initial state. The reason for that is the temperature influence on the waveguide mounting. Especially the metal radius template expands and, hence, increases the pressure on the coupling point. The result is an enlarged contact area, as it is depicted in Fig. 6.40, which leads to higher coupling rates. Hence, stability of the AOBC mainly depends on the mounting and later packaging of the waveguides.

A microscope image of the bus waveguide sample after 18 days of temperature cycling test reveals only minor damages to the waveguide. Figure 6.41 shows some small cracks on the waveguide surface along the direction of propagation, which are caused by the alternating mechanical stress at the coupling point. However, no significant influence on the optical power could be observed.

With aerosol jet printed waveguides, a second temperature cycle test is performed with the same test conditions. Figure 6.42 shows the results of the experiment, revealing the same alternating behavior as in the aforementioned measurement with the photolithographic waveguide samples. However, the overall coupling power—which is the optical output of the module waveguide—decreases slowly during the experiment until the connection is lost after seven days. Parallel to this, the remaining power in the bus waveguide increases, since a smaller amount of power is coupled to the module waveguide. Since the aerosol jet printed

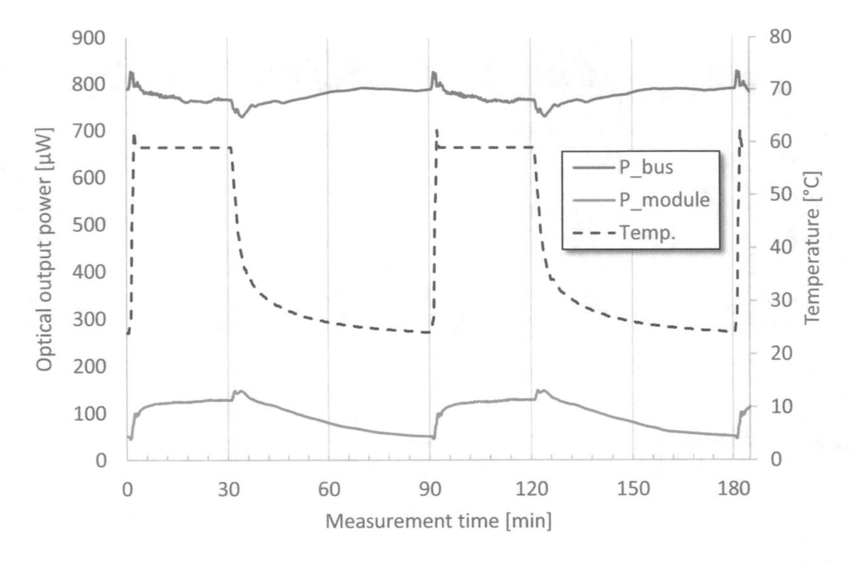

Fig. 6.39 Detailed illustration of two temperature cycles with the according optical output powers for a bus-to-module coupling

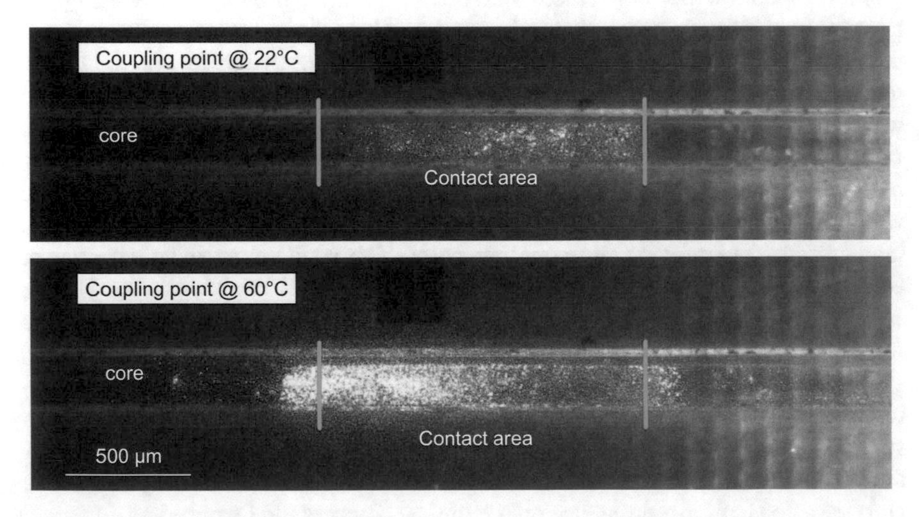

Fig. 6.40 Contact area increases at higher temperatures resulting in higher coupling ratios

waveguides have a circular segment cross section, they slowly slip away from each other until the coupling is interrupted.

Overall, the first tests on the long-term stability and reliability of the AOBC revealed its vulnerability to temperature changes. The alternating optical power mainly depends on the AOBC's package and has to be considered in future applications, since it could affect sensitive signals. Furthermore, the waveguide shape

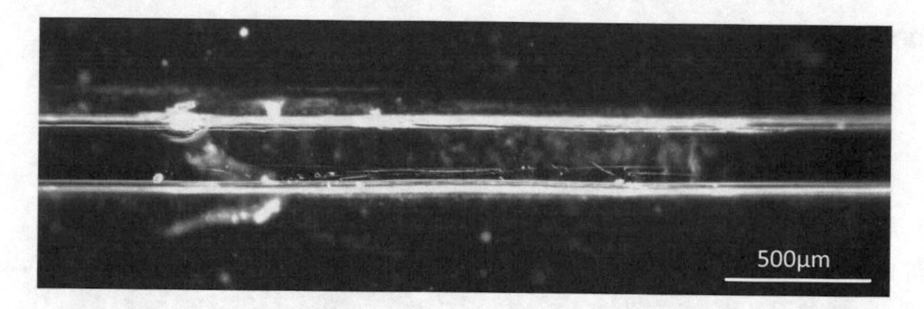

Fig. 6.41 Microscope image of the bus waveguide after 18 days of temperature cycling; despite the output power is still on the initial level, the waveguide shows some minor damages

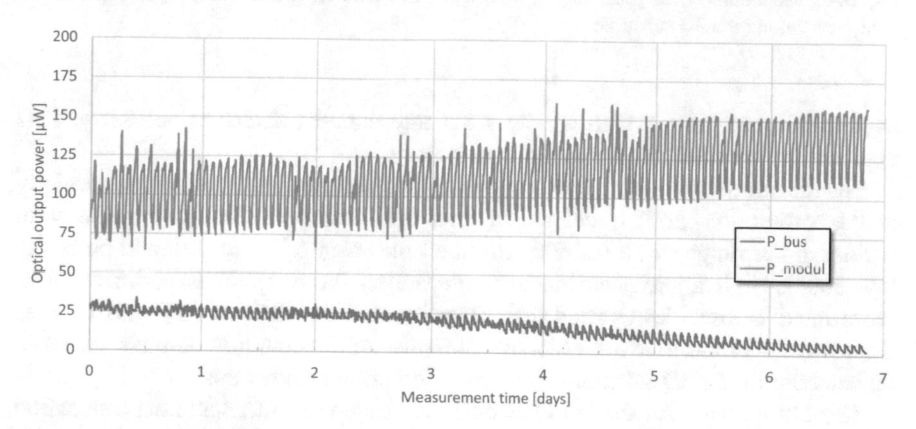

Fig. 6.42 Long-term stability of the bus-to-module coupling during a 60 °C/22 °C temperature cycling test with aerosol jet printed waveguides

has a major influence on the long-term stability. A flat top surface is preferable as it provides a better mechanical grip. Otherwise, a mechanical guidance has to be implemented, which prevents slipping of the coupling partners.

6.4 Technologies for Three-Dimensional Electro-Optical Interconnects

6.4.1 Packaging Demands and State-Of-The-Art for 3D-Opto-MID

After proving the functionality of the coupling concept in simulation and experiment, it is now necessary to present concrete solutions for the packaging of the asymmetric optical bus coupler. A schematic of such a package is depicted in Fig. 6.43, where the part with the bent waveguide is defined as module and the straight one as bus. As mentioned earlier, the defined radius is crucial for the

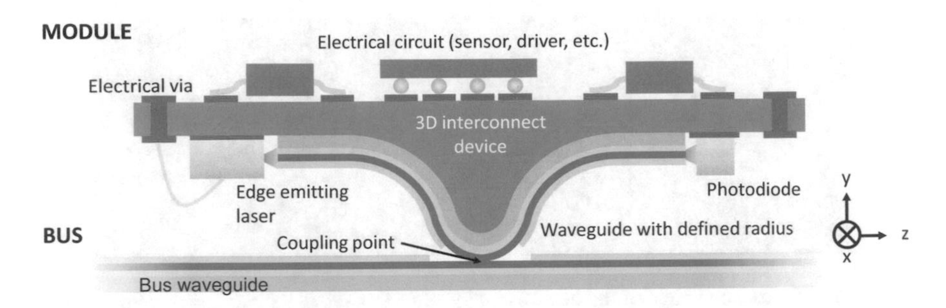

Fig. 6.43 Schematic of a packaging approach for the AOBC, where the module contains the bent waveguide on a 3D substrate

asymmetric coupling, which is why a 3D interconnect device is necessary. This explains the need for a 3D-Opto-MID assembly.

For the coupling itself, the two coupling partners need to be aligned especially in the x-direction, as it is defined in Figs. 6.10 and 6.43. For waveguides with widths in the range of (100…400) μm, the alignment tolerances should be in the low double-digit micrometer-range. Furthermore, the coupling experiments with the printed waveguides showed their fragility, as it is depicted in Fig. 6.44. It is therefore important to avoid excessive pressure on the coupling partners. An elastic behavior of the 3D substrate at the coupling point ensures this.

The last demand for the 3D-Opto-MID package is a sufficient heat dissipation of the substrate for the operation of the electro-optical converters. Lasers need a stable operating temperature, especially when they are directly modulated for data transmission. Otherwise, the operating point is shifted and the signal levels are pushed into an impermissible range. The following bullet points summarize the demands:

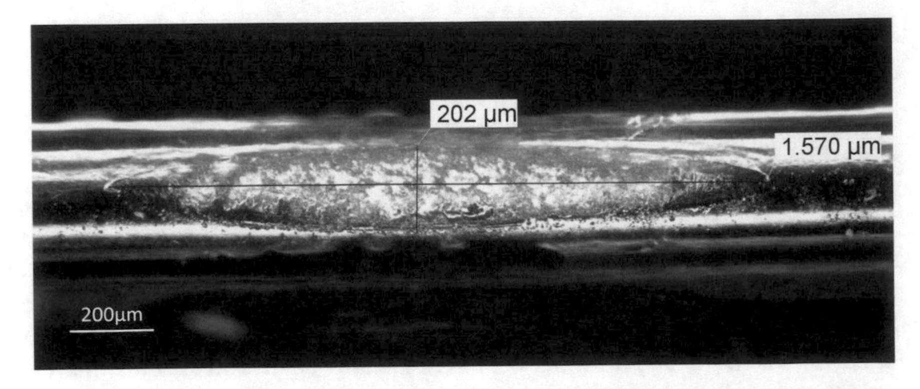

Fig. 6.44 Damaged aerosol jet printed waveguide after a coupling experiment with too much contact pressure

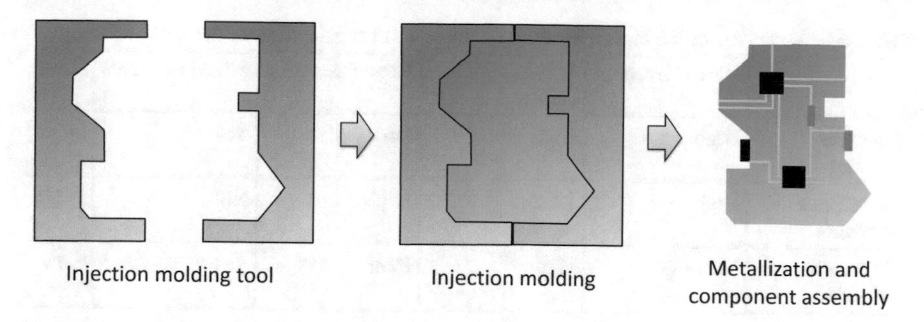

Fig. 6.45 Main process steps for the manufacturing of 3D-MID by injection molding [40]

- 3D-shaped wave form for the defined bending radius of the module waveguide
- Alignment structures with high accuracy
- Elastic behavior of the module substrate at the coupling point
- Good heat dissipation for a stable operation of the electro-optical converters

For the integration of the coupling principle, it is necessary to apply the waveguides to the assembly and to provide a defined radius for the asymmetrical coupling. Two technologies can be used for this purpose: Either the waveguides are directly printed in the desired form [37] or flexible waveguides are applied subsequently [38].

Besides the application of the waveguides, the fabrication of the 3D substrate is crucial for the realization of the coupling approach. Currently, most often injection molding is used for standard 3D-MID. It combines electronic and mechanical functions in one component. For this purpose, the structural assembly is molded from a thermoplastic material. The electric circuit is pre-structured with a laser (laser direct structuring) and metallized or a second, electrically conductive material is injection molded in addition (Fig. 6.45) [39, 40]. Discrete components can then be glued and soldered onto these substrates.

The major drawback of this technology is the heat dissipation of the polymer material. This is especially important for 3D-Opto-MID where electro-optical converters are used. The lasers need stable temperatures to work properly. While ceramic filler materials (up to 50% by volume) can improve the thermal behavior of injection-molded assemblies [41], it is still an order of magnitude behind ceramic substrates. Standard Al_2O_3 thick-film substrates reach 24.0 $W{\cdot}m^{-1}{\cdot}K^{-1}$ [42], while filled polymers for injection molding reach 0.3…4.3 $W{\cdot}m^{-1}{\cdot}K^{-1}$ [43]. Furthermore, the initial effort for an injection-molded substrate is very high, because of the necessary tools. While the fully additive manufacturing of polymer 3D-substrates by stereolithography or fused deposition molding can overcome this drawback, it also suffers from a poor heat dissipation.

Three-dimensional ceramic assemblies offer better thermal management. The injection molding of a composite of organic binders, solvents and ceramic particles is a technology for the production of ceramic structural assemblies with

Table 6.6 Overview of the fabrication technologies worth considering for 3D-Opto-MID

	Design freedom	Substrate quality	Heat dissipation	Mixed materials	Effort
Injection molding	High	High	Poor	Yes	High
3D ceramic structures	High	Poor	High	No	High
3D Printing of polymers	Very high	High	Poor	Yes	Low

excellent heat management. This composite is pressed into a mold similar to polymer injection molding. To form a fully cohesive ceramic matrix, the component is sintered afterward. This causes a volume shrinkage, which often leads to defects in the structure [44]. For metallization, a laser structures the substrate and copper, nickel or gold is chemically deposited. These layers are then bond- and solderable [45].

Furthermore, a fully additive manufacturing via stereolithographic, selective laser sintering, fused deposition molding or ink jet printing of ceramic components is possible. For that, the ceramic particles are hold together by a binder during the printing and sintered afterward [46]. However, common errors of this production method are visible and invisible delamination, poor layer bonding and assembly errors, up to the complete destruction of the assembly as it is removed from the building platform [47, 48]. Furthermore, the pixel size is $(40 \times 40)\ \mu m^2$ [46, 47], which is not accurate enough for waveguide alignment. In addition, the required elastic behavior at the coupling point cannot be achieved with ceramic-only interconnect devices. Hence, a mixed material substrate is necessary. However, because of the shrinkage, additional materials can only be applied after the sintering.

The short look into the state of the art for three-dimensional interconnect devices revealed that there is a variety of technologies for classical (electric-only) 3D-MID. However, for the special demands of electro-*optical* 3D-Opto-MID, all these methods have major disadvantages, as it is summarized in Table 6.6.

6.4.2 Polymer–Ceramic Hybrid Assembly for 3D-Opto-MID

Additive manufacturing is promising for the realization of the defined radius for the AOBC. On the other hand, a polymer-only substrate is not suitable for the electro-optical converters. Hence, this work presents a hybrid approach. For that, a standard thick-film alumina substrate carries the critical electro-optical converters and (optional) other electronic components. On this substrate, a three-dimensional polymer form, necessary for the waveguide coupling, is applied. Depending on the application, a flexible material can be used at the coupling point to avoid waveguide damage. Furthermore, the mechanical alignment- and fixing structures can be printed in the same step. This guarantees small tolerances between the waveguide guidance and the alignment structures, since they are fabricated in the same step. Figures 6.46 and 6.47 show a schematic of the hybrid packaging approach and a CAD model, respectively. While the bus waveguide can be mounted on any substrate or structure, the module with the defined bending needs to be a three-dimensional interconnect device. Hence, this work concentrates on the realization of the module.

Figure 6.48 illustrates the process steps for the fabrication of the polymer–ceramic hybrid package. Because several packages can be fabricated in parallel, it is a panel-level process. The fabrication of the electrical circuit uses standard thick-film technology, which is especially important for the e/o converters. Besides the heat dissipation of the ceramic, the thick-film pads are responsible for the height alignment of the laser and photodiode to the waveguide. Hence, an accurate screen-printing of the paste is crucial. The stereolithographic printing of the three-dimensional polymer part of the package can be optionally used to embed components for a higher package density or to protect the components from environmental influence [49]. After the 3D-print is cured and applied to the ceramic, the waveguides can be assembled to the substrate. After the waveguides are placed, the e/o converters can be visually aligned and assembled by a flip-chip bonder for direct butt coupling. Hence, the alignment of the laser and photodiode is fully passive. In the future, it is aimed for a direct conditioning and printing of the waveguides on the package without the use of an additional flexible foil. In this case, the e/o converters are mounted together with the other electric components and the waveguides are printed directly on the laser and photodiode.

In the stereolithographic printing method, the component hangs upside down on a platform, which is lowered into a resin tank. During the printing process, the platform is then gradually pulled out of the tank while a UV exposer cures the individual layers of the component. After the printing is finished, the component is removed from the build platform. For this work, an "asiga PICO HD27" printer is used, as it is depicted in Fig. 6.49. For curing after the actual printing, an ozone generator is used to generate the UV radiation. After the curing, the printed part can be applied on the thick-film substrate.

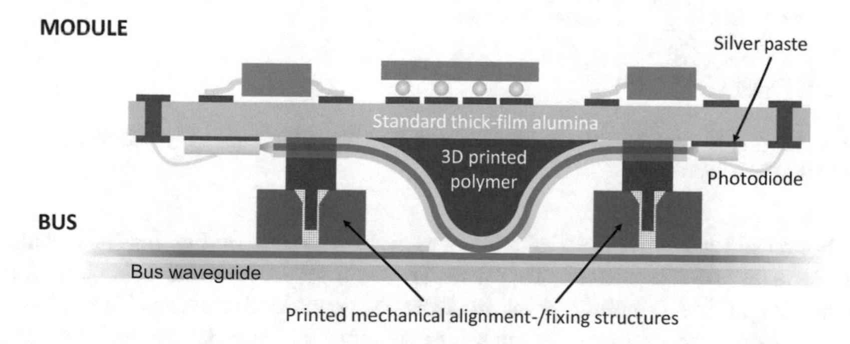

Fig. 6.46 Schematic of a polymer–ceramic hybrid assembly for the AOBC

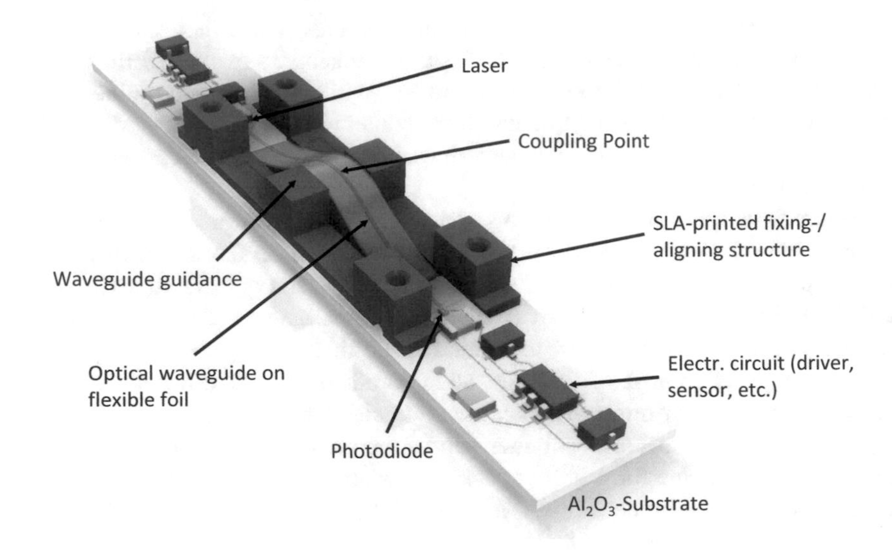

Fig. 6.47 CAD model of the module assembly

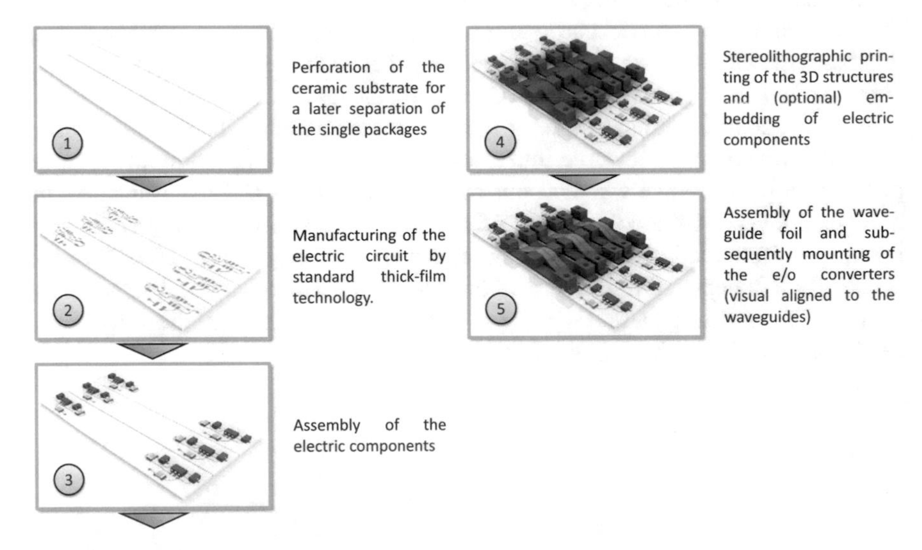

Fig. 6.48 Single steps for the manufacturing of a 3D-Opto-MID assembly with the polymer–ceramic hybrid approach

For printing the module assembly, as it is depicted in Fig. 6.47, "Formlabs FLFLGR02" (GR2) is used. This material has a slightly elastic behavior, which is needed at the coupling point to avoid waveguide damaging. The adhesion of the printed structure to the thick-film substrate is essential for the success of the assembly. For this reason, the adhesion of printed structures to the ceramic

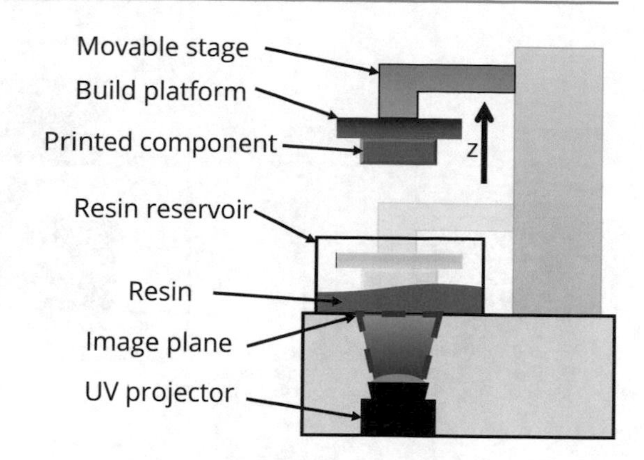

Fig. 6.49 Schematic of a stereolithographic printer

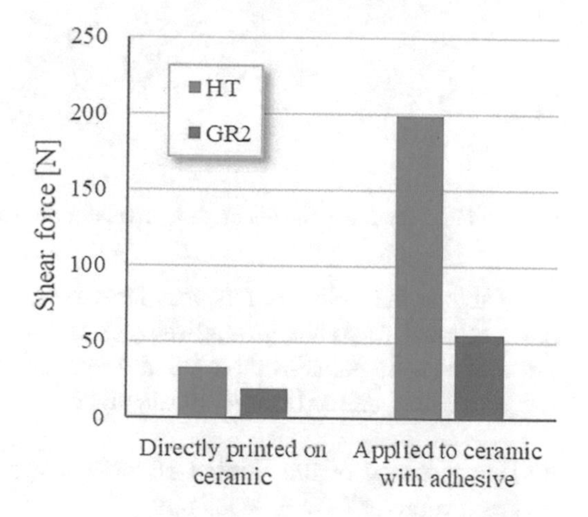

Fig. 6.50 Average shear force for cuboids with a (4 × 4) μm² footprint for two polymer materials

substrate is investigated first. To do this, cuboids with a (4 × 4) μm² footprint were either directly printed on the ceramic or applied to it afterward using an adhesive. After that, a "XYZtec Condor Sigma" shears off the samples and measures the force. For the flexible GR2 material, "Panacol Penloc GTN" is used as adhesive, while "Panacol Penloc GTI" is used for the HT polymer. Figure 6.50 shows the results for two different polymers, the flexible GR2 and the rigid "Formlabs FLHTAM01" (HT), which is later used for the alignment structures of the bus for the coupling experiments. The experiment reveals the poor adhesion of the polymers, if these are directly printed on the ceramic [50]. Since coupling and decoupling of the AOBC induce mechanical stress to the assembly, a strong adhesion between polymer and ceramic is desired, which is why the three-dimensional structures are glued on the ceramic.

A major advantage of the asymmetrical optical bus coupler is its simple alignment. The waveguides only need to be aligned in one lateral and one rotational direction, as shown in Fig. 6.10. Since the distance between the alignment

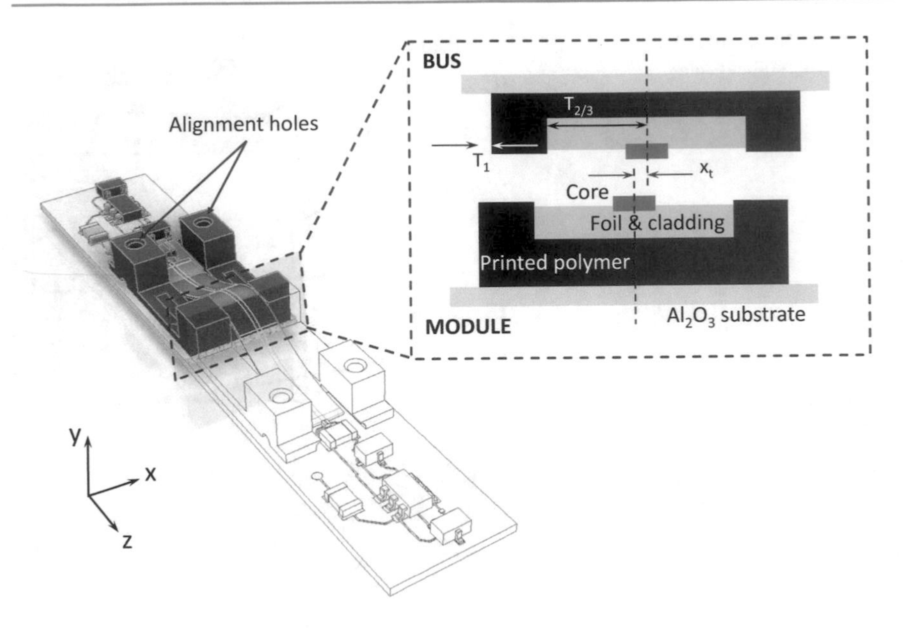

Fig. 6.51 Schematic of the tolerance x_t, crucial for the coupling

structures in the z-direction is very large compared to the width of the waveguide, the rotational tolerance around the y-axis can be neglected. Hence, the offset in x-direction x_t remains as the critical tolerance, as shown in Fig. 6.51. The following parameters define the overall tolerance:

- The accuracy of the printed alignment structures. This was measured with a mean value of $T_1 = \pm 14.25$ μm.
- The width deviation of the printed guiding groove for the waveguide, which is $T_2 = \pm 10.4$ μm if bus and module is considered together.
- The width tolerance of the waveguide substrate, which is defined by the laser with $T_3 = \pm 10$ μm if bus and module are again considered together.

The result is an overall tolerance of $x_t = \pm 34.65$ μm, which is mainly determined by the printed structures and thus by the resolution of the stereolithographic printer. In the presented case, the "asiga PICO HD27" has a minimum pixel size of (27×27) μm^2. Nevertheless, the obtained tolerance is sufficient for multimode waveguides in the width range from 100 μm, as it is the case for aerosol jet printed waveguides.

As shown in Fig. 6.48, the first three manufacturing steps are the preparation of the alumina substrate and the electrical circuit. The latter one depends on the actual application. For demonstration purposes, only a simple circuit is shown in this work. It includes connectors for the power supply, a laser driver, an operational amplifier for the photodiode and a status LED, indicating successful coupling.

First, a laser perforates the ceramic substrate for later separation. The next processing step is to screen print and burn the electrical circuits and pads of the front side and repeat the steps on the back side. The layer thickness after burning is approximately 10 μm, which is why additional conductive layers are printed until the desired pad height is reached to align the e/o converters to the optical axis of the waveguide. Figure 6.52 shows the result.

The assemblies are completed with the 3D-printed parts, the waveguides and the e/o converters. Figure 6.53 shows both the module and the bus assembly ready for coupling. The alignment of the two is realized by pins on the bus assembly, which are then connected to the corresponding holes on the module part. Because of the elastic behavior of the modules printing material (GR2), a tight press fit is achieved, as it is depicted in Fig. 6.54. With the fabricated AOBC in a 3D-Opto-MID package, the coupling can be successfully demonstrated. Figure 6.55 shows the fully passive coupling of the two parts with successful optical signal transmission via the interruption-free waveguide coupling indicated by the green status LED.

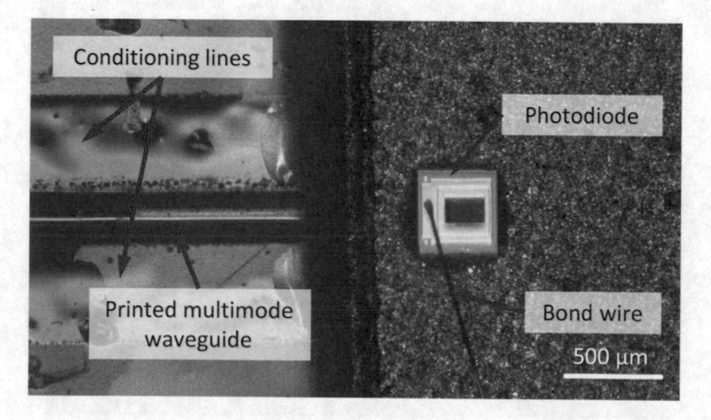

Fig. 6.52 Microscope image of the butt-coupled photodiode on the double-layer thick-film pad

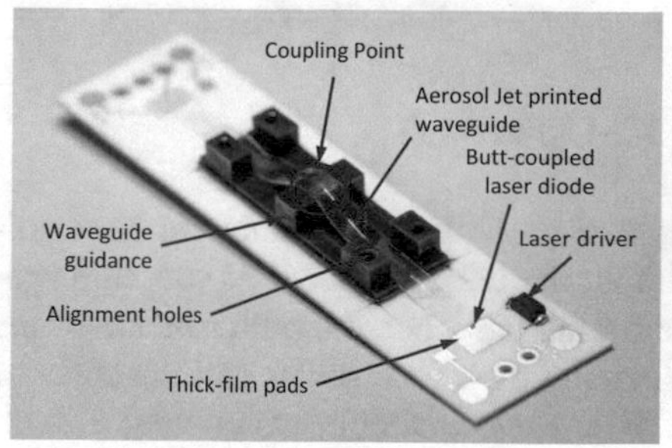

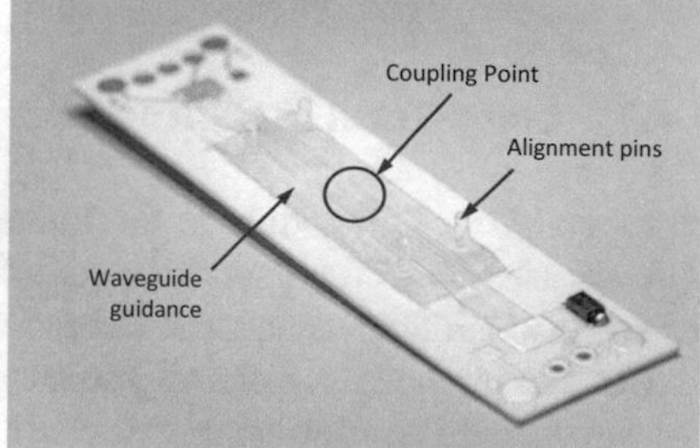

Fig. 6.53 Image (left) of the module and (right) of the bus assembly; the main part of the electrical circuit is located on the back side of the assemblies

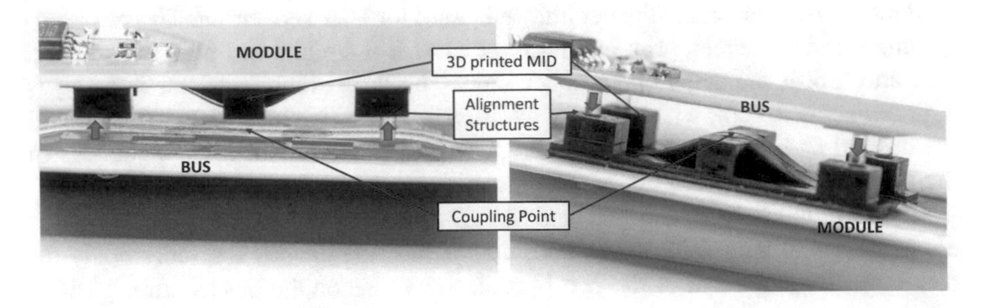

Fig. 6.54 Image of the coupling between bus and module using the printed alignment structures

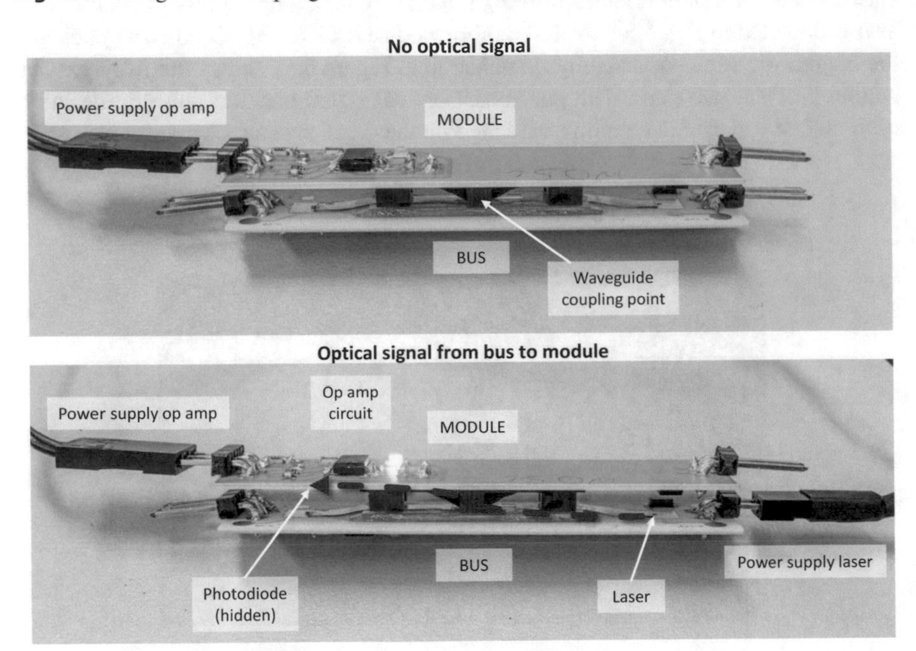

Fig. 6.55 Demonstration of the optical signal coupling from bus to module with a 3D-Opto-MID package, indicated by the green LED

6.4.3 Long-Term Stability of the 3D-Opto-MID

When coupling the module to the bus assembly, both components are mechanically stressed by the contact pressure. With an increasing number of coupling processes, this can lead to material fatigue and thus influence the performance of the AOBC. During coupling, the alignment pins and their receptacles and the cores of the waveguides are particularly stressed. When the assembly is coupled, a compressive load is applied to the alignment components and the waveguides, while the tensile load is applied only to the alignment elements during decoupling. The press fit of the pins in the receptacles can also lead to mechanical abrasion and thus to greater clearance in the positioning of the coupling elements.

To analyze the repeatability, 100 cycles of coupling and decoupling are performed. At the beginning of the measurement, the photocurrent when coupled from the bus to the module is 49.3 μA. Subsequently, the AOBC is manually coupled and decoupled 100 times and the current in the coupled state is measured in each case. Figure 6.56 shows the results of the repeatability measurement, which reveal a decreasing mean value of the photocurrent and an increasing variability. This indicates increasing damage to the assembly. [50]

In order to investigate the reliability of the coupler under different environmental conditions, one module element and one bus element each are aged in the coupled state according to Table 6.7. Meanwhile, the output voltage of the transimpedance amplifier on the receiver side of the coupler is recorded to monitor the function of the coupler.

During the HTS and TH aging, the voltage drops very soon to a few millivolt. The remaining signal is probably caused by scattered light at the photodiode. Figures 6.57 and 6.58 show the results for the HTS and the TH, respectively. At about 75 °C, the physical contact of the coupling waveguides is totally lost. Which is why the AOBC fails. This can be explained by the temperature expansion of the polymer structures at high temperatures, which is why the pressure at the coupling point is reduced until it is totally lost. Hence, a stronger mechanical coupling mechanism is needed to improve the behavior of the AOBC at higher temperatures. Nevertheless, for the HTS experiment, the coupling is restored after the samples reach lower temperatures again, which is also indicated by the green LED on top of the package. Hence, the assemblies are not damaged in particular by the high temperature. Even the initial voltage of 3.7 V is reached again, if the

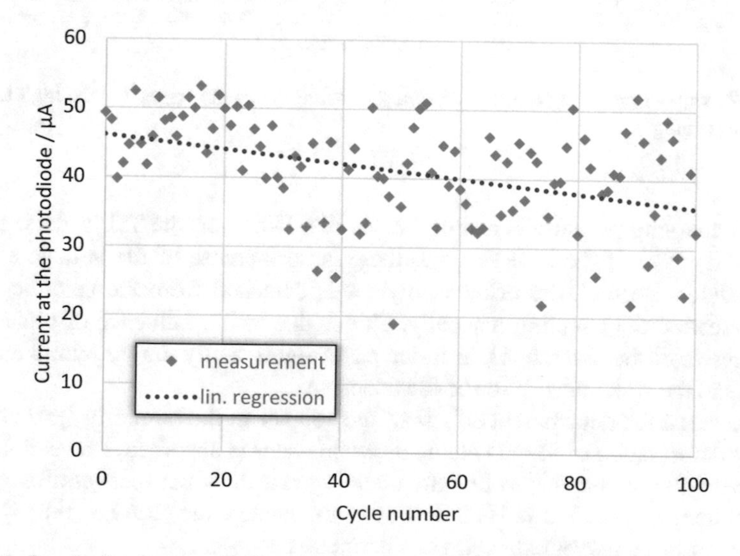

Fig. 6.56 Results for the repeatability of the coupling with the 3D-Opto-MID

Table 6.7 Conditions for the aging of the AOBC

Aging experiment	Conditions	
High temperature storage test HTS	Temperature	85 °C
	Humidity	<40%
	Duration	168 h
Temperature humidity test TH	Temperature	85 °C
	Humidity	85%
	Duration	168 h
Temperature cycling test TC	Min. temperature	− 40 °C
	Max. temperature	85 °C
	Dwell at T_{max} and T_{min}	10 min
	Temperature gradient	2°K/min
	Number of cycles	50

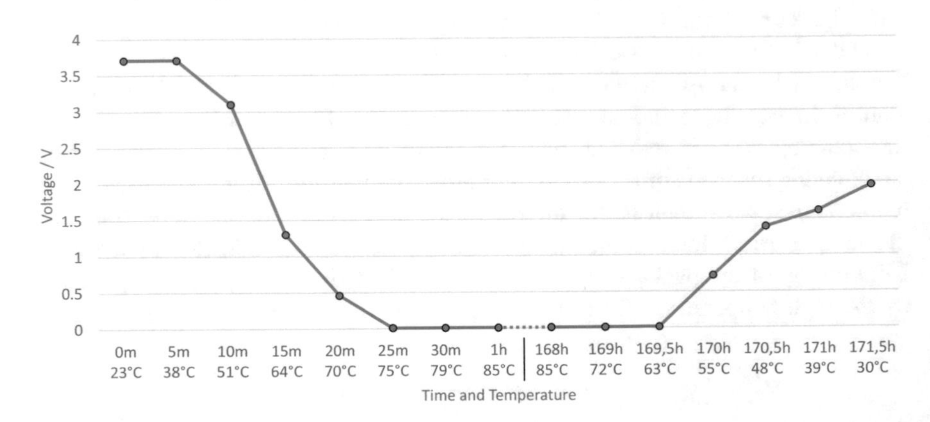

Fig. 6.57 Voltage progression at the receiving transimpedance amplifier for the high temperature storage aging

original coupling pressure is restored manually. However, the TH leads to an irreversible damage of the AOBC. First, there is no increase of the voltage at lower temperatures. Second, the printed polymer is detached from the ceramic, if one tries to restore the coupling manually. This is due to the influence of moisture on the adhesive of the assemblies, as it decreases significantly during humid temperature due to the water absorption of the adhesive.

Because of the detachment effects of the coupler in the humidity–heat chamber and the deformation of the 3D elements when water is absorbed, irreversible damage occurs to the coupler, so that the coupling rate does not recover after cooling to room temperature. While HTS thus does not damage the package, the ingress of moisture must be prevented, e.g., via a hermetic package.

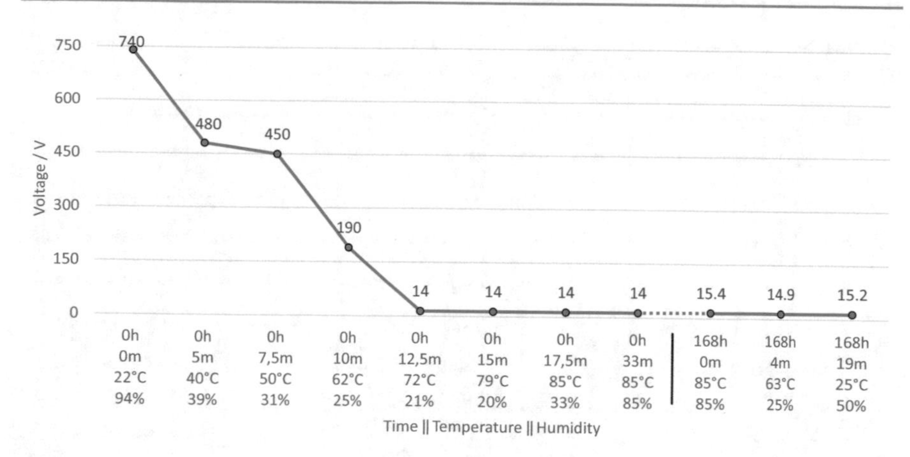

Fig. 6.58 Voltage progression at the receiving transimpedance amplifier for the temperature humidity aging

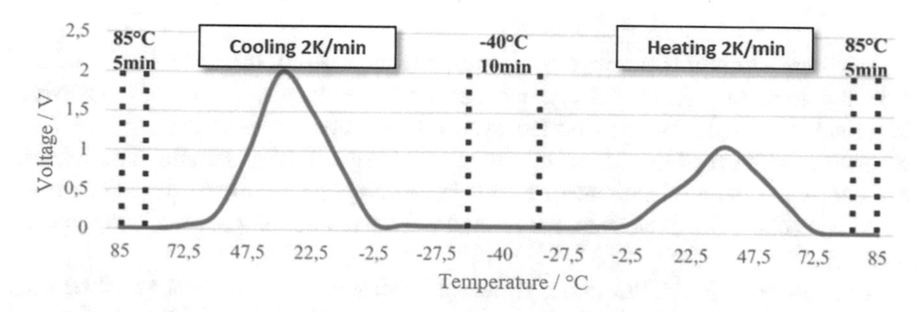

Fig. 6.59 Voltage progression related to the temperature during one cycle

The deformation influence of the 3D-printed elements on the coupling rate becomes obvious in the temperature cycling experiment. The measured voltage changes in correlation to the temperature change, as it is depicted in Fig. 6.59. It is revealed that there are two voltage peaks during one cycle, each around room temperature. A cycle starts at 85 °C. Then, the parts are cooled to −40 °C and again heated to 85 °C. It can be seen that the voltage peaks during cooling are higher than during the heating process. This could be because the printed wave of the module element is relatively massive compared to the rest of the assembly and methyl methacrylates have a rather low thermal conductivity of approx. 0.18 $W \cdot (m \cdot K)^{-1}$. Hence, the temperature inside the wave lags behind the temperature profile, causing it to expand even further during cooling and thus providing a higher contact pressure, while the delayed expansion during heating leads to reduced pressure and thus a smaller contact area of the waveguide cores to each other. [50]

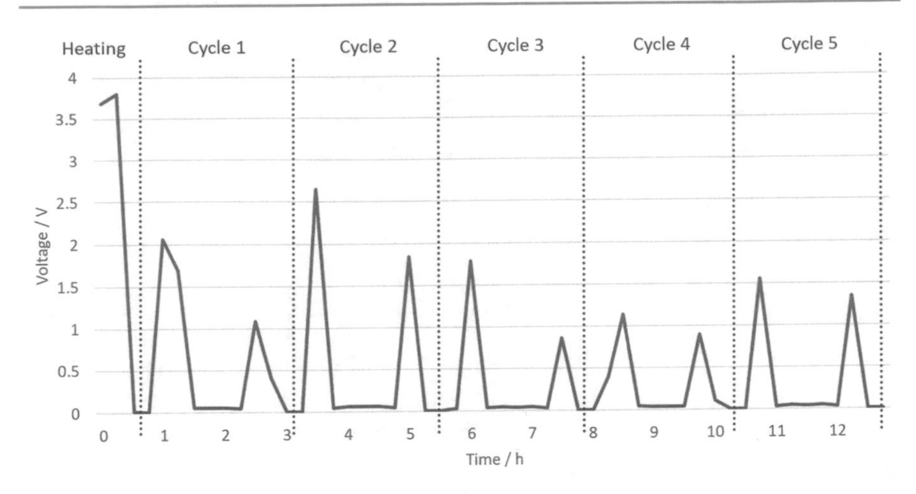

Fig. 6.60 Voltage progression during the first five temperature cycles

With an increasing number of cycles, the voltage at the receiving module steadily decreases. Figure 6.60 shows the first five cycles, where this behavior is already revealed. For the first ten cycles an average voltage peak of 1.33 V is measured, while for the last ten cycles only averaged 0.25 V remain. The reason for that is the material fatigue of all package components due to the alternating stress, especially the adhesives, the printed polymers and the polymer waveguides itself. [50]

The tests of the AOBC under different environment conditions revealed the weak spot of the 3D-Opto-MID package approach. While a failure of optical components and assemblies caused by high humidity is not unusual and can be solved by a hermetic housing, the loss of coupling at dry heat needs to be prevented. The dependence on an always sufficient coupling pressure is the main issue here. This could be solved by the use of other materials with a lower expansion coefficient, by an actively controlled coupling pressure or simply by a stronger mechanical fixing. Nevertheless, with the demonstrated 3D-Opto-MID package approach it is possible to build fully passive aligned asymmetric optical bus couplers with cost-effective and highly flexible manufacturing technologies. Hence, this could open the market for 3D-Opto-MID in the long term.

Acknowledgements We sincerely thank the Deutsche Forschungsgemeinschaft for funding the research group OPTAVER FOR 1660.

References

1. Lorenz, L.: Beiträge zur effizienten Kopplung von optischen Wellenleitern in der Gerätekommunikation. In: Bock, K., Wolter, K., Zerna,T. (Hrsg.) Dresden, TUDPress, Germany (2018)
2. Anderson, D., Beranek, M.: 777 opitcal LAN technology review. In: 48th IEEE Electronic Components and Technology Conference, Seattle (1998)
3. MIT Microphotonics Center: On board optical interconnection digest (2013)
4. Montero, D., Vázquez, C., Möllers, I., Arrúe, J., Jäger, D.: A self-referencing intensity based polymer optical fiber sensor for liquid detection. Sensors **9**(8) (2009)
5. Fischer, U.H.: Optoelectronic packaging. VDE Verlag, Berlin Offenbach (2002)
6. Coutandin, J., Groh, W., Herbrechtsmeier, P., Theis, J.: Verfahren und Vorrichtung zur Herstellung eines Sternkopplers aus Polymer-Lichtwellenleitern. Deutschland Patent Hoechst AG EP0403896 (B1) (1990)
7. Poisel, H., Puls, M., Trommer, G.: Directional coupler for multimode fibers. Europa Patent Messerschmitt-Bölkow-Blohm GmbH EP0346528 (A2) (1989)
8. Bamiedakis, N., Hashim, A., Penty, R., White, I.: A 40 Gb/s optical bus for optical backplane interconnections. J. Lightwave Technol. **32**(8), 1526–1537 (2014)
9. Dou, X., Wang, A., Lin, X., Chen, R.: Photolithography-free polymer optical waveguide arrays for optical backplane bus. Opt. Express **19**(15), 14403–14410 (2011)
10. Karioja, P., Tammela, S., Trevonen, A., Honakanen, S.: Passive fiber optic bus using bidirectional integrated optic bus access couplers. Opt. Eng. **34**, 2551–2559 (2005)
11. Nieweglowski, K., Rieske, R., Wolter, K.-J.: Demonstration of board-level optical link with ceramic optoelectronic multi-chip module. Proc. IEEE 59th Electronic Components and Technology Conference (ECTC) 1879–1886 (2009)
12. van Erps, J., Vervaeke, M., Debaes, C., Ottevaere, H., Hermanne, A., Thienpont, H.: Deep proton writing: A rapid prototyping tool for polymer micro-optical and micro-mechanical components. In: Rapid Prototyping Technology—Principles and Functional Requirements, InTech, pp. 339–362 (2011)
13. Hendrickx, N., van Erps, J., van Steenberge, G., van Daele, P.: Tolerance analysis for multilayer optical. J. Lightwave Technol. **25**, 2395–2401 (2007)
14. Lüngen, S., Tiedje, T., Nieweglowski, K., Lorenz, L., Charania, S., Killge, S., Bartha, J., Bock, K.: 3D optical coupling techniques on polymer waveguides for wafer and board level integration. In: Proc. of IEEE 67th Electronic Components and Technology Conference (ECTC), Orlando, FL, USA (2017)
15. Kagami, M., Kawasaki, A., Ito, H.: A polymer optical waveguide with out-of-plane brancing mirrors for surface-normal optical interconnections. J. Lightwave Technol. **19**(12), 1949–1955 (2001)
16. Jiang, G., Baig, S., Wang, M.: Polymer waveguide with tunable optofluidic couplers for card-to-backplane optical interconnects. In: Proc. SPIE 8991 Optical Interconnects, San Francisco (2014)
17. Rosenberg, P., Mathai, S., Sorin, W., McLaren, M., Straznicky, J., Panotopoulos, G., Warren, D., Morris, T., Tan, M.:Low cost, injection molded 120 Gbps optical backplane. J. Light. Technol. 590–596 (2012)
18. Tan, M., Rosenberg, P., Yeo, J., McLaren, M., Mathai, S., Morris, T., Straznicky, J., Jouppi, N., Kuo, H., Wang, S., Lerner, S., Kornilovich, P., Meyer, N., Bicknell, R., Otis, C., Seals, L.: A high-speed optical multi-drop bus for computer interconnections. In: 16th IEEE Symposium on High Performance Interconnects (2008)
19. Yariv, A.: Coupled-mode theory for guided-wave optics. IEEE J. Quantum Electron. **9**(9), 919–933 (1973)
20. Yang, J., Flores, A.: Array waveguide evanescent ribbon coupler for card-to-backplane optical interconnects. Optics Letters **32**(1) (2007)

21. Flores, A., Song, S., Yang, J., Liu, Z., Wang, M.: High-speed optical interconnect coupler based on soft lithography ribbons. IEEE J. Light. Technol. **26**(13) (2008)
22. Yang, J.: Optical waveguide evanescent ribbon coupler. Patent US 7142748 B1, 28 November 2006
23. Olsen, C., Trewhella, J., Fan, B., Oprysko, M.: Propagation properties in short lengths of rectangular epoxy waveguides. IEEE Photonics Technol. Lett. **4**(2), 145–148 (1992)
24. Nieweglowski, K., Rieske, R., Sohr, S., Wolter, K.-J.: Design and optimization of planar multimode waveguides for high speed board-level opitcal interconnects. In: IEEE 63th Electronic Components and Technology Conference (ECTC), Las Vegas, NV, USA (2013)
25. Kaufman, K., Terras, R., Mathis, R.: Curvature loss in multimode optical fibers. J. Opt. Soc. Am. **71**(12), 1513–1518 (1981)
26. Snyder, A., Love, J.: Reflection at a curved dielectric interface—electromagnetic tunneling. IEEE Transactions on Microwave Theory and Techniques **23**(1), 134–141 (1975)
27. Makino, K., Ishigure, T., Koike, Y.: Waveguide parameter design of graded-index plastic optical fibers for bending-loss reduction. J. Light. Technol. **25**, 2108–2114 (2006)
28. Marcuse, D.: Field deformation and loss caused by curvature of optical fibers. J. Opt. Soc. Am. **66**(4), 311–320 (1976)
29. Papakonstantinou, I., Wang, K., Selviah, D., Fernandez, A.: Transition, radiation and propagation loss in polymer multimode waveguide bends. Opt. Express **15**, 669–679 (2007)
30. Schermer, R., Cole, J.: Improved bend loss formula verified for optical fiber by simulation and experiment. IEEE J. Quantum Electron. **43**(10), 899–909 (2007)
31. Loosen, F., Backhaus, C., Zeitler, J., Hoffmann, G.-A., Reitberger, T., Lorenz, L., Lindlein, N., Franke, J., Overmeyer, L., Suttmann, O., Wolter, K.-J., Bock, K.: Approach for the production chain of printed polymer optical waveguides-an overview. Applied Optics, Bd. 56, Nr. 31, pp. 8607–8617 (2017)
32. Brenner, K. H., Singer, W.: Light propagation through microlenses: a new simulation method. Appl. Opt., AO (Applied Optics), Bd. 32, Nr. 26, pp. 4984–4988 (1993)
33. Lorenz, L., Nieweglowski, K., Wolter, K.-J., Loosen, F., Lindlein, N., Bock, K.: Optical beam propagation and ray tracing simulation of interruption-free asymmetric multimode bus couplers. J. Microelectron. Electron. Packag. **14**, 1–10 (2017)
34. Lorenz, L., Nieweglowski, K., Wolter,K., Bock, K.: Two-stage simulation for coupling schemes in the device communication using ray tracing and beam propagation method, 2018 7th Electronic System-Integration Technology Conference (ESTC), pp. 1–6 (2018)
35. Lorenz, L., Nieweglowski, K., Al-Husseini, Z., Neumann, N., Plettemeier, D., Wolter, K.-J., Reitberger, T., Franke, J., Bock, K.: Asymmetric optical bus coupler for interruption-free short-range connections on board and module level. J. Light. Technol., Bd. 35, Nr. 18, pp. 4033–4039 (2017)
36. Lorenz, L., Nieweglowski, K.: Vorrichtung und Verfahren zum Testen eines zu überprüfenden Lichtwellenleiters. Deutschland Patent 102017204034.B3, 10 März 2017
37. Reitberger, T., Hoffmann, G.-A., Lorenz, L., Wolter, K.-J., Overmeyer, L., Franke, J.: Integration of polymer optical waveguides by using flexographic and aerosol jet printing. In: 12th International Congress Molded Interconnect Devices (MID), Würzburg, Germany (2016)
38. Lorenz, L., Ackstaller, T., Bock, K: Stereolithographic printed polymers on ceramic for 3D-opto-MID. In: Proc. SPIE 11349 3D Printed Optics and Additive Photonic Manufacturing II (2020)
39. Jürgenhake, C., Falkowski, T., Dumitrescu, R.: Classification of MID-prototypes. In: 12th International Congress Molded Interconnect Devices, Würzburg, Germany (2016)
40. Ackstaller, T.: Entwicklung einer opto-elektrischen dreidimensionalen Baugruppe, TU Dresden, Institut für Aufbau- und Verbindungstechnik, Dresden (2019)
41. Hörber, J., Franke, J., Ranft, F., Heinle, C., Drummer, D.: Thermisch leitfähige Kunststoffe für kostengünstige Fertigung und erweiterte Funktionalität in der MID-Technologie. Produktion von Leiterplatten und Systemen (PLUS) **12**, 2870–2886 (2010)

42. Songhan Plastic Technology Co. Ltd.: CeramTEc Rubalit 708S Datasheet: http://www.look-polymers.com/pdf/CeramTec-Rubalit-708S-Alumina-96.pdf. Accessed: Jan. 2020
43. Soltani, M., Liu, Y., Zimmermann, A., Kulkarni, R., Barth, M., Groezinger, T.: Experimental and computational study of array effects on LED thermal management on molded interconnect devices MID. In: 13th International Congress Molded Interconnect Devices, Würzburg, Germany (2018)
44. Hsu, F.-J., Huang, C.-T., Lee, R., Liao, D.: Study on the mechanism of the surface defect of the ceramic injection molded (CIM) insulator. In: AIP Conference Proceedings 2065 (2019)
45. Ermantraut, E., Zimmermann, A., Müller, H., Wolf, M., Hahn-Schickard, W., Ninz, P., Kern, F., Gadow, R.: Laser induced selective metallization of 3D ceramic interconnect devices. In: 13th International Congress Molded Interconnect Devices, Würzburg, Germany (2018)
46. Schwentenwein, M., Homa, J.: Additive manufacturing of dense alumina ceramics. Int. J. Appl. Ceram. Technol. **12**(1) (2015)
47. Scheithauer, U., Schwarzer, E., Moritz, T., Michaelis, A.: Additive manufacturing of ceramic heat exchanger: Opportunities and limits of the lithography-based ceramic manufacturing (LCM). J. Mater. Eng. Perform. **27**, 14–20 (2018)
48. Götze, E., Postler, K., Buschulte, S., Zanger, F.,Schulze, V.: Limits of ceramics in the 3D-MID with additively produced aluminum substrate. In: 13th International Congress Molded Interconnect Devices, Würzburg, Germany (2018)
49. Tiedje, T., Lüngen, S., Schubert, M., Luniak, M., Nieweglowski, K., Bock, K.: Will low-cost 3D additive manufactured packaging replace the fan-out wafer level packages? In: Proc. of IEEE 67th Electronic Components and Technology Conference (ECTC), Orlando, FL, USA, (2017)
50. Lorenz, L., Hanesch, F., Nieweglowski, K., Hamjah, M.-K., Franke, J., Hoffmann, G.-A., Overmeyer, L., Bock, K.: Reliability of 3D-Opto-MID packages for asymmetric optical bus couplers. In: IEEE 71st Electronic Components and Technology Conference (ECTC), San Diego, CA, USA (2021)

7 Feasibility of Printed Optical Waveguides Over the Entire Process Chain by OPTAVER

Gerd-Albert Hoffmann, Alexander Wienke, Stefan Kaierle, Ludger Overmeyer, Mohd-Khairulamzari Hamjah, Thomas Reitberger, Jochen Zeitler, Jörg Franke, Lukas Lorenz, Karlheinz Bock, Carsten Backhaus, Florian Loosen and Norbert Lindlein

The integration of the knowledge of different engineering disciplines to build a three-dimensional optical short-range interconnect system is the core of this book. For this purpose, the physical and technological principles of optical technologies have to be considered. The focus here is on optical transmission technology. Attenuation, the propagation of optical signals, the construction of printed optical waveguides and their manufacturing processes play a special role. Likewise, product development methods of technical systems are a related field; since up to now, especially for 3D-Opto-MID, no adequate methodology exists. Analogies to mechatronic systems and their sub-processes do exist, but these have to be evaluated and expanded with regard to the new optomechatronic components.

A separate procedure that differs from conventional development methods is just as important as the challenges that must be placed on modeling systems, the design and the production of the 3D-Opto-MID. From these defined requirements, it is possible to derive a concept for a 3D optomechatronics CAD software (OMCAD) in Chap. 2, which contains the essential steps for creating these products. In particular, the circuit synthesis and the layout process of the physical

G.-A. Hoffmann (✉) · L. Overmeyer
Institut für Transport- und Automatisierungstechnik, Leibniz Universität Hannover, Garbsen, Germany
e-mail: g.hoffmann@lzh.de

L. Overmeyer
e-mail: ludger.overmeyer@ita.uni-hannover.de

A. Wienke
Production and Systems Department, Laser Zentrum Hannover E. V., Hannover, Germany
e-mail: alexander.wienke@lzh.de

S. Kaierle
Laser Zentrum Hannover e. V.,, Hannover, Germany
e-mail: s.kaierle@lzh.de

J. Franke et al. (eds.), *Optical Polymer Waveguides*,
https://doi.org/10.1007/978-3-030-92854-4_7

development steps from the import of the logical circuitry to the laying of the conductor structures represent the greatest challenge. This is especially true since 3D-Opto-MIDs are applied on complex surface structures.

For this reason, various path planning strategies are considered that can be applied to 3D-Opto-MID components. On the one hand, partially automated 2D routing algorithms can be applied to 3D components through the use of unwinding methods. On the other hand, 3D path planning methods, which have already been presented and discussed in other publications, play an important role, which could also be proven in the context of this work.

In the prototypical implementation presented, it is shown how a software tool for fulfilling the most important functionalities, in connection with the necessary validation functions, can look. Thus, not only manual layouts of electrically characterized components can be modeled, as in the state of the art, but also spatially complex optomechatronic circuits. In addition to path planning, the synthesis of logical circuit diagrams with 3D design, as well as the placement of components on spatially shaped substrates, is of great importance. A significant extension of this functionality is provided by validation options, which on the one hand make it possible to observe production-relevant design rules, but on the other hand also ensure the functionality of the optical conductors. This is made possible by a specially designed interface to a non-sequential raytracing simulation system, which checks the previously planned layouts of the optical conductors.

M.-K. Hamjah · T. Reitberger · J. Zeitler · J. Franke
Institute for Factory Automation and Production Systems, Friedrich-Alexander-Universität Erlangen-Nürnberg, Erlangen, Germany
e-mail: mohd-khairulamzari.hamjah@faps.fau.de

T. Reitberger
e-mail: reitberger.thomas@gmx.de

J. Zeitler
e-mail: jochen.zeitler@neotech-amt.com

J. Franke
e-mail: joerg.franke@faps.fau.de

L. Lorenz · K. Bock
Institut für Aufbau- und Verbindungstechnik der Elektronik, Technische Universität Dresden, Dresden, Germany
e-mail: lukas.lorenz@tu-dresden.de

K. Bock
e-mail: karlheinz.bock@tu-dresden.de

C. Backhaus · F. Loosen · N. Lindlein
Institute for Optics, Information and Photonics, Friedrich-Alexander-Universität Erlangen-Nürnberg, Erlangen, Germany
e-mail: carsten.backhaus@fau.de

F. Loosen
e-mail: florian.loosen@fau.de

N. Lindlein
e-mail: norbert.lindlein@fau.de

The question how three-dimensional printed waveguides with a circular segment cross section can be simulated optically is discussed in Chap. 3. For waveguides whose smallest dimension (10–50 μm) is at least one order of magnitude larger than the wavelength (850 nm), the geometric approach using ray tracing has proven to be suitable. To represent a surface defined by a three-dimensional printed waveguide in a mathematical model in the ray tracing algorithm, a new approach is necessary. By means of the BARc algorithm, a methodology could be found to allow an efficient simulation of up to several million rays propagating in a highly three-dimensional waveguide. The results of the simulation are very useful for the analysis of polymer optical waveguides (POW), not only because of the attenuation value obtained, but also because of the intensity distribution in the whole waveguide. With a stable running and easy-to-use simulation, a wide variety of effects can now be systematically investigated, such as roughness of the surfaces, three-dimensional curves of the POWs as well as macroscopic defects (droplets and inclusions). After measuring the roughness on printed waveguides, it is found that the influence on the attenuation is <0.1 dB/cm and thus not a limitation for the optical system. For curves, minimum radii can be determined, which are used as information in the design of the printing path on three-dimensional functional parts. However, the avoidance of macroscopic defects is found to be decisive for optimal data transfer. So-called droplets and inclusions result from the application of too much or too little material in the aerosol jet printing process. With simulations, it could be shown that sizes typical for the OPTAVER process lead to a high attenuation value. However, if none of these imperfections are present, attenuation values as low as 0.2 dB/cm from measurements can be confirmed by simulation. While the attenuation can be determined by ray tracing, however, the intensity distribution of the simulation differed from the measurement. This can be attributed to the fact that interference effects are not considered in simulations with ray tracing. For this reason, a wave-optics approach to simulate POWs is also applied: the wave propagation method (WPM). By a numerical solution of the Helmholtz equation using the angular spectrum of plane waves, light is propagated along the waveguide axis by dividing the waveguide into small layers. However, before this can be used to simulate POWs, instabilities and errors in the simulation method have to be investigated. The use of the fast Fourier transform (FFT) algorithm having periodic boundary conditions results in unphysical behavior, since light which leaves the simulation field at one side enters it on the other. This is solved by using absorbing boundary conditions of the simulation field, which lead to an absorption of light at the edges. With this modification, it is shown that WPM is ideally suited to simulate POWs. A reflection of a Gaussian beam within a waveguide or even partial transmission and reflection can be simulated directly with the native algorithm. To represent the POWs within the algorithm, it is only necessary to know the refractive index distribution in each layer, so that a complicated surface model as in ray tracing does not arise. With WPM simulations a very good agreement of the speckle-like intensity distribution from simulation and measurement is achieved. But, it can also be verified by averaging the intensity distribution along the waveguide axis over some distance, which eliminates the speckle, that this averaged intensity distribution coincides quite well with the intensity distribution obtained by ray tracing. Finally, Chap. 3 discusses the data exchange between the optical simulation tool and the computer-aided design (CAD) tool. As mentioned earlier, results from the optical simulation are incorporated into the printing design. However, there is also an exchange in the other direction, where the printing path and surface normals from the CAD tool are integrated into the optical simulation tool.

In Chap. 4, the local wetting control of film substrates using flexographic printing is investigated in order to enable the additive manufacturing of optical waveguides in a subsequent process step. In this way, process-specific advantages are exploited in order to achieve increased performance of the optical structures in combination with other processes. After identifying a suitable active principle for printing wetting control by examining the state of the art, the current application areas of functional flexographic printing and its process flow are described. To this end, the system technology used is first described and the characterization, design and selection of printing tools and materials are presented. These are subsequently used to experimentally analyze the printing of conditioning lines and thus develop an understanding of the resulting structural geometry. This is supplemented by modeling of the printing stamp deformation using FEM analysis, which allows an explanation of the process characteristics. In order to evaluate the usability of the conditioned film substrates, the self-alignment of core material on the film surface is first investigated in general. For example, prior conditioning increased the achievable aspect ratio (waveguide height/waveguide width) from 0.039 to 0.167 compared to the aerosol jet process without conditioning lines. This increase results in better compatibility with existing optoelectronic systems and components. To enable the application of the optical waveguides on three-dimensional surfaces, spatial expansion and characterization are performed using thermoforming. The predicted increase in optical performance of optical waveguides fabricated in this way compared to direct printing is demonstrated by determining an average attenuation for aerosol jet printed optical waveguides on conditioned foil substrates.

Conditioning of film substrates by flexographic printing shows that wetting control for the application of optical waveguides is possible. In the future, the locally resolved wetting behavior could also be transferred to other substrate materials and thus optimize not only the application of optical waveguides but also other functional structures, e.g., electrical conductors and components. The goal may be to achieve higher resolution, maximize achievable layer thicknesses, or enable new deposition processes for production. Subtractive functionalization by means of ablative manufacturing processes, e.g., laser material processing, can also take over the role of changing wetting properties and thus achieve an even higher resolution than can be realized with the flexographic printing process.

The second part of Chap. 4 deals with the functionalization of the flexographic printing plates used, which can be achieved in various ways. The focus here is on laser-based functionalization, with the emphasis on the introduction of microstructures. For this reason, fundamental considerations are first made regarding the selection of a suitable laser. On the one hand, printing-form specific parameters such as the chemical composition and the resulting trans- or absorption properties have to be considered. On the other hand, different physical parameters such as the different absorption mechanisms depending on the used pulse duration have to be considered. Furthermore, ablation tests of the printing form material are carried out before microstructures are introduced into the actual printing form. First of all, it is investigated which surfaces have what kind of influence on the printing result. It is found that the structuring of the stamp surface has the greatest influence on the print image. On the other hand, the transition area between the stamp surface and the stamp flank is critical for the resulting edge waviness. Depending on the size of the microstructures in the stamp face, different manipulations of the printed image can be achieved. With structure sizes in the range of 30 μm, for example, one can selectively adjust the material distribution within a conditioning line, while ablations in the range of 100 μm width already lead to gaps in the printed image. Ultimately, the flexibility of laser processing allows a flexographic printing form to be processed in such a way that it is perfectly matched to the requirement. No less important, however, is the entire printing process itself, since, for example, the infeed of the individual rollers to each other also has a very high influence, which can even be higher than the microstructures introduced. This means that in order to carry out successful printing, the printing form production, the printing form processing and the printing process must be suitably coordinated with each other.

In Chap. 5, state of the art in POWs fabrication by using aerosol jet printing (AJP) technology is presented. The fabrication covers a complete process step from selecting reliable POW materials, studying the optimum AJ process parameters and strategies in POWs printing on 2D/3D substrate surfaces.

AJ printed POWs are characterized for their geometrical, defect, mechanical, lifetime performance and optical quality analysis. Analysis of the geometrical profile is essential to draw a significant theoretical geometry to be used in the computer-aided design of electro-optical assemblies (Chap. 2) and three-dimensional simulations of optical multimode waveguides (Chap. 3) of this book. Theoretically, a circle segment cross section of POW is produced. For the optimum optical function, the aspect ratio of height to width (~1:5) and contact angle of ~ 45° to 60 ^{O} are desired to be obtained. The aspect ratio and contact angle requirements can be realized by printing the POW onto PMMA foil substrates with already applied conditioning lines. A thorough investigation of the various POW materials resulting in J + S 390,119 material can produce significant optical quality, and the material is stable with the AJ printing system by adapting temperature control of the material during printing. The sleeve temperature of 45 °C and dwelling time of 30 min is required during the printing process.

Optical quality measurement of the printed POW shows attenuation rates of up to 0.2 dB/cm is achieved. The waveguide can transmit up to 10 Gbit/s and likely become a game-changer to the metallic conductor. For the printing strategies, multi-layers printing without the intermediate UV-curing process produced desired qualities of the printed POW. The range of layer is varied in between 6 and

10 layers depending on the selected value of the AJP parameters (e.g., sheath gas, atomizer gas, exhaust gas, printing distance and printing speed). As an application, AJ printed POW is used in an asymmetric optical bus coupler, which allow light to be transferred from a module-POW to a bus-POW. Details of the mentioned application are given in Chap. 6 of this book.

The key advantage of the AJ printed POWs is the capability of the process to be directly printed on the 2D and 3D conformal surfaces, which significantly reduced the cost for new/customization production.

Despite that, the OPTAVER process result is possible to be used in optoelectronic sensing functions such as fiber Bragg grating (FBG) pressure sensor, interferometry sensor and photonic strain sensors, to name a few. This sensor enables to sense any physical changes such as temperature, tensile strain, pressure, torsion, or any other parameters, especially for medical health monitoring and equipment/device performance monitoring.

The central question for Chap. 6 is: How to connect optical multimode waveguides without interruption? Optical bus systems are a possible solution for this. Such systems are needed in device communication for the coupling of 3D-Opto-MID.

Within OPTAVER, a novel coupling concept (AOBC: asymmetrical optical bus coupler) is developed. By coupling the two cores at the side surfaces, it is possible to realize a bidirectional connection. This also allows the coupling rates to be adjusted by adjusting the overlap area through changing the contact pressure. The most important innovation of the presented principle is the possibility of asymmetric coupling, depending on the coupling direction. This makes it possible to couple less power from the bus into the module than in the opposite direction. This ensures that there is enough power along the bus to provide the signal for all connected modules. On the other hand, it ensures that a module signal can also be reliably detected at the end of the bus. This functionality is missing in known coupling principles and has so far prevented purely optical bus systems from gaining acceptance in extended structures.

Asymmetric coupling is made possible by bending one coupling partner while keeping the other straight. This results in a shift of the power maximum within the bent core to the outer radius. Since the coupling takes place there, more power is available at the coupling point, which can overcouple from the bent module waveguide into the bus. On the straight bus waveguide, on the other hand, this has no effect, which is why asymmetrical coupling occurs. To prove this, a simulation is first carried out using the newly developed algorithm from Chap. 3. The simulation confirmed the principle of the asymmetrical bus coupler. Using the collected design rules, the experimental investigation is connected within the scope of this work. Thus, a coupling without waveguide interruption at variable coupling rates can be obtained, which is moreover solvable and without power loss when not coupled. However, it is crucial to prove the asymmetric coupling, which can be achieved. Only then it is reasonably possible to use the coupling principle in bus coupling for device communication applications. Finally, we are able to successfully demonstrate the suitability of the AOBC for data transmission by transmitting signals at 12.5 Gbps. Finally, we are able to integrate the AOBC into a novel mixed-material package, which expands the field of 3D-MID by optical functionality to 3D-Opto-MID. The 3D-Opto-MID setup consists of stereolithographically printed three-dimensional structures for guiding the optical waveguides, assembled on standard thick film substrates. The latter have excellent electrical and thermal properties, which are particularly important for the optical transducer devices. The aging tests of the connected bus and module assemblies reveal a loss of coupling at high or low temperatures. Because of the thermal deformation of the 3D polymer elements, the waveguides lose contact and the coupling fails. Hence, we have to redesign the alignment/fixing structures in future works. Another option would be the use of CTE-adjusted polymers, so the mechanical load during temperature variation can be reduced. In case of the temperature humidity test, the package is irreversibly damaged. A hermetic housing can solve this issue.

Over the project duration of the research group OPTAVER, the state of knowledge in the field of additively manufactured optical waveguides has been extensively expanded. In the context of four dissertations as well as 10 journal articles and 39 contributions in conference proceedings with quality assurance in combination with oral and poster presentations on national and international conferences, the scientific knowledge about the entire process chain is published and subsequently cited about 150 times in numerous publications by other researchers. This process chain ranges from simulation and design, through manufacturing of the individual components, to integration into the overall assembly. The newly developed approach for manufacturing polymeric optical multimode waveguides using functionalized tools in production machines enables both two- and three-dimensional optical waveguide printing. This answers the research question if multimode optical polymer waveguides can be printed on spatially shaped surfaces. Especially, the latter case is beyond the current industry standards. Another research question was: Can sophisticated data transmission properties be achieved? The additively manufactured waveguides allow for data transmission with low losses of down to 0.2 dB/cm, achieving a transmission rate of more than 10 Gbit/s and thus meets the challenges set. By means of the asymmetric detachable bus coupler, different coupling ratios depending on the direction (bus to module or vice versa) are achieved at the same node. With this new coupling scheme, several nodes can be connected in a physical optical bus network without any bypasses. Thus, the question if printed waveguides can be contacted and signal branches be realized is answered. Another research question was: "Can the precise geometry and functions to be realized be modeled in a multi-physical design system?" For the design and modeling of such three-dimensional optomechatronic assemblies (3D-Opto-MID), a CAD-supported design environment including optical component libraries is developed in OPTAVER, which takes into account the additive manufacturing of the optical waveguides by means of design guidelines. In a feedback loop within the CAD tool, it is possible to spatially simulate optical losses of the assembly by means of a radiation optical wave propagation method and thus to estimate the performance of the optical path already at the time of design including production process characteristics. Simulation results answer the research question if the influences of production processes on functional properties can be predicted in a process model. This is a major step toward a holistic design tool for the development of electro-optical assemblies, which could be a key enabler for optical technologies in short-range connections. Thanks to the results of the OPTAVER research group, the realization of lightweight and powerful optical data transmission lines on mechanical structures, e.g., door/roof modules, seats, side planks in automobiles, aircraft, etc., is thus possible. In addition, energy-efficient networking of sensors can be implemented on large structures such as wind turbine or aircraft wings, automotive bodies. At the same time, there is an opportunity to realize a resource-saving substitution of copper-based information networks in mobile devices. The successful cooperation of the interdisciplinary research units within the research group has created fundamental scientific knowledge and will thus enable the transfer of science to industry in further research projects in the future.

We sincerely thank the Deutsche Forschungsgemeinschaft for funding the research group OPTAVER FOR 1660.

Printed in the United States
by Baker & Taylor Publisher Services